TABLE OF CONTENTS

INTRODUCTION

The approach to drawing presented in this book is one I have used for a number of years in the teaching of life drawing and anatomy classes. It is aimed at students in a myriad of disciplines (animation, game art, concept design, comics, GED, etc.), and so does its best to remain consistent in the emphasis of many artistic fundamentals. In addition, the drawing process presented here can be treated as applicable to different artistic ventures. For example, the thought process outlined can be an aid in understanding sculpture, modeling, painting, etc. Thinking outside the immediate subject of drawing and training in the thinking process described will help you prepare for a number of different artistic avenues that require the same basic skill set.

The approach covered here is primarily concerned with the use of line, development of form, and the simplified design of anatomy — the basics of being able to convincingly invent a figure that exists in space. While contour, shading, and expression are important elements in this process, they are not at the forefront of this particular method.

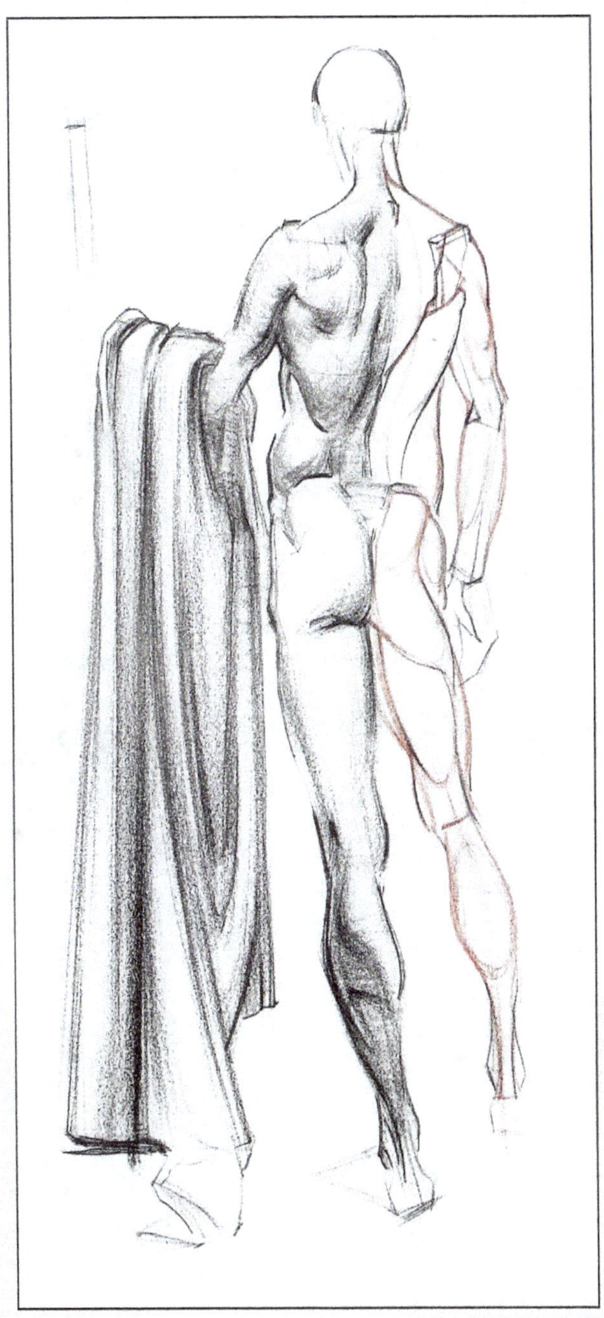

Through teaching this subject over a period of time, I have tried to assemble different technical elements to produce a consistent, beneficial result in student learning. However (before you leave screaming), I consider the approach outlined here as an open, changeable thinking/working process, meant at some point for the reader to personalize. It is my hope that there will be aspects to the process you disagree with, or deem to not be as important. After internalizing the information, I suggest altering the approach to more clearly reflect your ideas: such as reorganizing chapters, leaving some chapters out - or even adding something of your own! So, learn the drawing method outlined here for what it has to offer, and what I consider to be the essential elements of drawing the figure. But keep in mind that it isn't a belief system, or claim to any absolutes -- it is meant to help someone get started. After learning what you can from it, make it yours.

As we begin, keep in mind that each chapter builds upon the next. This approach should also apply to your drawings as you make them. Have discipline in your working process, understand how one step leads into the next, you should improve more quickly.

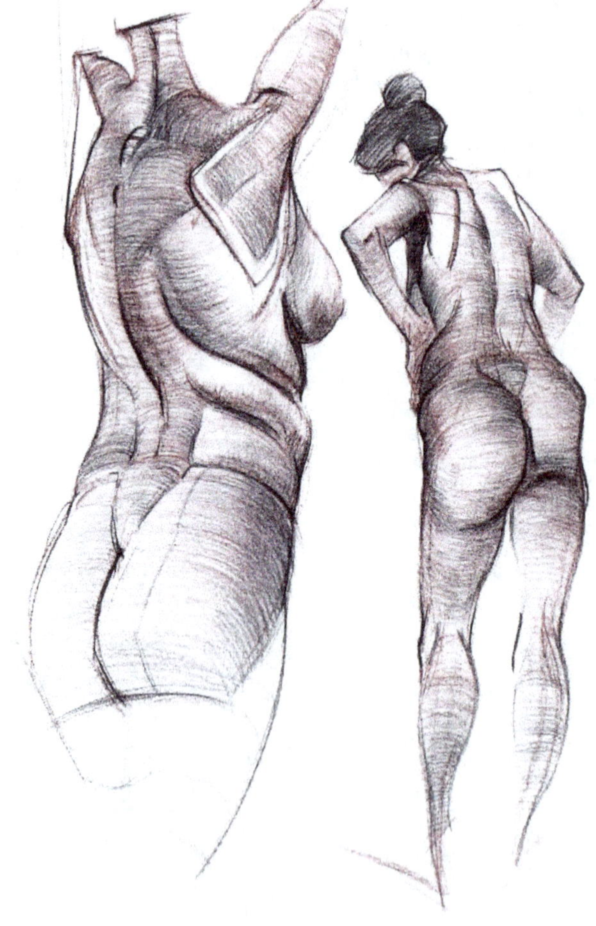

Remember, a major emphasis in this book is not on drawing the figure, but using the figure as an excuse to train oneself in the use of various formal principles in a myriad of artistic applications.

My goal is that this book can be a beneficial resource not only for drawing the figure, but also an introduction to the figure that facilitates knowledge and technical skills that are applicable to many other pursuits.

GESTURE DRAWING

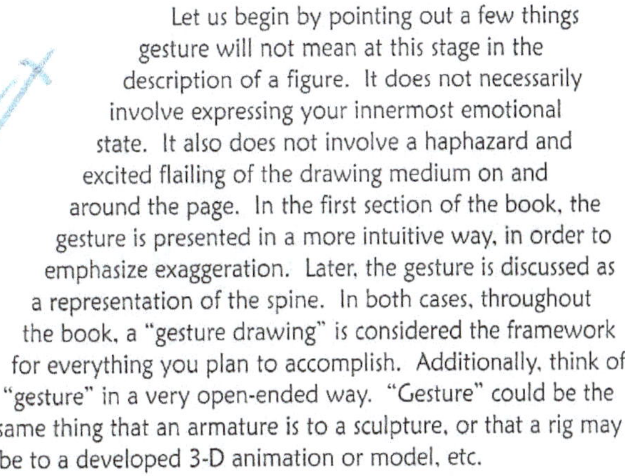

Let us begin by pointing out a few things gesture will not mean at this stage in the description of a figure. It does not necessarily involve expressing your innermost emotional state. It also does not involve a haphazard and excited flailing of the drawing medium on and around the page. In the first section of the book, the gesture is presented in a more intuitive way, in order to emphasize exaggeration. Later, the gesture is discussed as a representation of the spine. In both cases, throughout the book, a "gesture drawing" is considered the framework for everything you plan to accomplish. Additionally, think of "gesture" in a very open-ended way. "Gesture" could be the same thing that an armature is to a sculpture, or that a rig may be to a developed 3-D animation or model, etc.

At this early stage, the focus will be on communicating an idea to a viewer or audience. In order to communicate an idea effectively, you want to start by distilling everything seen into only the essential qualities of the figure/ character in front of you (or in your imagination). Through this drawing process, the goal is to take your attention outside of drawing the figure and onto the basic mechanics that allow that figure to manifest. By following this rationale, you will increase your whole artistic skill set, while learning to organize that skill set in a way that can produce a figure.

This chapter is the most important to the continued development of the book, and should be something studied continuously. It also begins the drawing process. It is important to understand that this drawing process is one for designing the figure from imagination (or life) with an emphasis on thinking structurally. My hope is that it remains generic enough to allow the addition of other influences, styles, etc.

At this stage, your goal is limiting the artistic means needed to build a concentrated sense of intention. Try only making lines that have a meaning, or that you could explain as intentional to the development of your drawing.

When developing a gesture drawing, it is important to be aware that you are describing the eight parts of the body.

These eight parts include:

- Head
- Spine
- Arms (2)
- Pelvis
- Rib Cage
- Legs (2)

The essential elements you will describe using these eight parts include a sense of story and composition. Giving the pose a "sense of story" means communicating a unique sense of positioning or attitude. Every person has a specific way of holding himself or herself when moving. By exaggerating the "story," you give your viewer a compelling image to experience. When creating a gesture drawing, this involves developing your figure's proportions and giving your figure a sense of balance and weight.

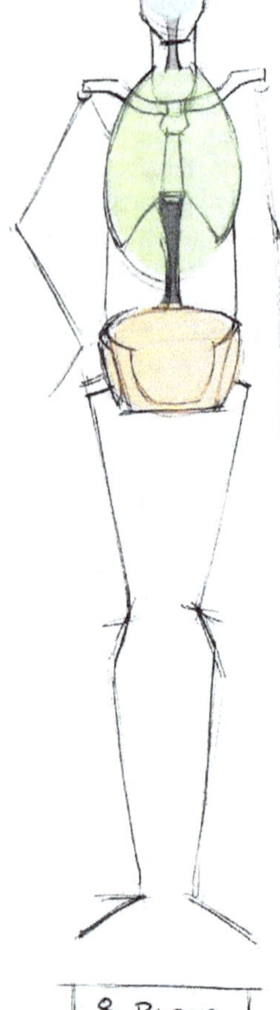

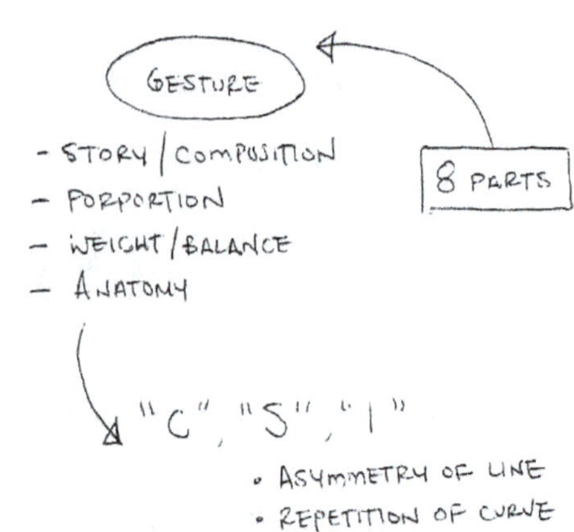

The lines most crucial to showing a figure are the "C" curve, the straight (line), and "S" curve. These lines will continuously reappear throughout the book. In this drawing process, you will never use any other type of line.

TIP When checking the proportion of the figure, try to avoid doing any slow, methodical measuring. Instead, base the proportions off of what looks correct after establishing the figure from head to foot. If it looks incorrect, change it — the drawing is still at an early stage when correction is easy. The downside to slowly measuring out the figure is that it stiffens the poses. Focus on the activity — the proportion can be corrected later.

The most important thing to keep in mind while drawing the figure is that the human from is essentially a balancing act.

This illustration is a diagram of the figure from the side and from the front.

In the side view, the head is suspended out over the rib cage by the forward angle of the neck. The neck and head are in turn balanced by the rib cage as it pushes at the opposite angle.

The pelvis moves opposite to the tilt of the rib cage, and the legs stabilize the body in the shape of a large "S".

The side view shows us that the skeleton is designed in a way that naturally balances the figure.

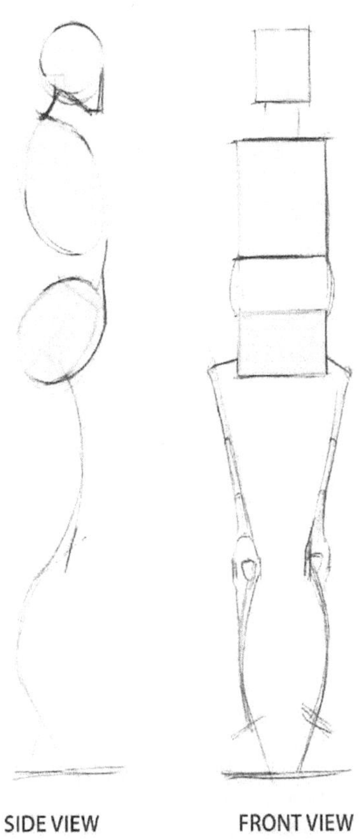

SIDE VIEW   FRONT VIEW

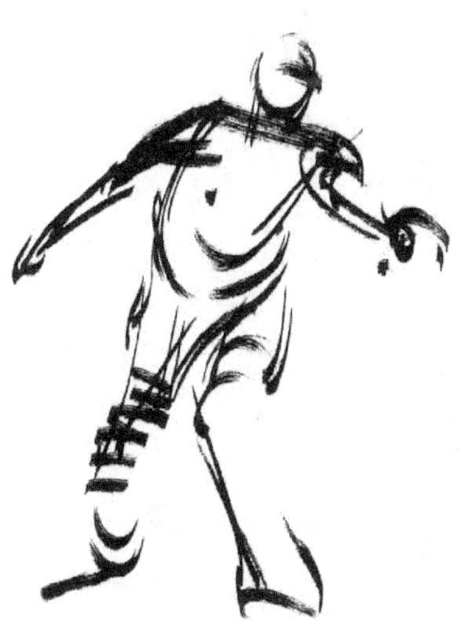

The diagram above right shows how the figure is balanced between hard and soft forms. The head, rib cage, and pelvis are all large areas of bone balanced between softer areas of muscle and flesh.

In a later chapter, we will study the active and passive groups of anatomy that create this form and balance.

In order to keep this natural quality of the human form a constant in our drawings, we need a use of line that continually emphasizes visual ideas of balance and movement.

ASYMMETRY

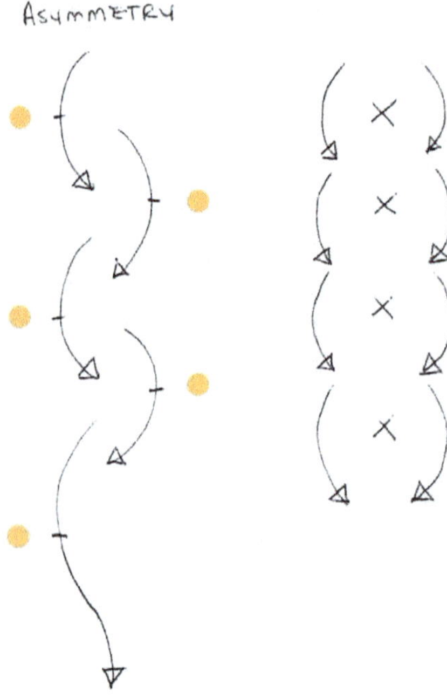

Beginning with only a "C" or "S" curve, the main focus is on positioning one of the curve's apexes higher than the one that follows.

The asymmetrical use of line (shown on the left) is the main line use to be emphasized when developing a gesture drawing. By keeping the high points of the curves slightly offset, the eye is forced to move through them. This gives you the ability to have a great deal of control over where the viewer's eye goes and how quickly. This is one way of dealing with composition at a very early stage of the drawing.

Avoid line use (shown on the right), which, instead of playing the curves off one another, uses mirroring or parallels. This approach closes off the form visually and does not allow for a flow between forms. Furthermore, the diagram on the right does not emphasize a natural sense of balance and movement, which are paramount qualities in describing the figure.

TIP: In order to keep the two examples separate, try remembering that the asymmetrical lines give the viewer's eye a pinball-like experience — always bouncing the attention to a line into another direction. The symmetrical curves stiffen that experience into a snowman-shaped appearance – generally, we don't associate snowmen with a great deal of excited movement.

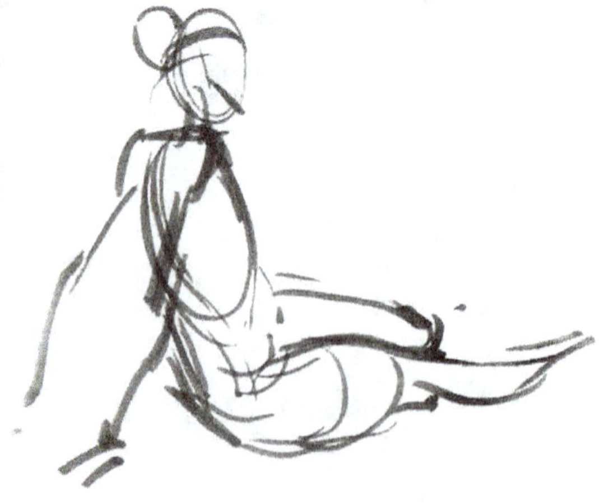

In addition to using asymmetry, the second quality of curve used is that of repetition. Any time a similar curve or shape is repeated twice or more, it provokes a visual movement.

In the diagram to the right, study how the three "C" curves placed next to one another start to push the eye from left to right.

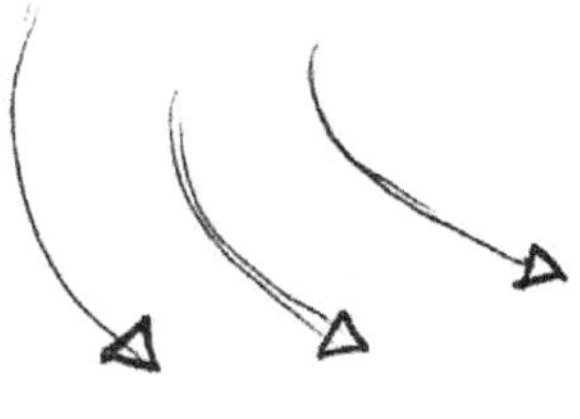

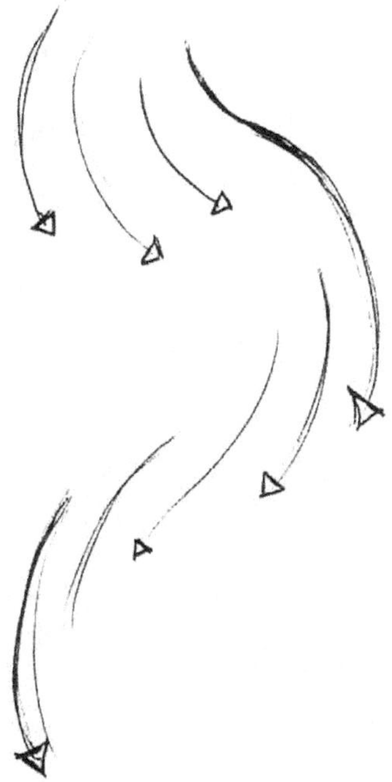

Using asymmetrical curves in addition to repeating curves gives your gesture drawings a solid sense of composition, fluidity, and timing.

In the diagram, notice how repeating curves cause the eye to slow down as it moves through the dominant asymmetrical curves.

Depending on the different combinations of line used, different visual experiences and speeds can be developed.

Fast and slow visual movements are a very important quality in the design of the figure at the gesture stage. Try slowing down the eye (emphasizing repeating lines to produce more side-to-side motion) in more complex areas (areas of intersection: midsection, shoulders, hips, knee, elbow) and speeding it up along the length of forms (such as asymmetrical lines creating a faster push downwards).

By playing one thing against another, you will keep your designs as appealing and life-like as possible. Also, you present the viewer with an experience closer to how we actually see — we scan at different speeds, lingering longer in some areas and quickly glossing over others. Rarely do we view everything before us at a consistent, steadied pace.

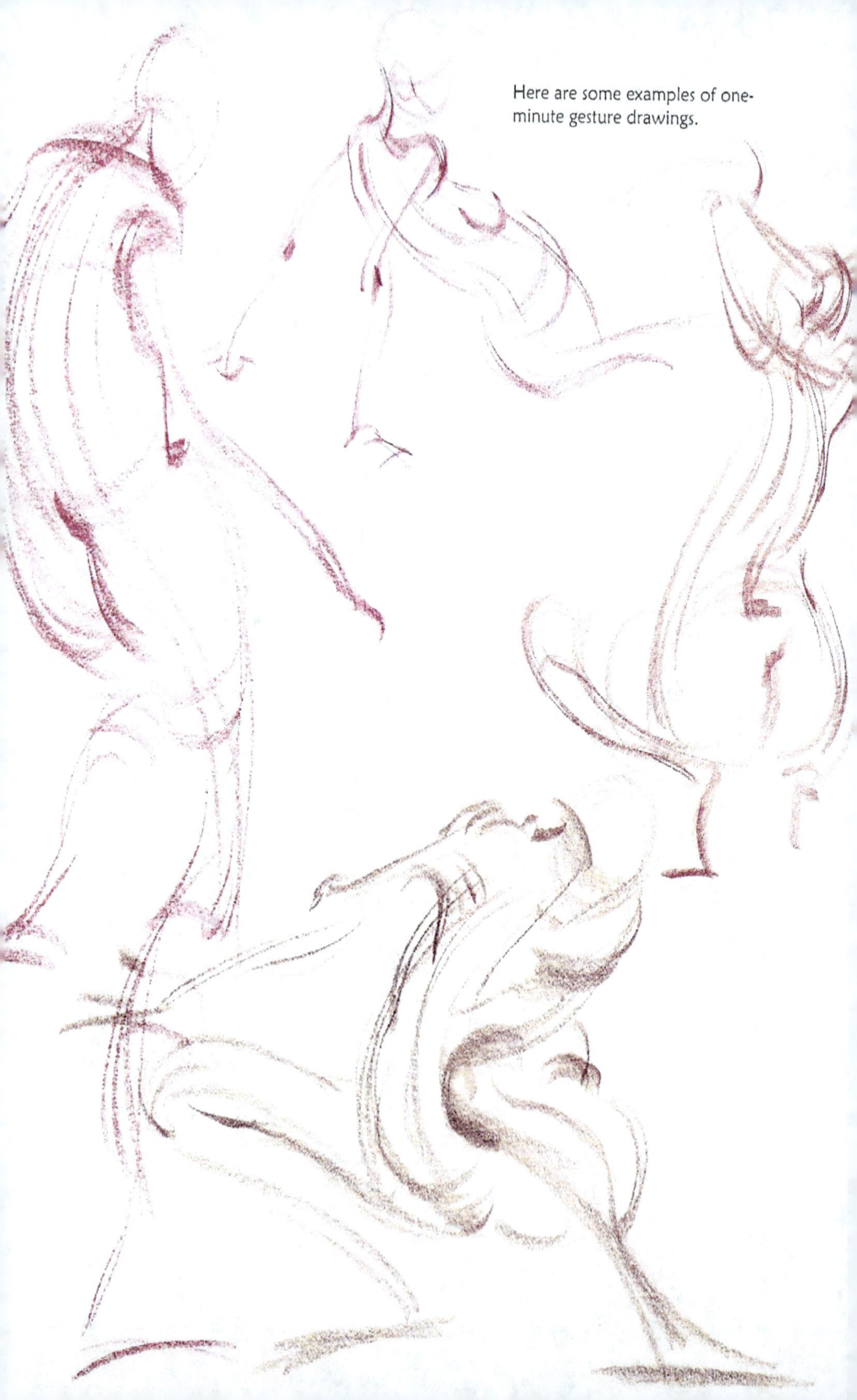

Here are some examples of one-minute gesture drawings.

Analyze the drawings on these pages for the ideas discussed so far. At this point, the eight parts of the body are indicated in an exaggerated activity. They are summarized into relationships using the straight, "C" curve, and "S" curve. The curves are used asymmetrically to play with a dynamic sense of timing and balance.

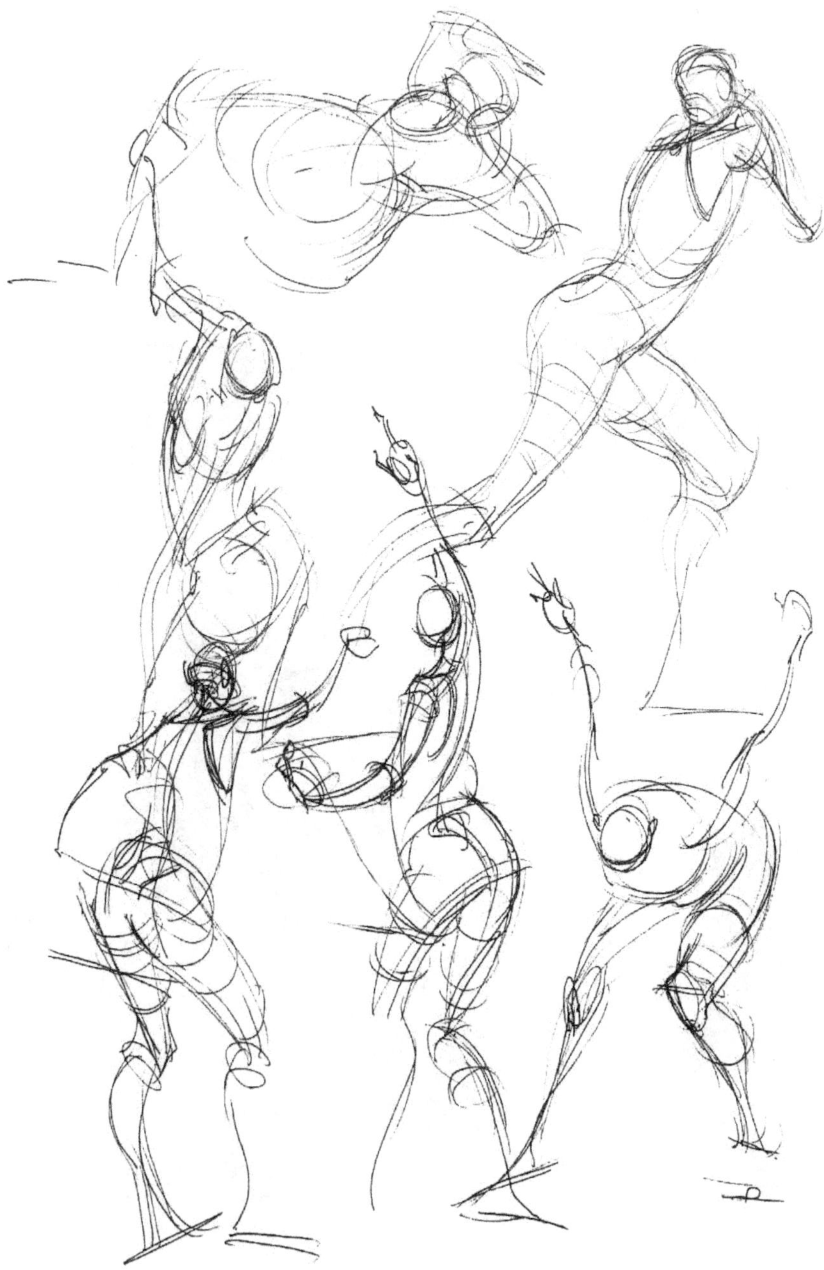

The last type of curve used in a gesture is wrapping lines. In a quick sketch, wrapping lines are curves that move across and around a form to indicate perspective.

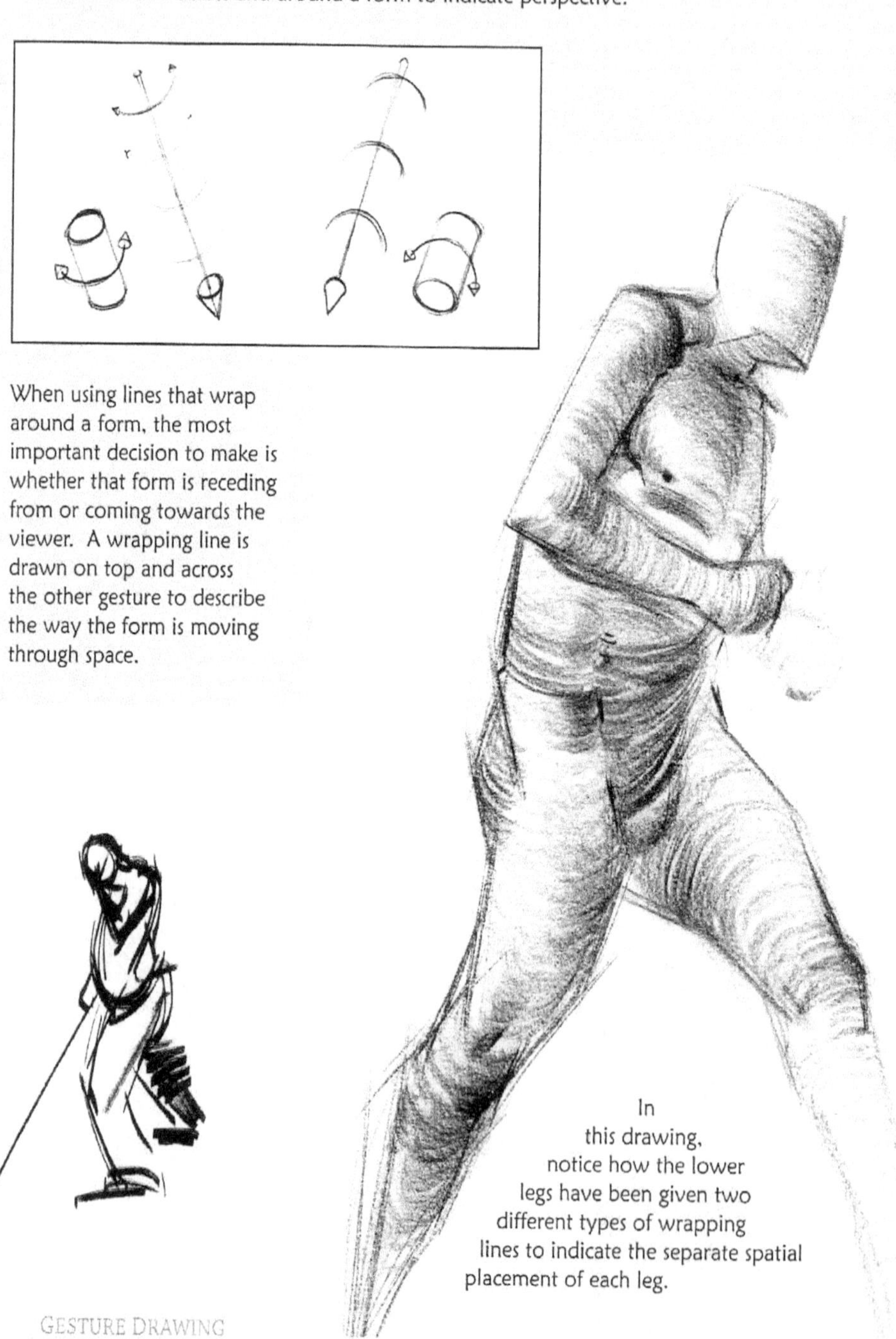

When using lines that wrap around a form, the most important decision to make is whether that form is receding from or coming towards the viewer. A wrapping line is drawn on top and across the other gesture to describe the way the form is moving through space.

In this drawing, notice how the lower legs have been given two different types of wrapping lines to indicate the separate spatial placement of each leg.

After using wrapping lines, the last step in creating a gesture drawing is to include the shapes of the head, rib cage, and pelvis.

When doing this, keep in mind that including these shapes will be a powerful tool in showing proportion, weight, and balance. At this point, keep the shape of the head very simple as a sphere. The rib cage should be shown as a conservative egg-shape that is standing up, while the pelvis is an oval laying on its side.

Refer to the diagram at the beginning of this chapter for an illustration of the shapes.

 Try to think of wrapping lines as rubber bands or string tied all the way around a form. The point of this exercise is to never draw a straight line across your drawing. From now on, only use lines that travel around an imagined surface. This will develop a short hand of form/perspective for you and for the viewer.

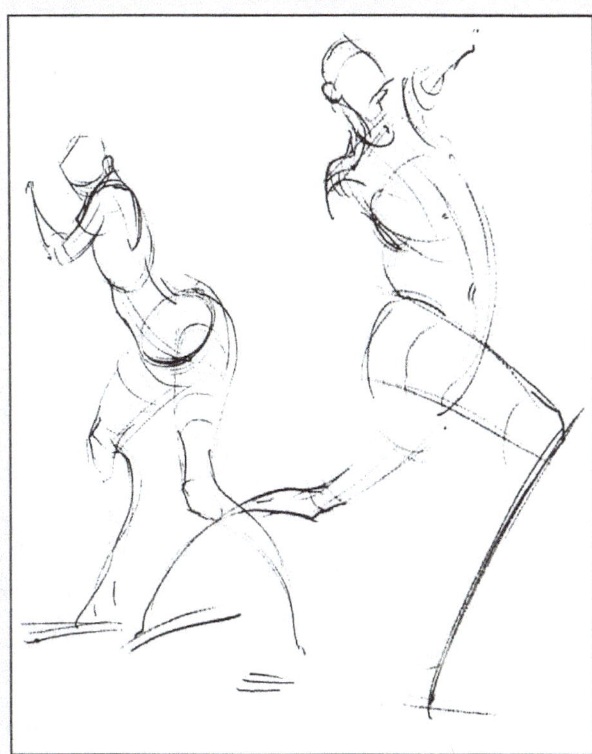

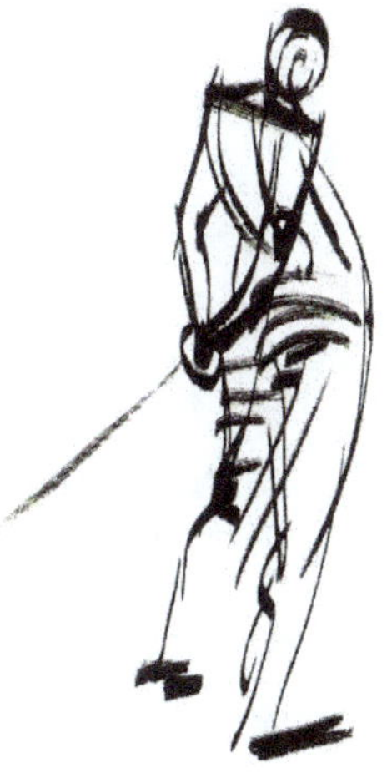

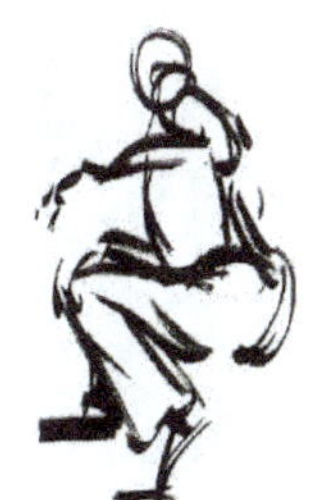

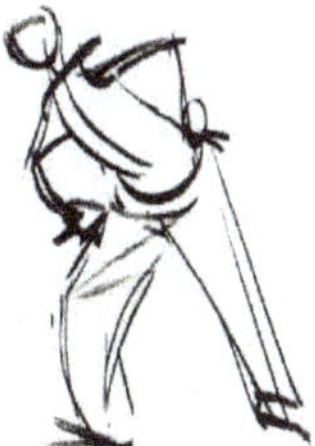

Similar to the diagram to the right, all of the
wrapping lines are volumetric contours, or lines
that travel across the surface of a form from side
to side. As a form changes direction spatially,
the lines will reflect that change.

However, keep in mind that you will never be
using a straight line. Using a straight line will,
at this point, start to become a reference to a
shape and begin to fatten out your drawings.

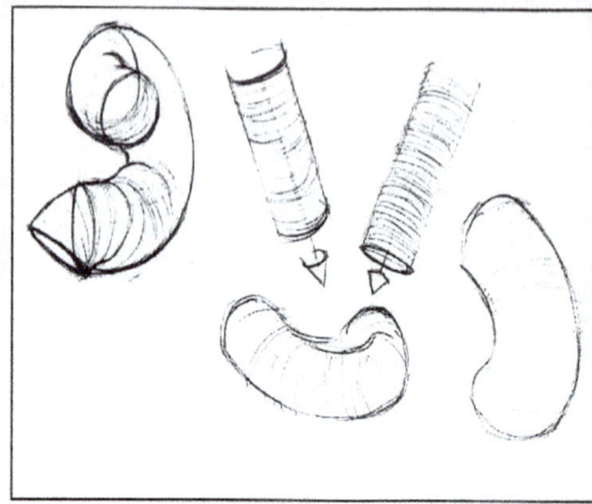

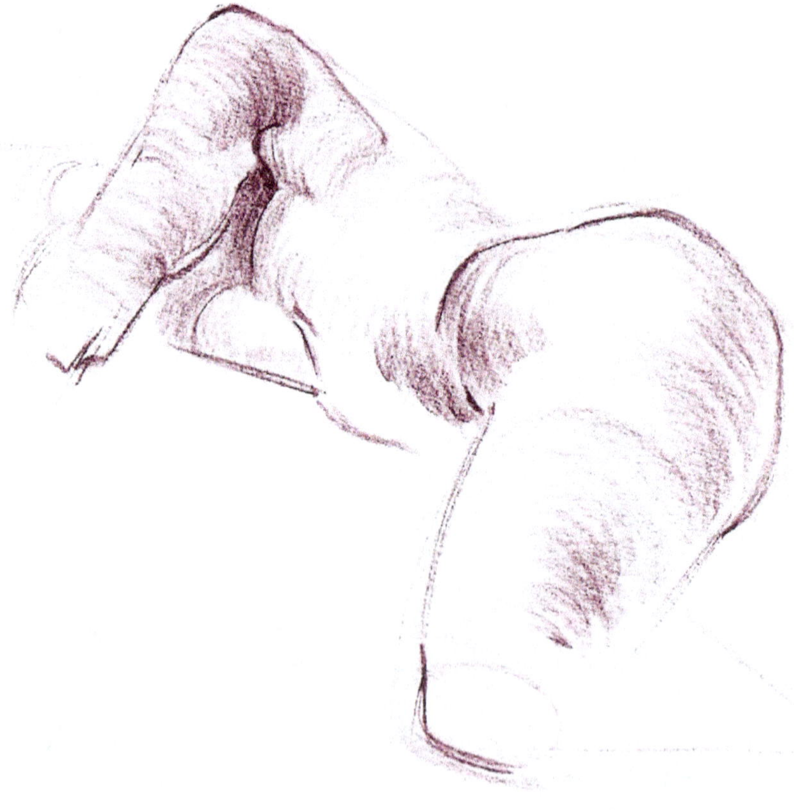

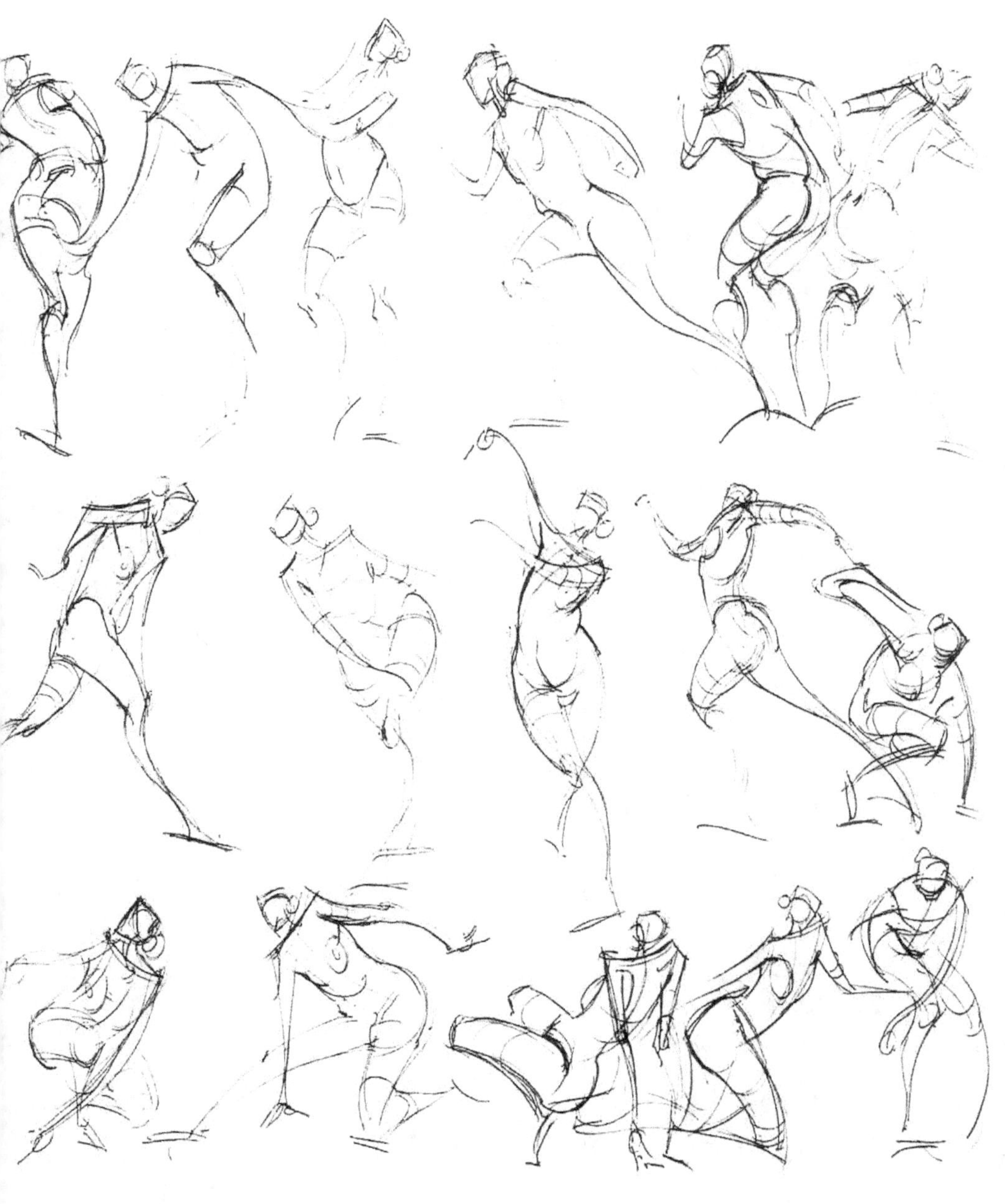

Developing the gesture involves considering the whole movement and relationship of the eight parts of the figure. The most important of these parts is the spine.

The spine is responsible for the organization and balancing of the three major masses (head, rib cage, and pelvis), as well as the arms and legs. This section describes how the spine influences the figure, and how that influence is shown in a gesture drawing. This section also explains the initial design of the three major masses based on the influence of the spine.

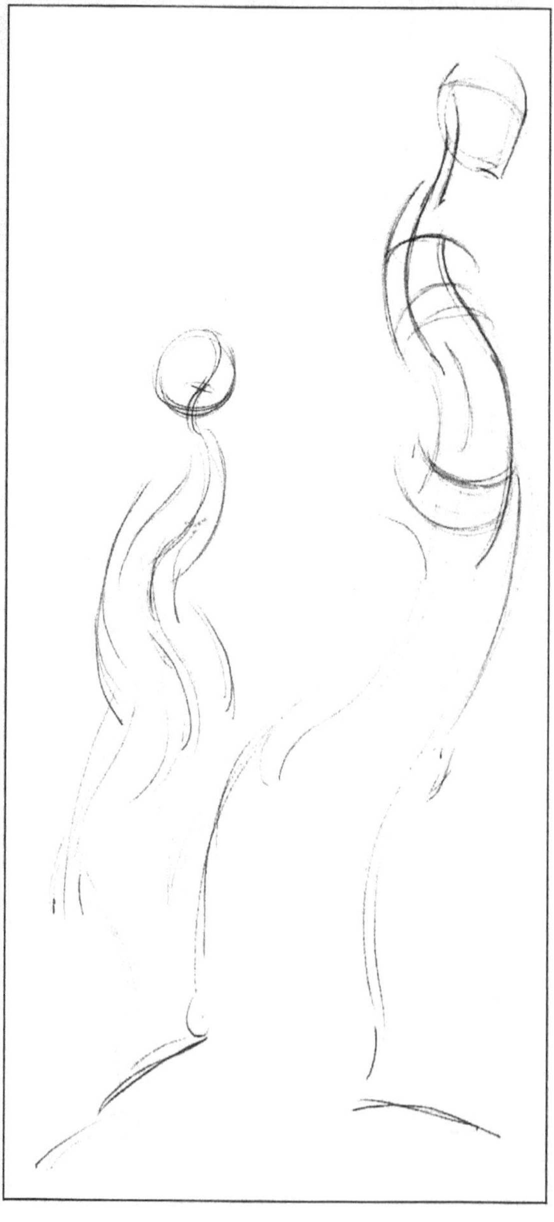

After becoming more intuitive with the use of line and curve, consider those same elements in a more concrete relationship to the movements of the spine.

Remember, the goal is to organize your mark-making in a way that communicates the natural designs of the figure.

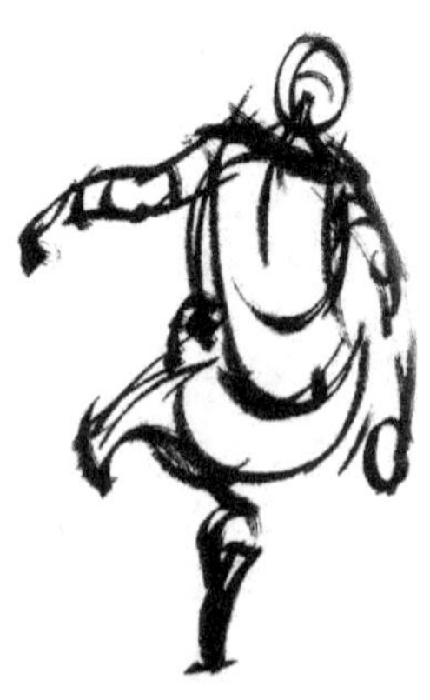

The diagram below shows four different illustrations of the spine, from a back three-quarter view. The spine is primarily an "S" curve in design -- the complexity is that the "S" needs to be thought of dimensionally.

BACK THREE-QUARTER VIEW

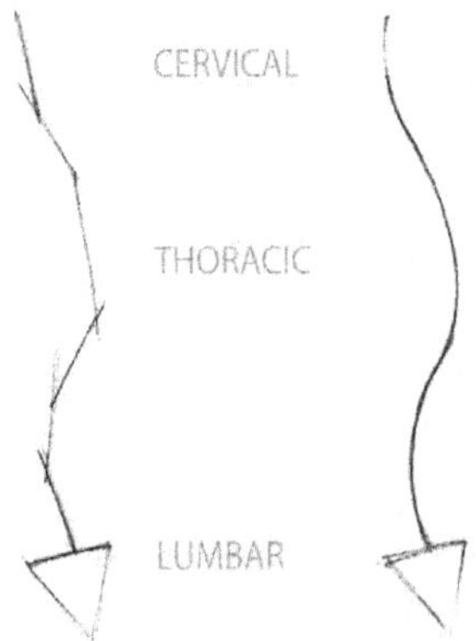

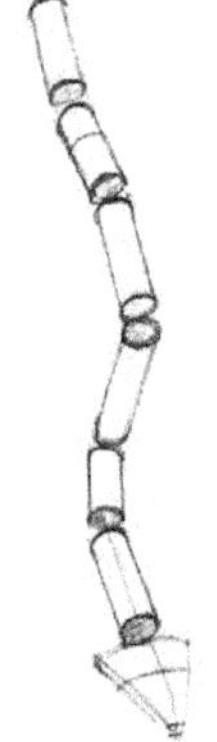

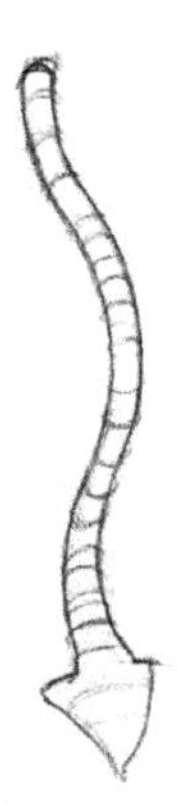

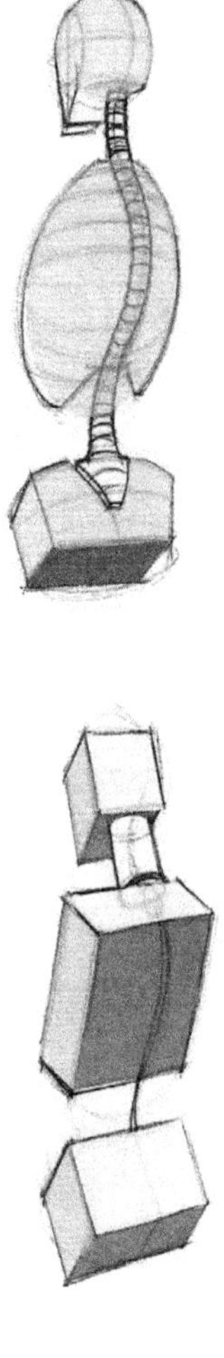

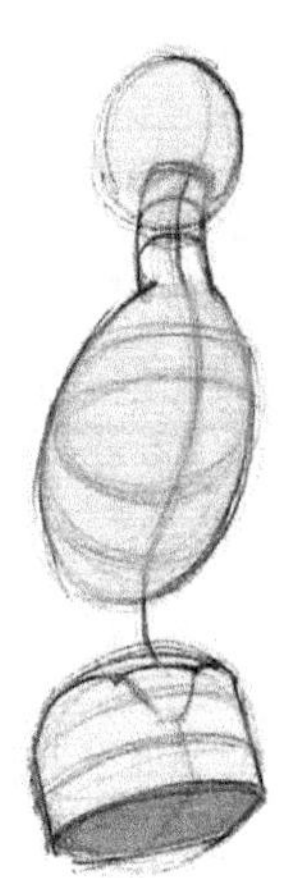

The first two drawings on the left show the design of the spine using only line.

The first drawing is done using only straight lines, illustrating the direction changes in the three areas of the spine: the cervical (neck), thoracic (upper and lower rib cage), and lumbar (lower rib cage and pelvis). Starting from the bottom triangle, notice that the lumbar section of the spine moves forward and away from the viewer's eye. Next, the direction of the spine changes and leans the opposite direction. As it moves further into the thoracic section, the rib cage again changes direction as it moves up and towards the neck. The thoracic section then moves into the cervical area of the spine.

The second drawing from the left shows how an "S" curve illustrates this complex movement in a simple fluid line.

The two drawings on the right illustrate the spatial position of the spine.

The first of these two (second drawing from the right) is similar to the first drawing on the left with the added element of perspective. Notice that the same 2-D directional changes are taking place, but now include the cylinders constructed on top to clarify the spine's snaking through space.

The last drawing on the right uses "S" curves to depict a more fluid design for the spine, using ellipses to delineate the perspective and surface changes.

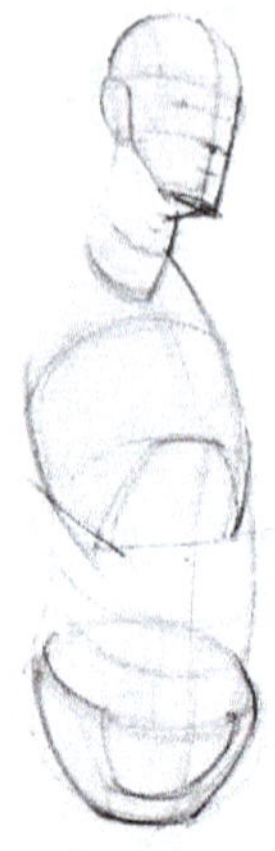

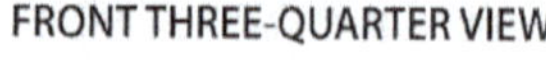

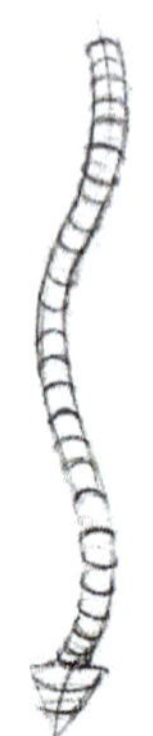

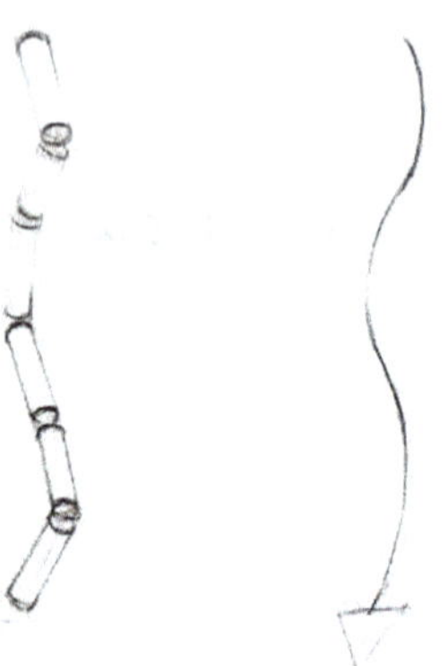

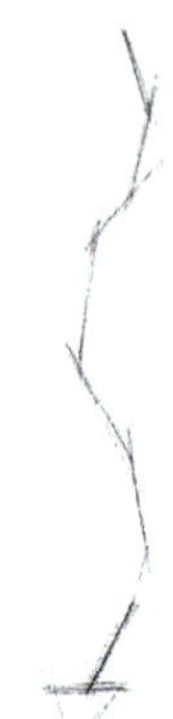

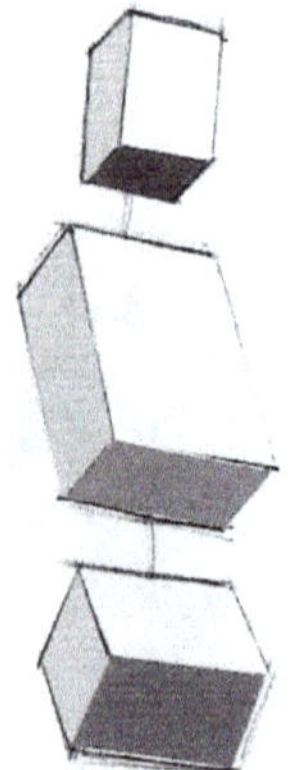

The diagram above shows the spine as if seen on a figure from a front three-quarter view. The same types of lines have been used as in the first illustration: straights, curves, cylinders, and a more organic shape. Compare the front view to the back view on the previous page. Notice that all of the movements detailed on the back view are now reversed in this front view.

In the illustration below, the same lines have been used to show the spine in profile.

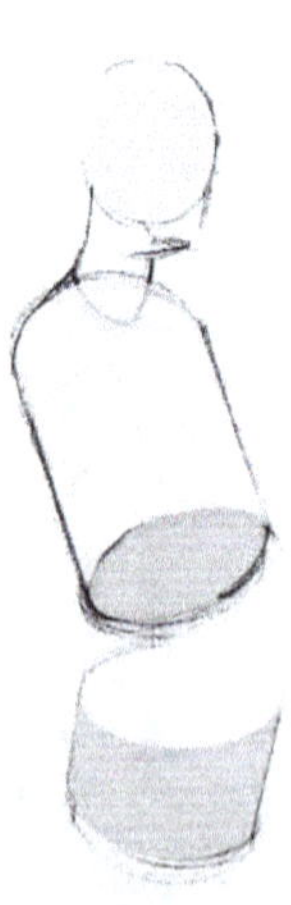

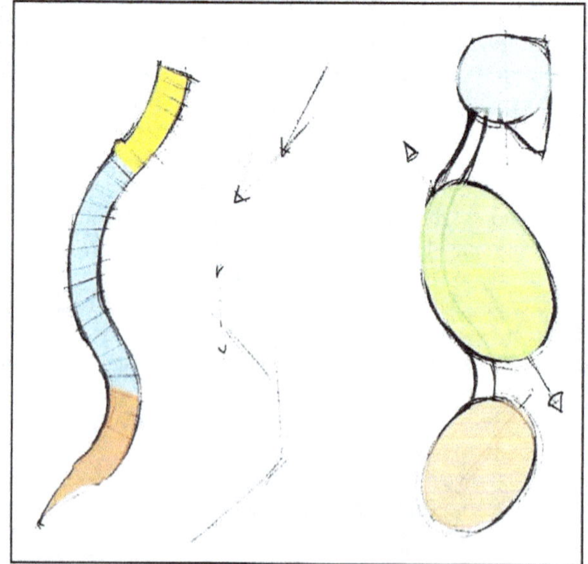

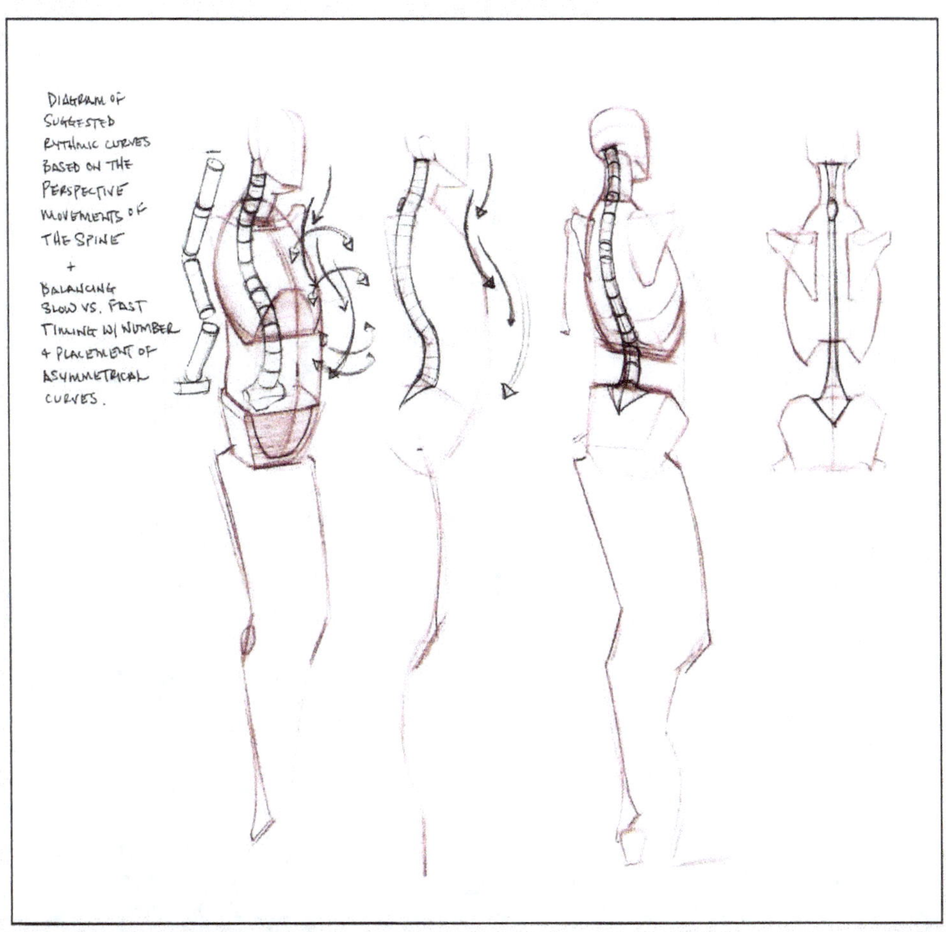

After some experience with gesture drawing, you will start to notice that passages or areas in the figure can be handled in a very formulaic way. For example, the same lines are always used to directionally express the movements of the spine. The diagrams above were done to illustrate the importance of trying to see with X-ray vision into the spine as a starting point in explaining the figure. Additionally, the first two figures have the gesture lines added to show the influence of the spine on their position and direction (design).

Always try and understand what the spine is doing — most everything in the figure can be explained as a consequence of it.

After developing your figure's pose as a gesture drawing, you will next give a more concrete description of the major masses: the head, rib cage, and pelvis. Manipulating the figure's center of gravity in an exaggerated manor is essential in creating an interesting pose. On top of the gesture, add a sphere for the head, an egg shape for the rib cage, and a horizontal egg shape for the pelvis.

The goal of using the center of gravity is to force an awareness of how the figure stands upright, while creating the ability to exaggerate positions.

Following ideas of balance, you can design a 2-D lean for the rib cage that is off the symmetrical center. (Of course, unless the figure is in a seated position, a pose using an object to remain upright, or if the majority if weight rests on the arms.)

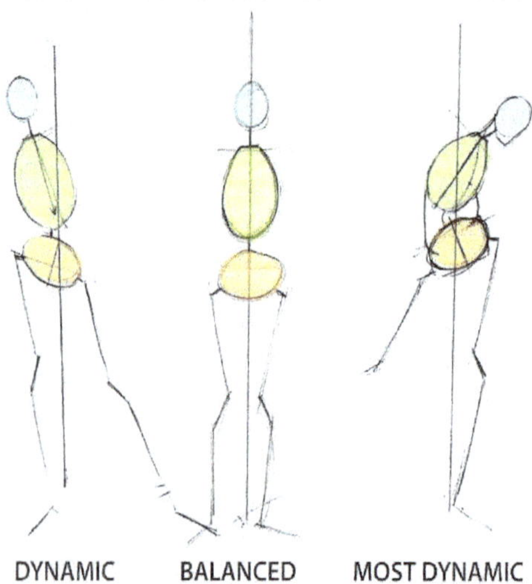

A common mistake when drawing the figure is keeping these shapes balanced and straight (center drawing). Notice that the shapes of the major masses all have an equal and balanced relationship to the center of gravity (shown as a vertical line).

Creating a dynamic pose involves creating a sense of tension with the figure's center of gravity. Just as our initial gesture lines create a sense of movement with an imbalance in the placement of line, you should flirt with the idea of imbalance when drawing the shapes of the head, rib cage, and pelvis.

On the left and right, notice how the major masses move around the center of gravity without lining up on it. The last pose is the most dramatic in its distribution of the masses in relation to the center of gravity.

Keep in mind that a balanced pose is no better or worse than using an out-of-balance pose. What matters is that you are able to build the correct position to match your story/ intention. Remember, though, that because of the spine, there is always some counter-balancing of the shapes of the three major masses.

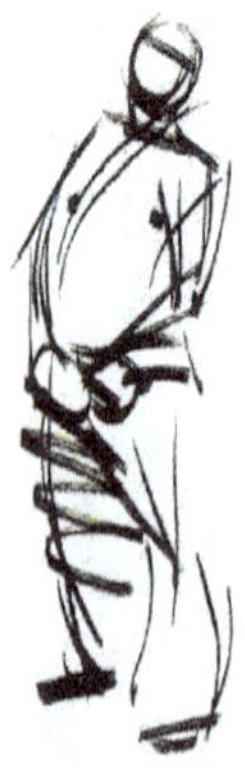

After identifying the center of gravity, the next step is to lay in the three major masses: the head, rib cage, and pelvis.

Because the head is a more complex form addressed in a later chapter, for now keep it as a simple sphere shape. When placing the shapes of the rib cage and pelvis, make sure they are consistent with the spine and the balance of the gesture.

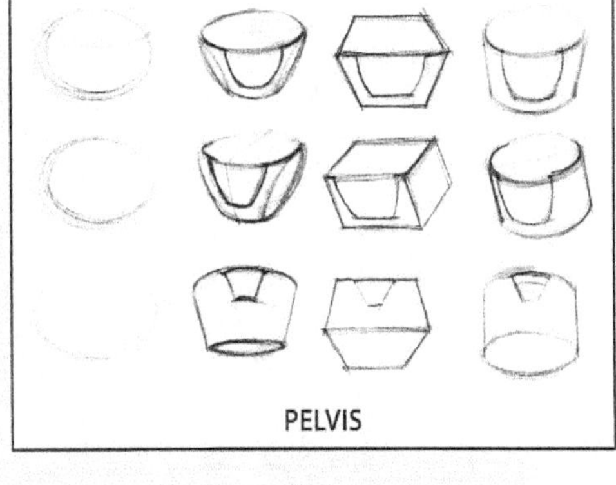

PELVIS

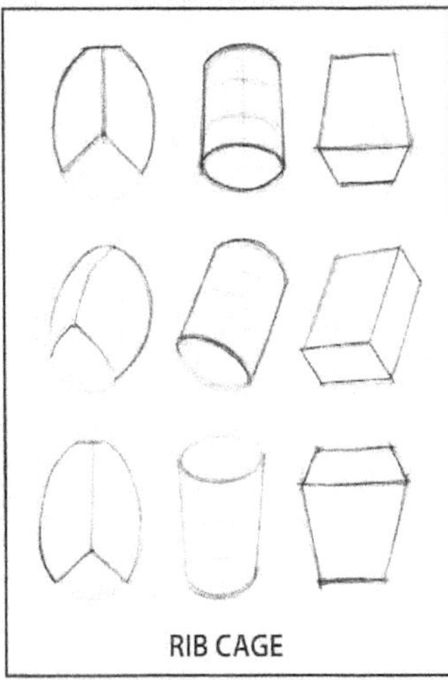

RIB CAGE

Before describing the shape of the pelvis or rib cage, look for the line of its tilt (2-D position/lean). An easy way to find this is to look for the weight-bearing leg. When the majority of weight is positioned on one leg, it usually causes this large area of bone to raise, dropping the other side. Draw this line of tilt and then place the shape on top. Options for the pelvis and rib cage are shown in the diagrams on this page.

At this early stage in the drawing, use the egg shape — which can then be used to develop more complex forms.

There are hundreds of different configurations for the creation of a pose, and each one is governed by the desired effect and context of a given story. The following exercise will help you create a sense of impending action, and is an exercise I generally give to students who are stuck making stiff symmetrical positions. While this exercise isn't the solution to how every pose should be thought through, it is one tool to use when thinking about the mechanics of the figure, and how these mechanics can be used.

Stiff, symmetrical poses, while good for a suggestion of power, strength and/or immobility, often lack a sense of lyricism and exaggeration. In an effort to push towards these more dynamic attributes in a pose, I ask my students to strive to create an "about to ..." quality, which is a pose or position in their drawing that is somewhere in mid-action, mid-step, etc. The "about to ..." effect is an engagement in the suspended interest or outcome of the figure. Stable, symmetrical positions keep the action in stasis; the action has either not begun, or it has ended. An "about to..." position engages viewers by making them anticipate the outcome of the action, hopefully wanting to fill in the rest of the story.

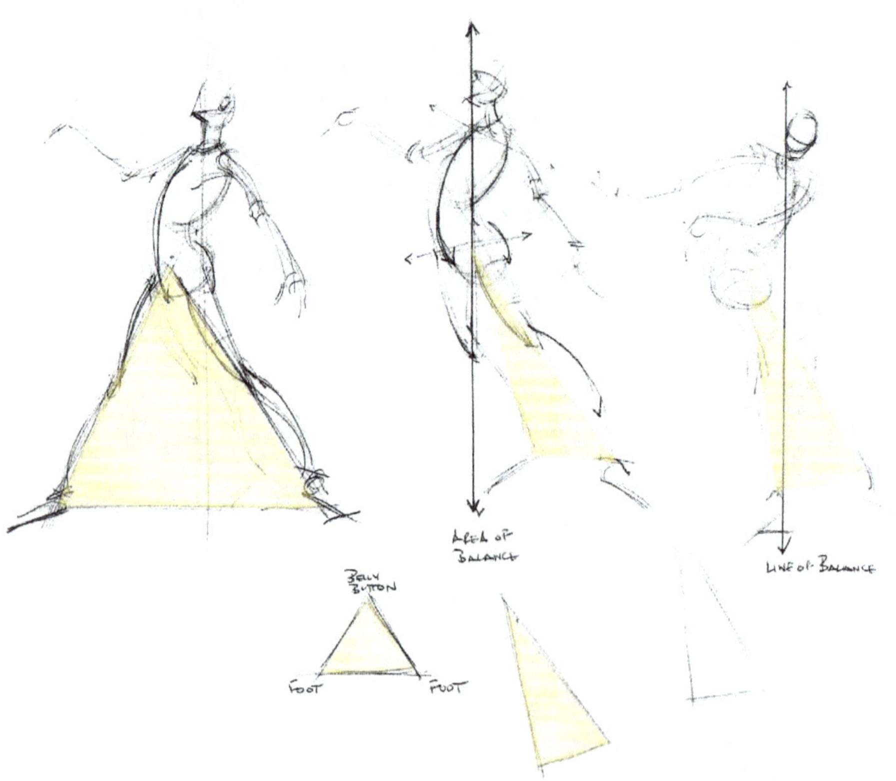

The difference between a stable pose and one in mid-action is
determined by how weight is distributed and balanced. While
this approach can be used to analyze most positions, here it is
demonstrated with the standing figure. Keeping in mind the
prior notes on the center of gravity, build a triangle between the
feet and either the belly button or nose. In poses that are very
stable, the triangle mostly appears very stable at the bottom.
Notice that in exaggerated positions, or out of balance poses, the
triangle looks more irregular.

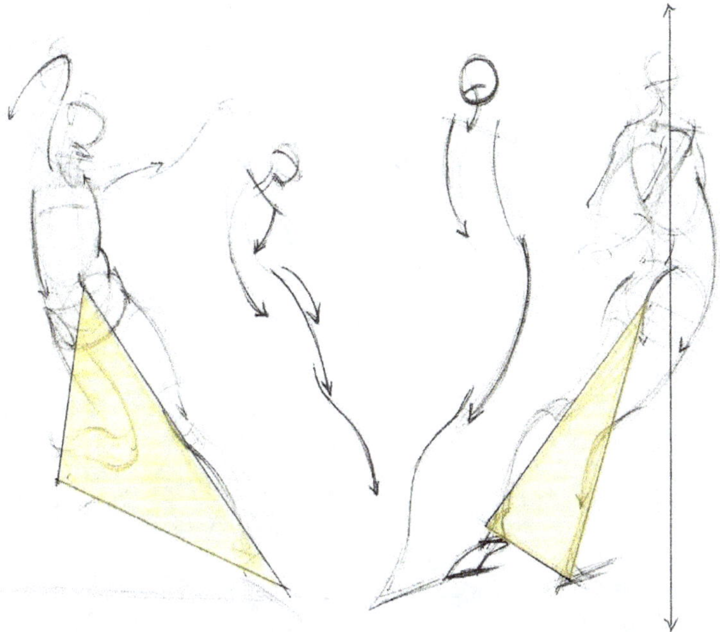

When developing a pose with these concerns in mind, use the
same approach discussed thus far. Begin with the head, working
the gesture lines down through the weight-bearing leg. This
organization of lines from the head to leg should be on a
diagonal line, which, judging from the center of gravity, looks
out of balance. When adding the second/supporting leg, place it
near the line of gravity to complete the out-of-balance posture.
This simple thought given to a figure's placement will create the
"about to ..." quality, engaging your viewer in the anticipation of
the potential outcome of the drawing's narrative.

ECONOMY of LINE

Economy of line is
yet another way to
clarify themes relating
to gesture. Read
through the diagrams
for suggestions on the
economical use of the
drawing medium.

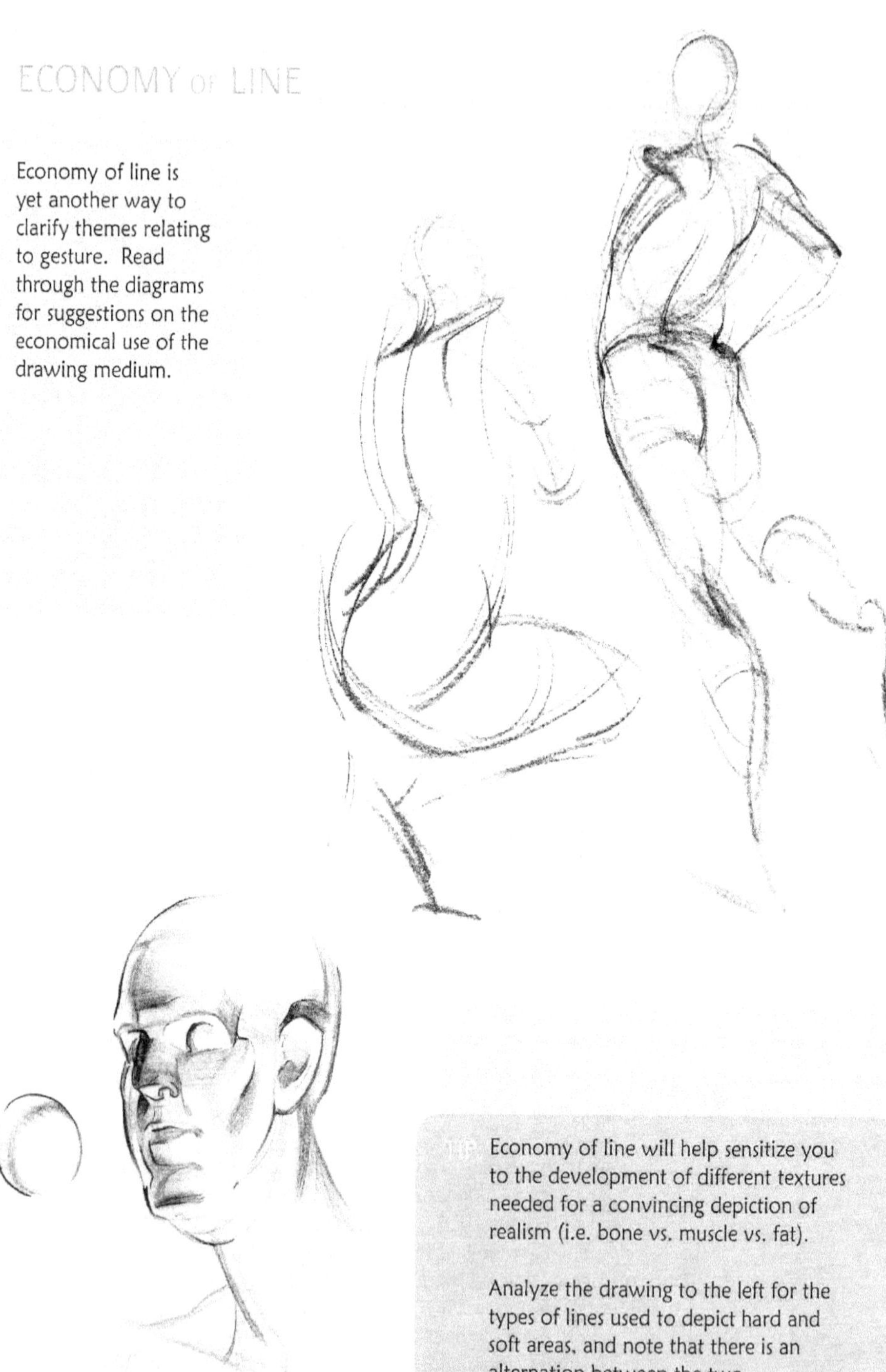

Economy of line will help sensitize you
to the development of different textures
needed for a convincing depiction of
realism (i.e. bone vs. muscle vs. fat).

Analyze the drawing to the left for the
types of lines used to depict hard and
soft areas, and note that there is an
alternation between the two.

"Economy of Line"

— Describing the difference between muscle, fat, + bone.

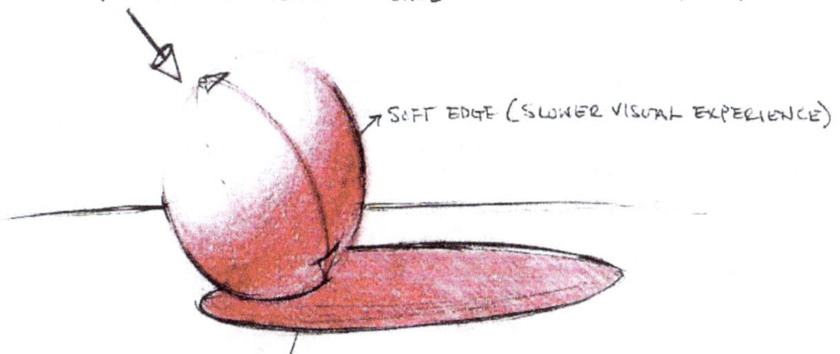

(One line created by turning the wrist while drawing. At first the lead is pointed at the surface then spun to run the broad side parallel to the paper.)

— Thin sharper lines describe abrupt changes in plane or light. These can generally be seen describing plane changes on a box or cast shadow edges.
— These types of line can be used for the description of hard, angular surfaces on the body, landmark points or bone.

— Softer edges gives a slower visual exp to the viewer
— This is the same as the gradation + edge describing the sphere
— This quality of line is reserved for smoother, softer (more sphere like) areas — muscle + fat

"Creating a sense of story" in your gesture drawings can mean a number of different things. Gesture can be the way we recognize moods through body mechanics, the innate ability to recognize your best friend from 20–30 feet away, or just simply being able to read the body as a type of communication. When studying gesture drawings, it will be a common exercise to exaggerate these positions until you become more comfortable with articulating a wide range of expressions. Once the ability to develop the exaggerated is achieved, the more natural subtleties of expression will be much easier to create.

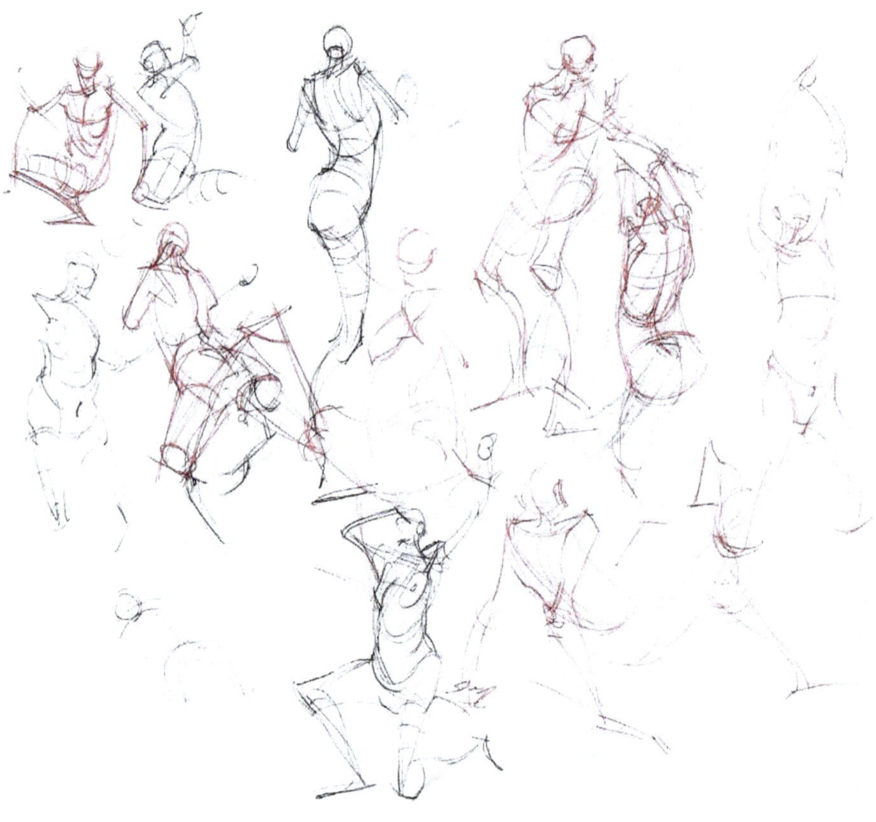

Remember that the figure is a machine in constant relationship with balance and imbalance — not just in the design of bones and muscle, but also of movement. Think of the natural activity of walking as an example. In order to walk, run — to move at all — we must throw ourselves out of balance, and with the next step catch it again. The reason all of our design elements are focused on asymmetry, balance, movement, etc. is essentially because we are describing a machine moving through a series of controlled falls.

So far, you may have noticed that there has been no discussion of measuring the figure or proportion. In this particular approach, there is an emphasis on achieving proportion through overall quick assessment of size. Work through the gesture lines from the head to the foot, then take a moment to decide if what you've done looks correct. This is not to say that this approach is better than another (because, ultimately, all should be considered); however, this approach allows for the emphasis to be placed on capturing the feeling of movement and position. One of the negative aspects of measuring is that, at times, it tends to produce static, stiffened poses with very little fluidity.

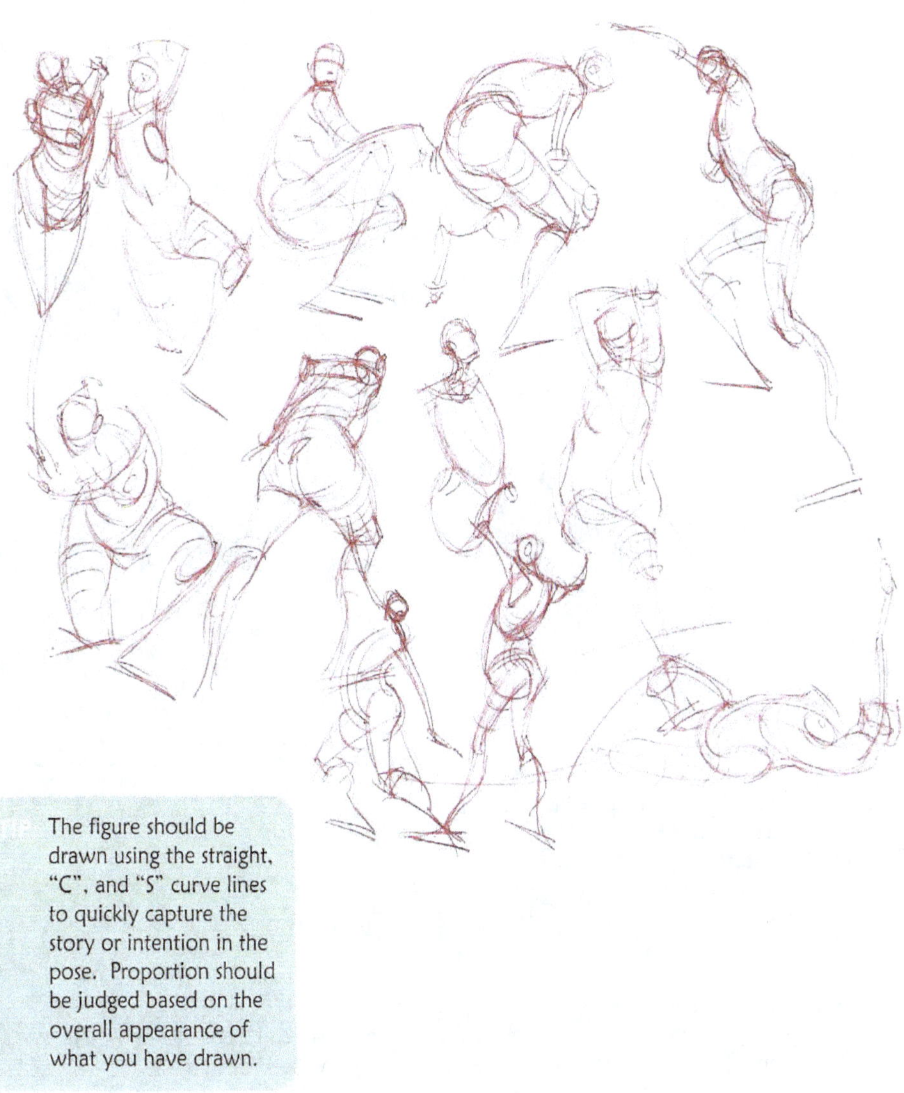

The figure should be drawn using the straight, "C", and "S" curve lines to quickly capture the story or intention in the pose. Proportion should be judged based on the overall appearance of what you have drawn.

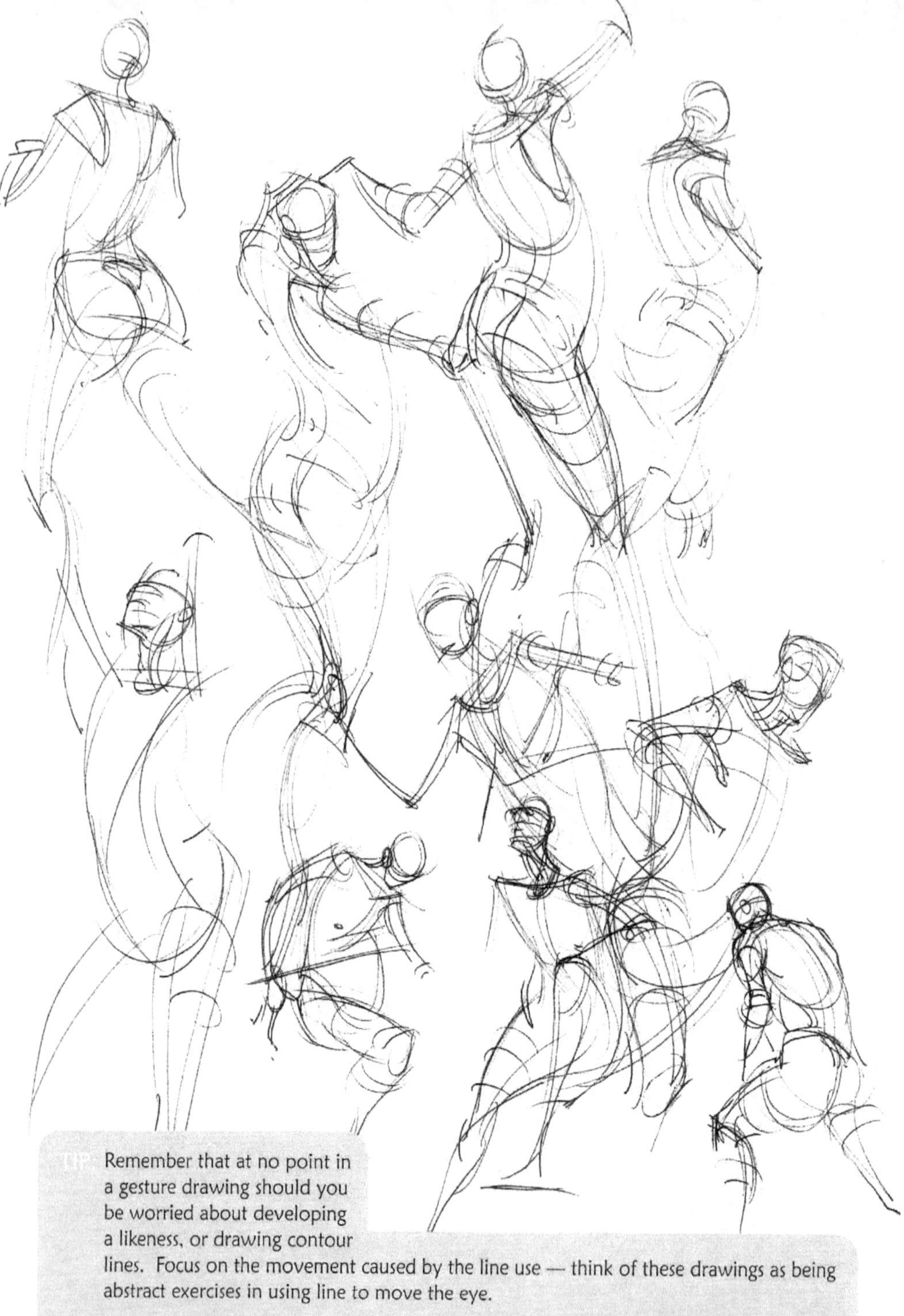

Remember that at no point in a gesture drawing should you be worried about developing a likeness, or drawing contour lines. Focus on the movement caused by the line use — think of these drawings as being abstract exercises in using line to move the eye.

Figure drawing is not so much about making something that looks like a person, but instead developing incredibly complex artistic practices and skills. Learning these skills will allow you to easily create any variety of figure or character desired.

Consider the gesture as the your animated way of capturing the lyricism of the entire figure. Do your best to keep the fluidity of the gesture, but still include the mechanics (skeleton, anatomy, perspective) in order to give believability to the overall figure.

 The next chapter will discuss using the gesture as a framework for developing a functional design for the skeleton. Adding the landmarks is the first step into a rigorous demonstration of how that gesture is possible. Regardless of the chapters and information to come, it is crucial to begin with a gesture.

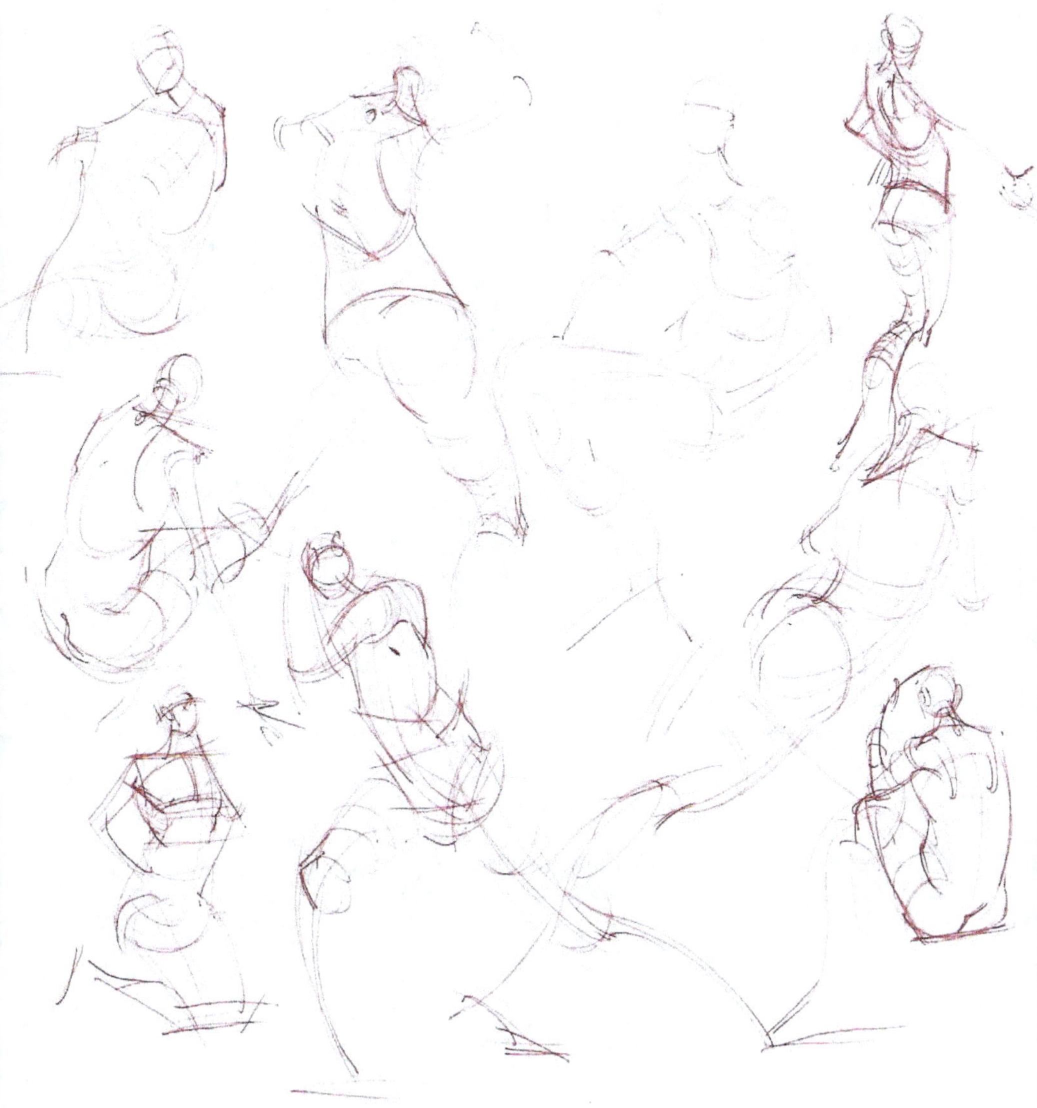

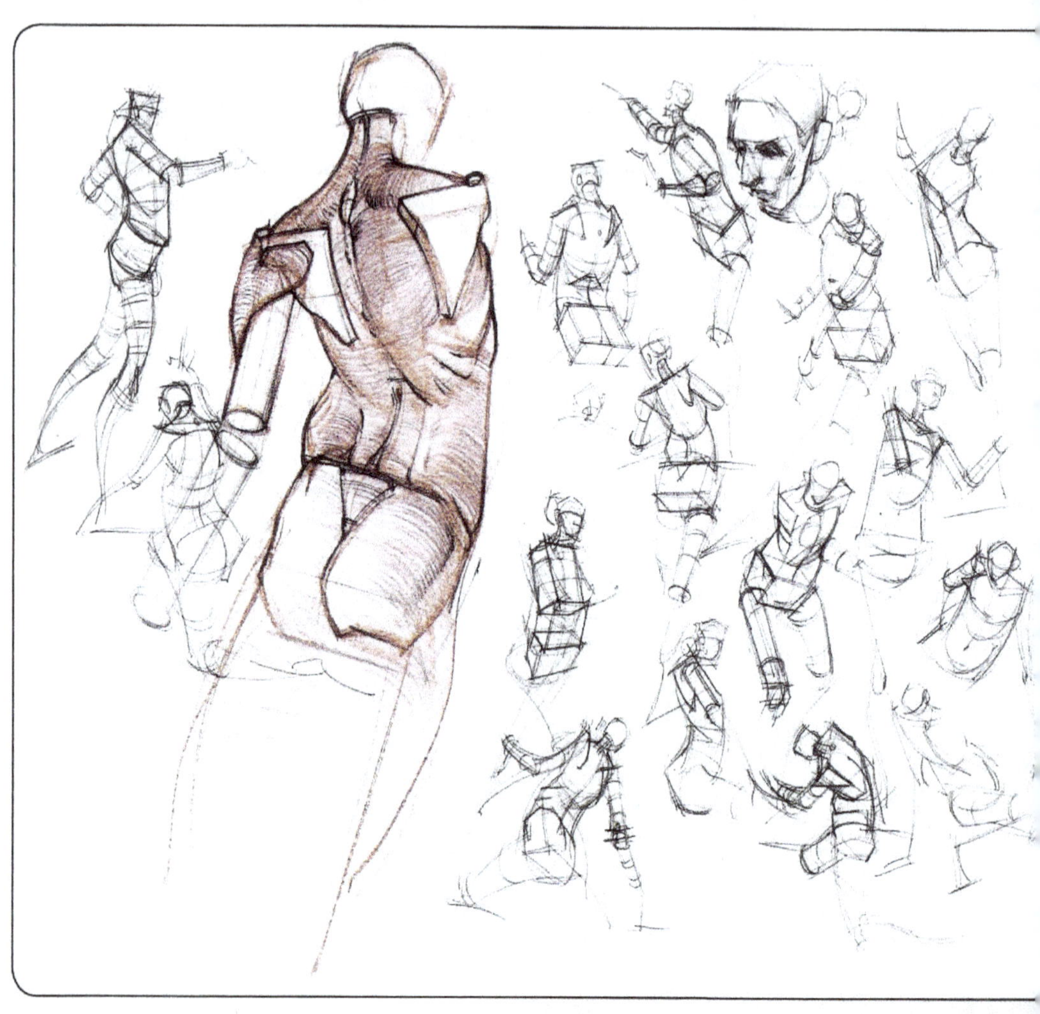

It is important to always build up these drawings in stages, and to always start with a gesture prior to this step.

LANDMARKS

Looking for the skeleton is the second stage in developing your figure drawings. This step is meant to give your drawings the look and feel of weight provided by the skeleton, as well as be a transitional stage in developing volume.

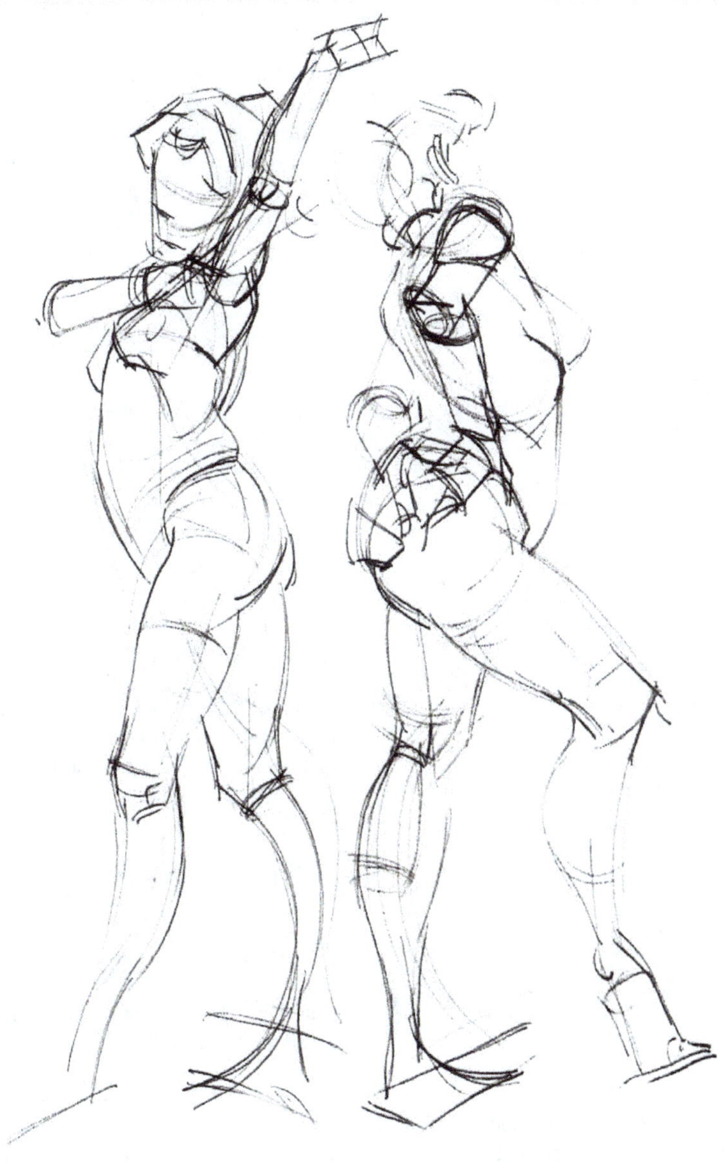

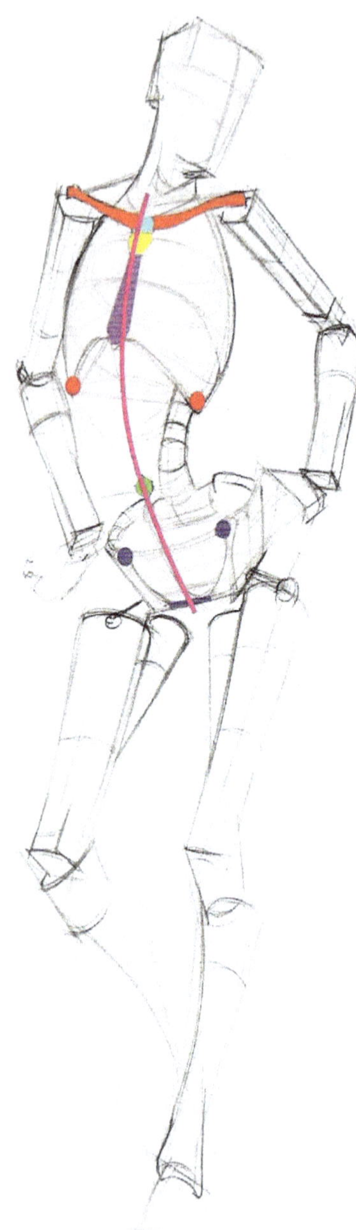

The skeleton can be used to look at the figure symmetrically. In a full-frontal or back view, a line down the center of the skeleton splits the figure into two equal halves (Examples A and E on the opposite page). Landmarks give us this line of symmetry.

The landmarks we need are color-keyed in the drawing to make their identification easier. All of these landmarks are areas of bone that visibly push through the flesh. For the time being, we are concerned with the landmarks of the rib cage and pelvis. Keep in mind that these are simplified designs based on knowledge of the skeleton.

The "pit" of the neck at the bottom of the throat.

The clavicles. Shape-wise, the clavicles resemble the handlebars of a bicycle, or a simplified bow. These two bones act as levers, enabling the arms to move around and away from the rib cage. The orientation of the clavicle will change depending on the position of the arm.

The manubrium. This is an area of bone that the two clavicles pivot from.

The sternum. This is a bone that fuses the bones of the rib cage together in the front. With the addition of the shape of the manubrium, these two bones resemble a neck tie.

The ends of the thoracic arch of the rib cage.

The belly button.

The ends of the iliac crest of the pelvis and the bottom of the pubic bone.

Remembering these areas helps to give your drawings the feel of an active skeleton. Observing the tilts across these points reveals the distribution of weight. These landmarks also help give the figure volume, perspective, and aid in the placement of anatomical shapes.

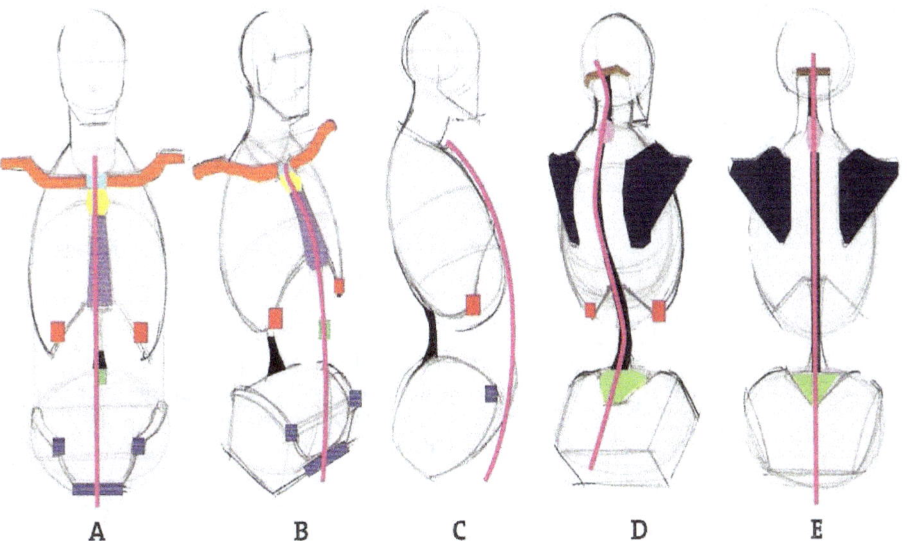

THE BACK

The drawing above shows the landmarks of the back. These include:

- The base of the cranial notch.

- The spine. The spine flows from the bottom of the cranial notch all the way down to the pelvis, ending at the sacrum.

- The sacrum.

- The seventh cervical vertebrae. This is a very pronounced area of bone towards the lower portion of the neck.

- The scapulae. The scapulae are two free-floating bones, which guide and aid the movement of the arms.

Examples B, C, and D show the positions of the landmarks as the figure starts to move through space. Notice that the line of symmetry on the three-quarter, side, and three-quarter back has remained, but now begins to favor, or move closer to, one side of the figure. Where the line of symmetry had previously divided the figure into two equal parts, now it helps to align the shape of the landmarks and show a turn. As the flat view (shown in the two drawings at top) becomes a slightly angled view, the rib cage and pelvis are shown with an interior corner. This interior corner will be used to show the perspective by allowing the rib cage and pelvis to be turned into a box.

The line of symmetry will always be a "C" curve when the rib cage and pelvis are facing the same direction. When the rib cage and pelvis are twisting, the line of symmetry will always be an "S" curve.

VOLUME

This diagram details the process of how to use your knowledge of the landmarks to show volume.

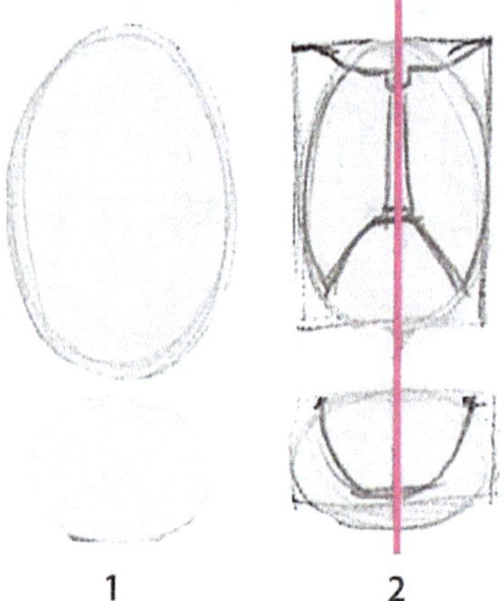

1   2

The first drawing shows the shape of the rib cage and pelvis in the gesture stage.

The second drawing shows how to begin developing the landmarks. This is a full-frontal pose, so all of the landmarks are shown symmetrically. The problem with this type of view is that it is very flat, emphasized in this drawing by the box drawn around the rib cage and pelvis. In making drawings that show form and volume, try to avoid focusing on shapes, such as the boxes, that only have two points (outside to outside).

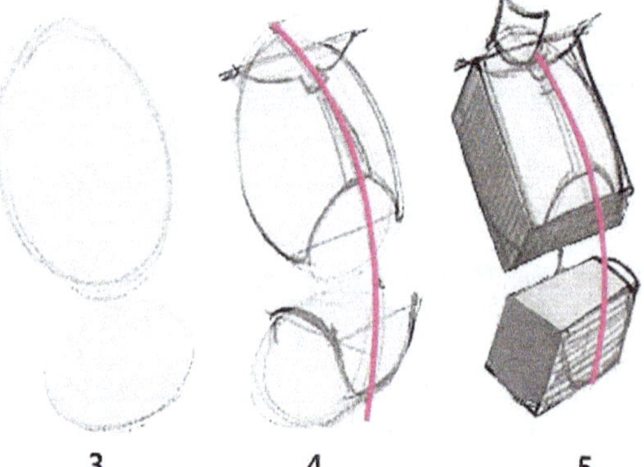

3   4   5

The third drawing shows the landmarks in a slightly-rotated view. Notice that the line of symmetry (found by placing the landmarks) favors the same side of the form as the direction that the figure is turning. For example, as the figure turns to the right, the line of symmetry moves closer to the right side of the form.

The fourth drawing shows the separation of front and side planes.

The fifth drawing uses landmarks to find the line of symmetry, keeping the front flat plane. A side plane has been added to reinforce the idea of the figure is turning in space. Note that all of the planes are based off of the landmarks. For example, the front plane of the figure is based off of four points: the ends of each clavicle and the two ends of the thoracic arch.

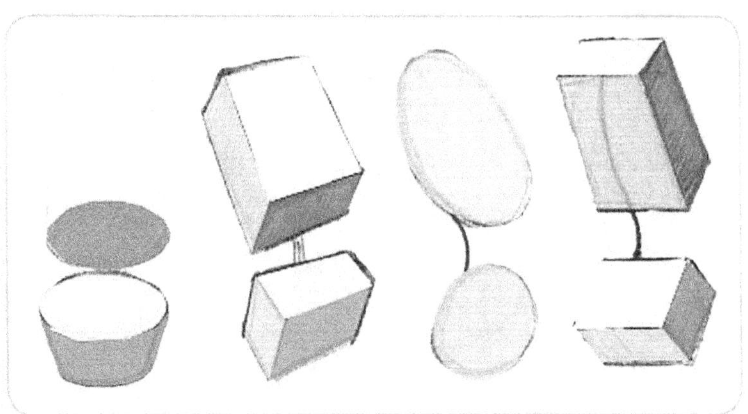

This illustration shows the perspectives you will want to emphasize based on the view.

On the left, this front view shows that the rib cage should always be pushed back in space to show the viewer the underside of the rib cage. To balance the rib cage, and reflect the spine's influence, the pelvis is seen from above.

The three-quarter view (second from the left) shows the addition of the side plane.

The side view (third from the left) shows the rib cage balancing on the pelvis.

The back view shows the reverse of the front. When drawing the back, always show the rib cage from above and the pelvis from beneath.

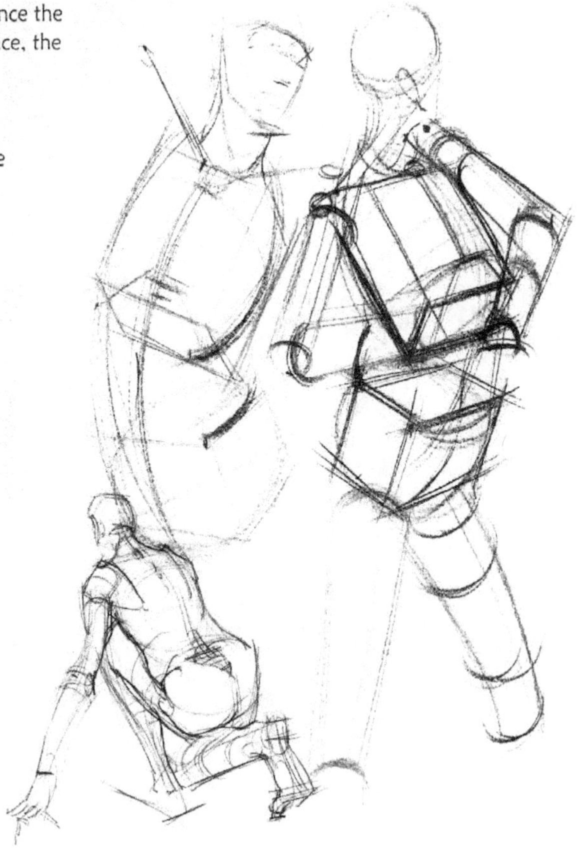

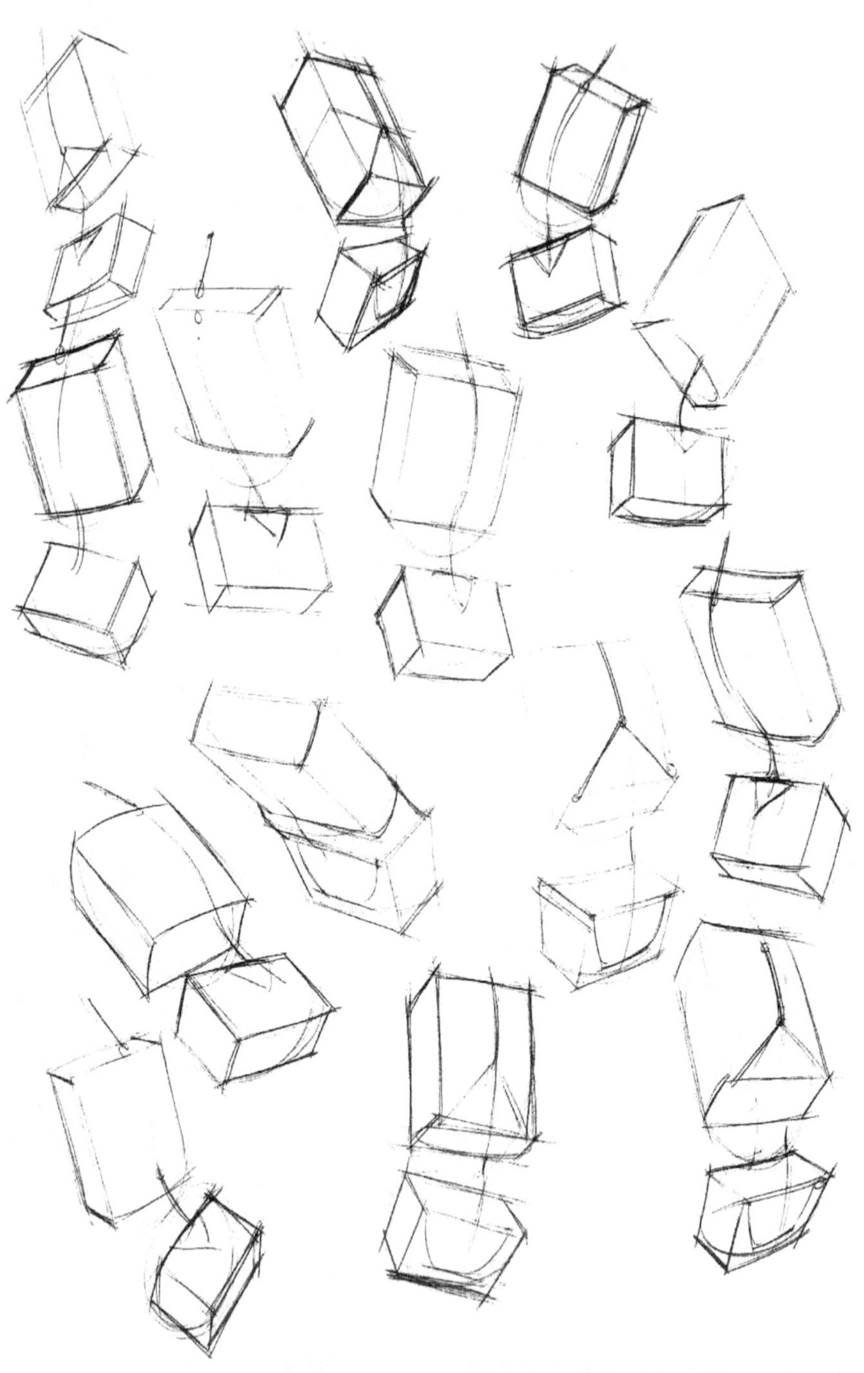

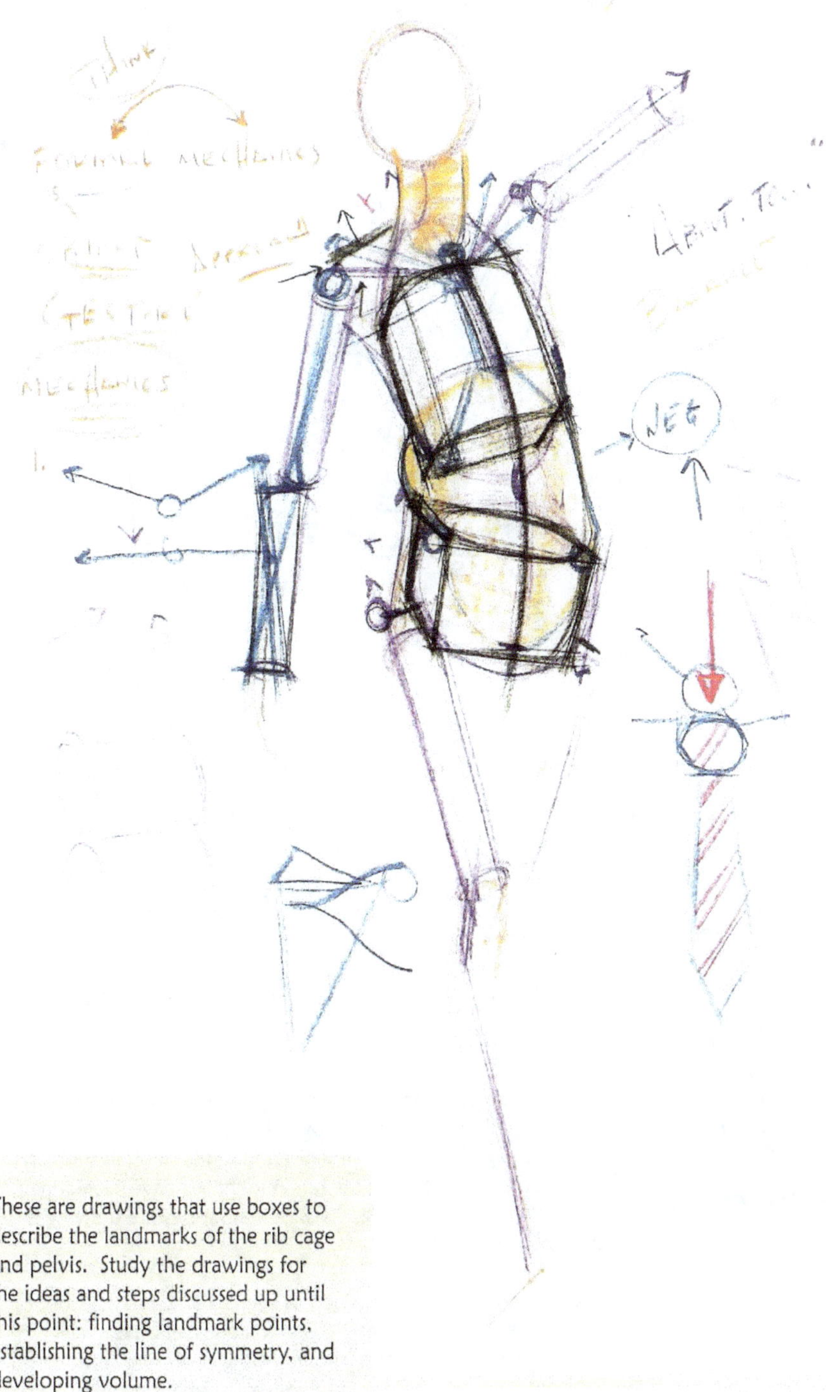

These are drawings that use boxes to describe the landmarks of the rib cage and pelvis. Study the drawings for the ideas and steps discussed up until this point: finding landmark points, establishing the line of symmetry, and developing volume.

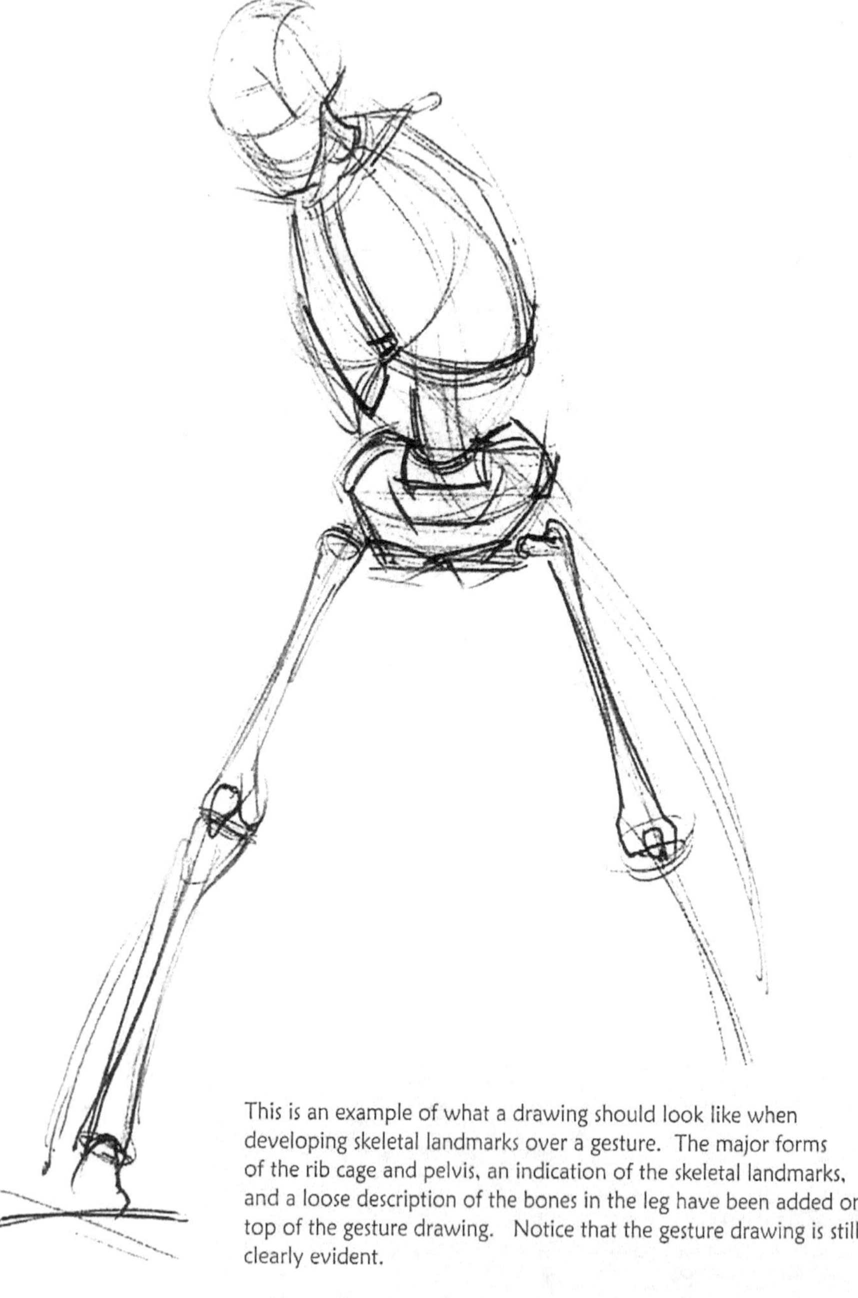

This is an example of what a drawing should look like when
developing skeletal landmarks over a gesture. The major forms
of the rib cage and pelvis, an indication of the skeletal landmarks,
and a loose description of the bones in the leg have been added on
top of the gesture drawing.   Notice that the gesture drawing is still
clearly evident.

The main focus during this step has been to preserve the story of
the gesture and to begin to indicate the distribution of weight. In
your drawings, the skeleton should help give a more concrete
explanation of the mechanics of weight distribution.

WEIGHT DISTRIBUTION

It is easier to determine how to handle areas of intersection by thinking of the figure mechanically. The drawing in the upper right is a simplified sketch of the distribution of weight-distribution and balance. Sometimes doing a small sketch like this before you begin your drawing can help solve problems at a very early stage.

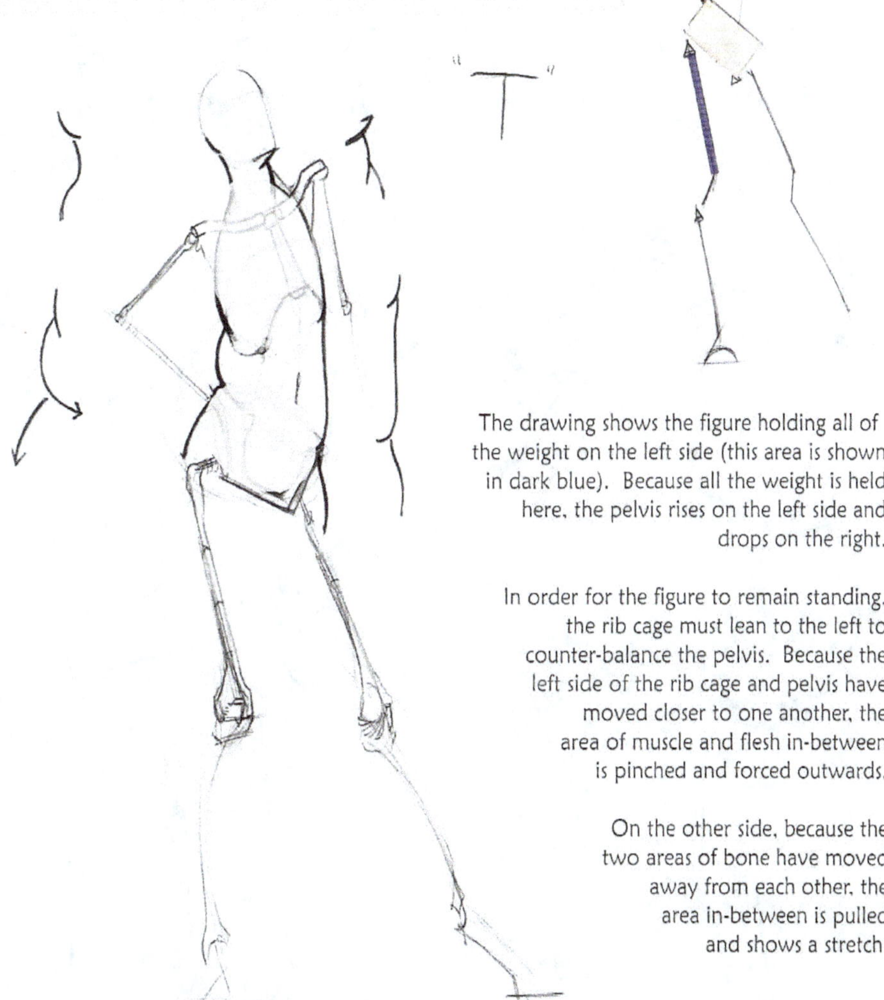

The drawing shows the figure holding all of the weight on the left side (this area is shown in dark blue). Because all the weight is held here, the pelvis rises on the left side and drops on the right.

In order for the figure to remain standing, the rib cage must lean to the left to counter-balance the pelvis. Because the left side of the rib cage and pelvis have moved closer to one another, the area of muscle and flesh in-between is pinched and forced outwards.

On the other side, because the two areas of bone have moved away from each other, the area in-between is pulled and shows a stretch.

When showing a pinch or a stretch, it is extremely important to develop a sense of space by using overlaps. This tool for showing recession and depth is commonly referred to as a "T" overlap. A "T" overlap should clearly show one line moving behind or in front of another line, much like the letter T. The illustration above shows how the "T" overlap is used when dealing with a pinch and stretch. This will be explained in more detail in the Connections section of this chapter.

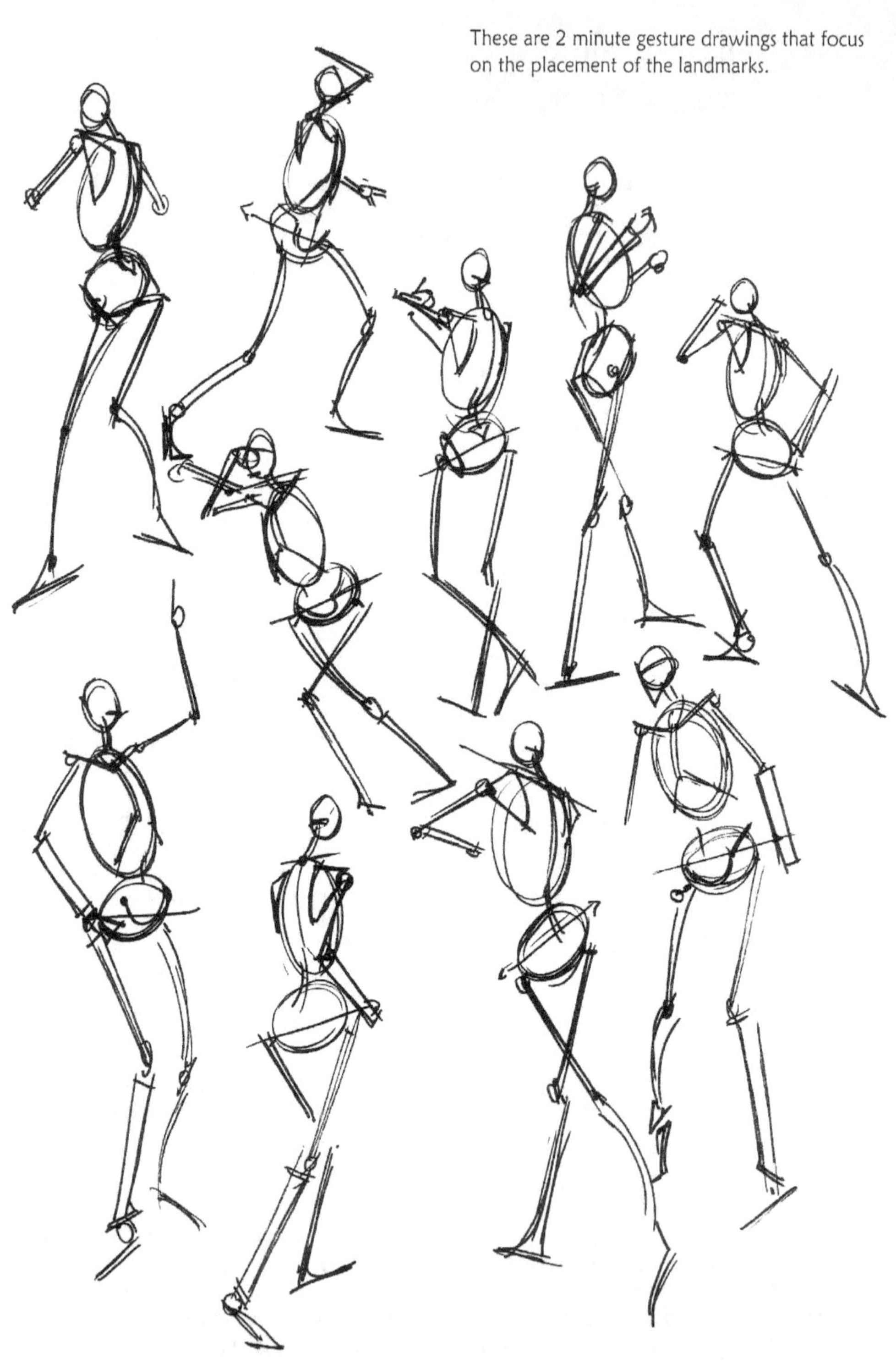

These are 2 minute gesture drawings that focus
on the placement of the landmarks.

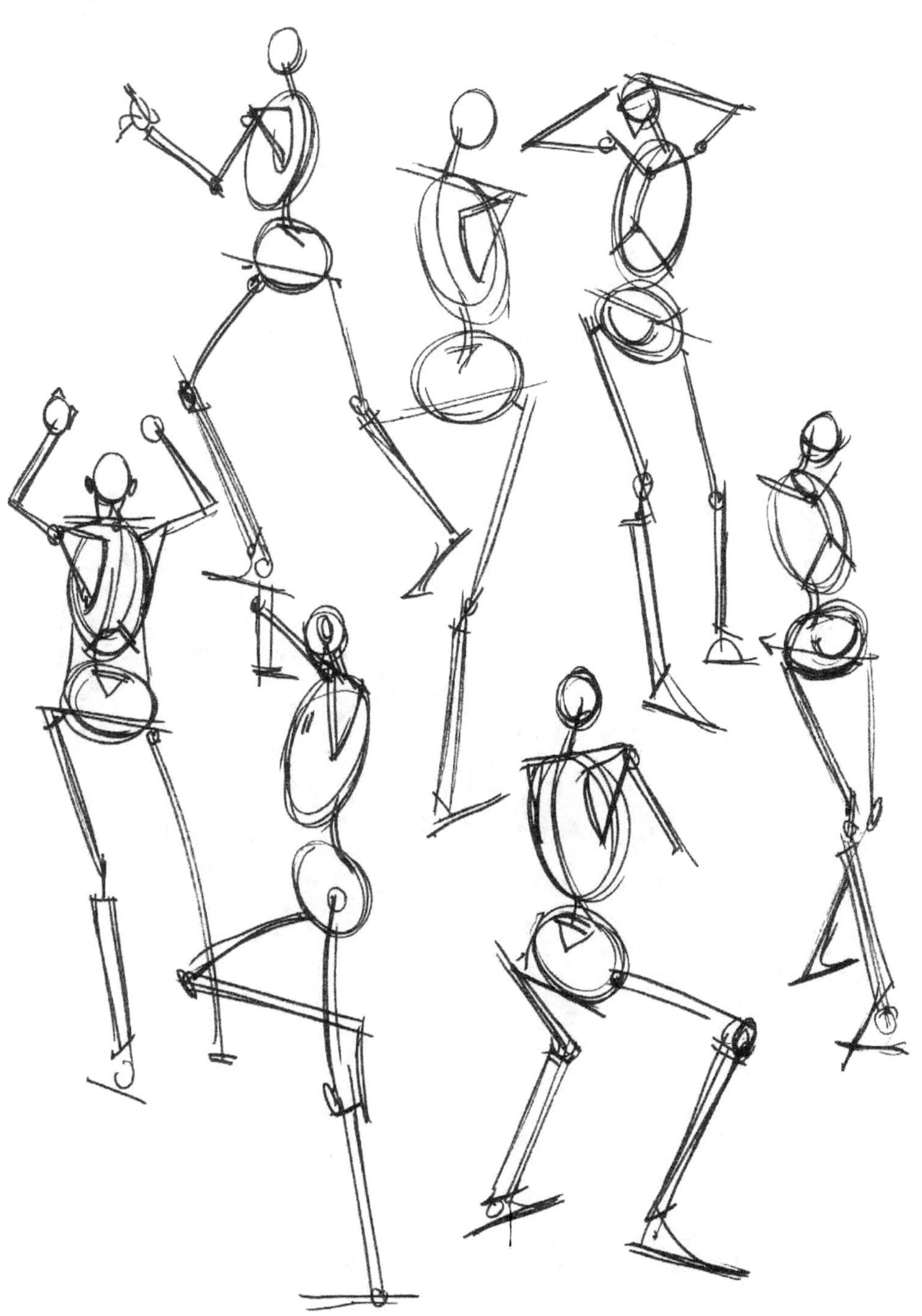

When showing depth using a "T" overlap, you
need to be especially clear about what form goes
in front or behind another. The diagram to the
right shows a demonstration of this idea, using
spheres. Notice that every time a form comes in
contact with one that is either in front or behind
it, they meet at a "T".

If you were drawing a figure foreshortened
or in a reclining pose, this would be a tool to
emphasize in a drawing. Because of the large
amount of information needed in describing a
figure, the "T" overlap will constantly be used to
help organize the major and minor forms.

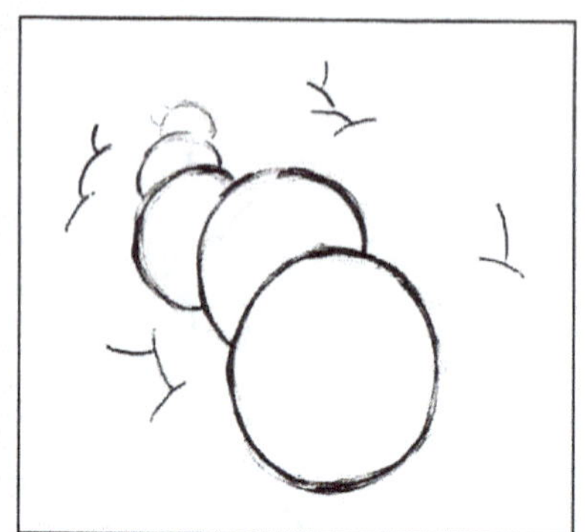

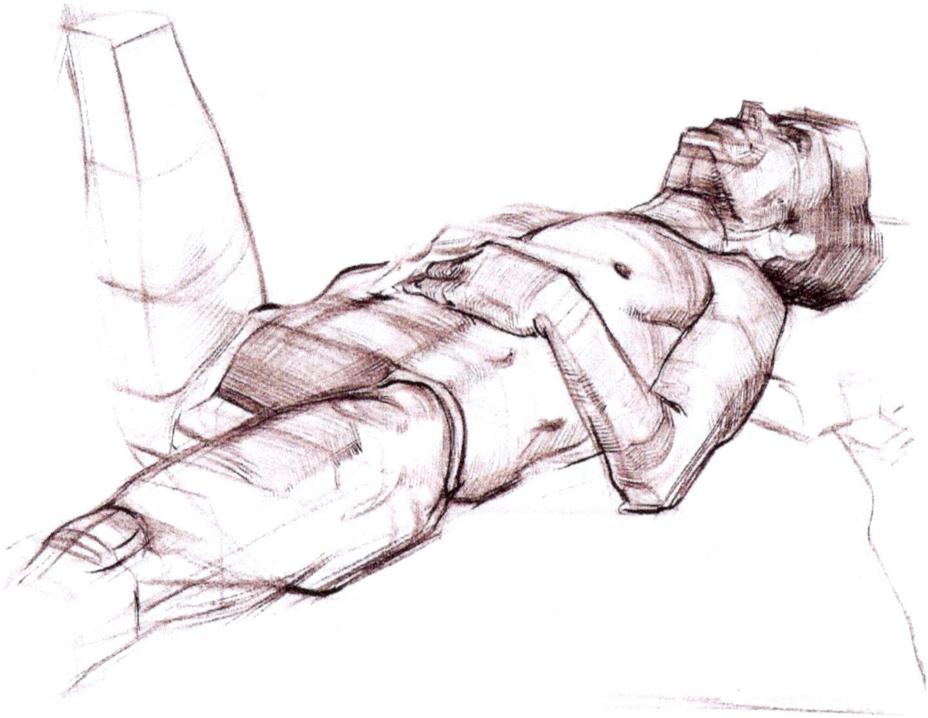

The "T" is used in many types of images and is a main tool for showing recession and depth
in landscapes. Study how the "T" overlaps are used to place one form in front of another
causing the eye to experience a sense of depth.

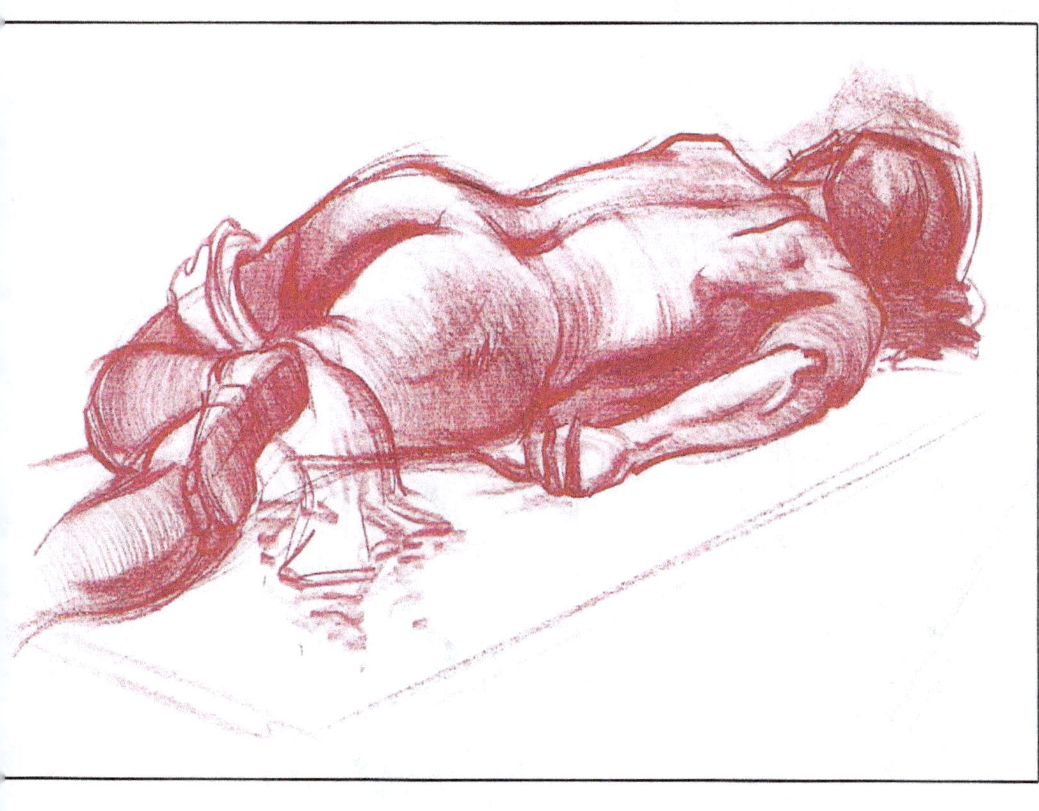

CONNECTIONS

Having established the skeletal landmarks and the figure in perspective, the next step deals with the connection and design of these forms.

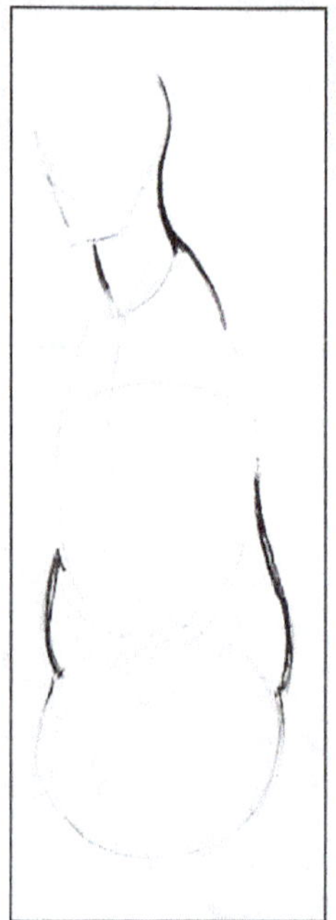

Notice, on the example to the left, that the head, rib cage, and pelvis are all connected with either an "S" or "C" curve. This use of curves will help tie the forms together consistently with the gesture drawing.

By limiting the line use, you design a fluid relationship to the page based on the negative space left behind. By only using these curves, you will have a more controlled and simplified understanding of how to describe the movement of the three parts against one another.

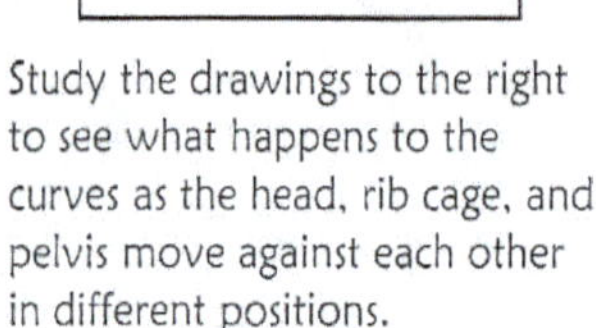

Study the drawings to the right to see what happens to the curves as the head, rib cage, and pelvis move against each other in different positions.

Notice in both drawings that depending on the movement, the connection between the rib cage and pelvis is either stretching or pinching.

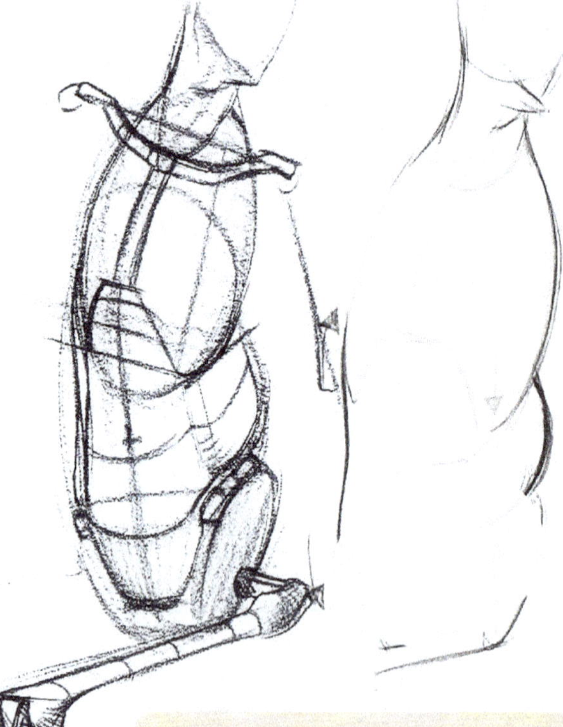

Connections are important areas in the figure to emphasize curves that play off of the harder shapes of bone. This maintains and exaggerates ideas of natural balance discussed in the gesture chapter.

This stretch or pinch is still a variation on a "C" or "S" curve.

Observe how the "C" curve becomes suggestive of the flesh between the two areas of bone compressing, and the "S" more clearly communicates an elongation or stretch between the two areas of bone as they move away from one another.

This use of curve will take place at every intersection on the body - the neck, midsection, between the pelvis and legs and the rib cage into the arms.

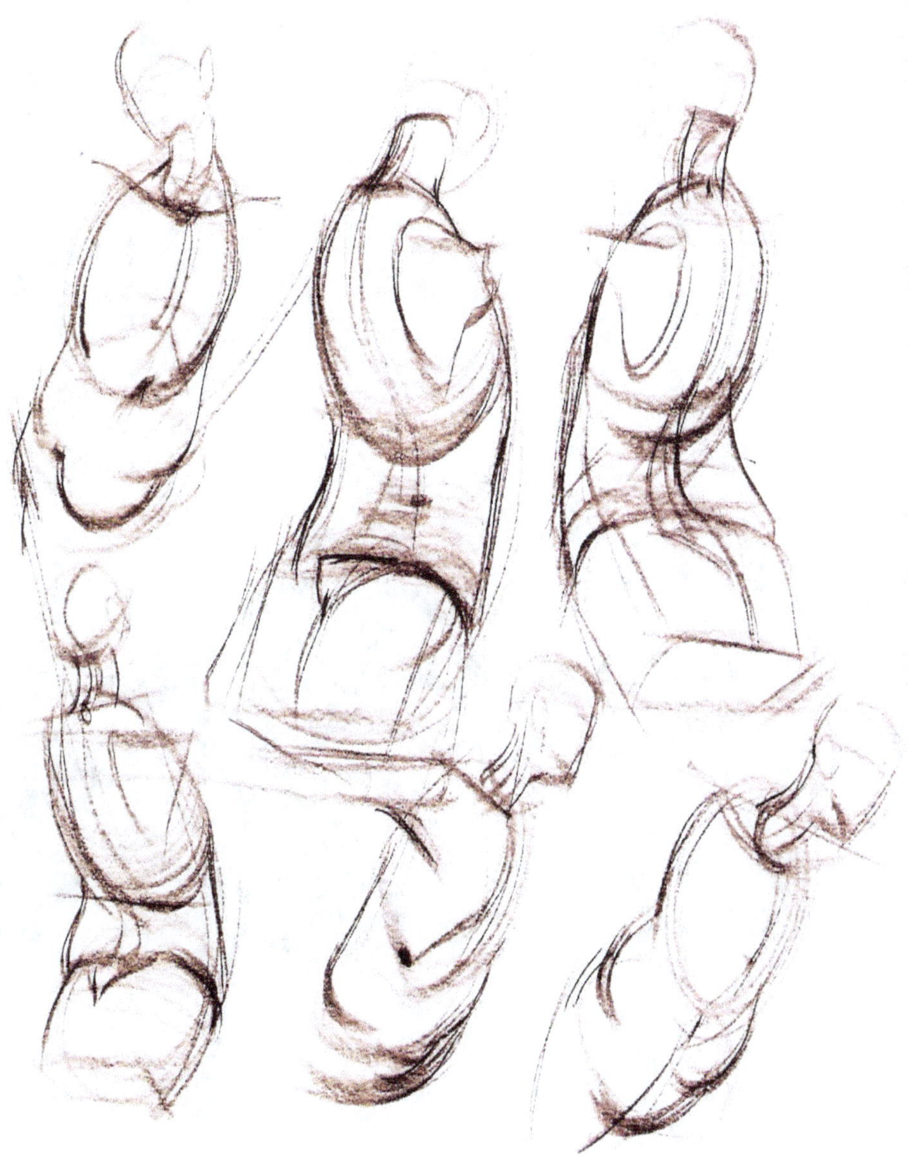

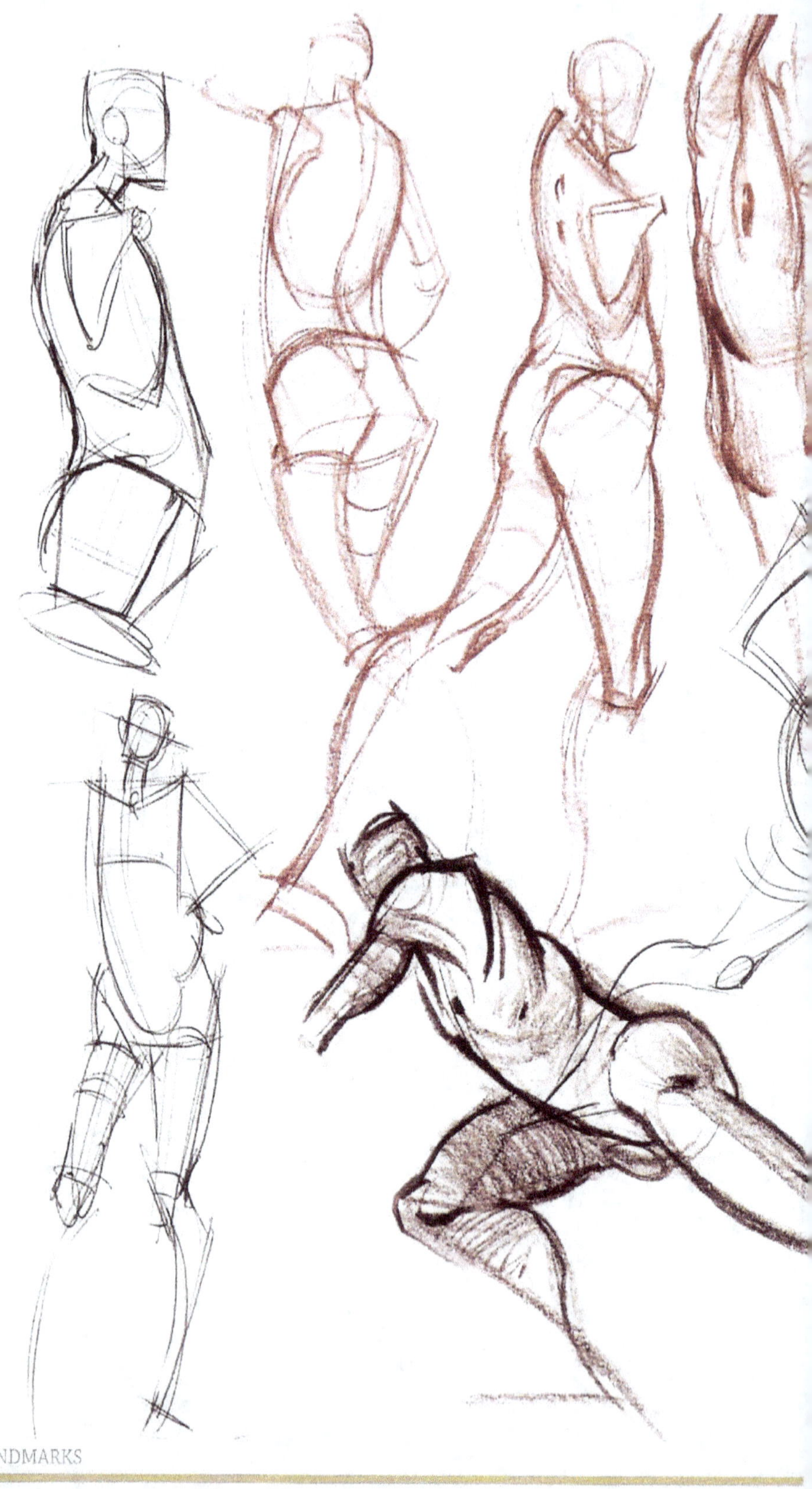

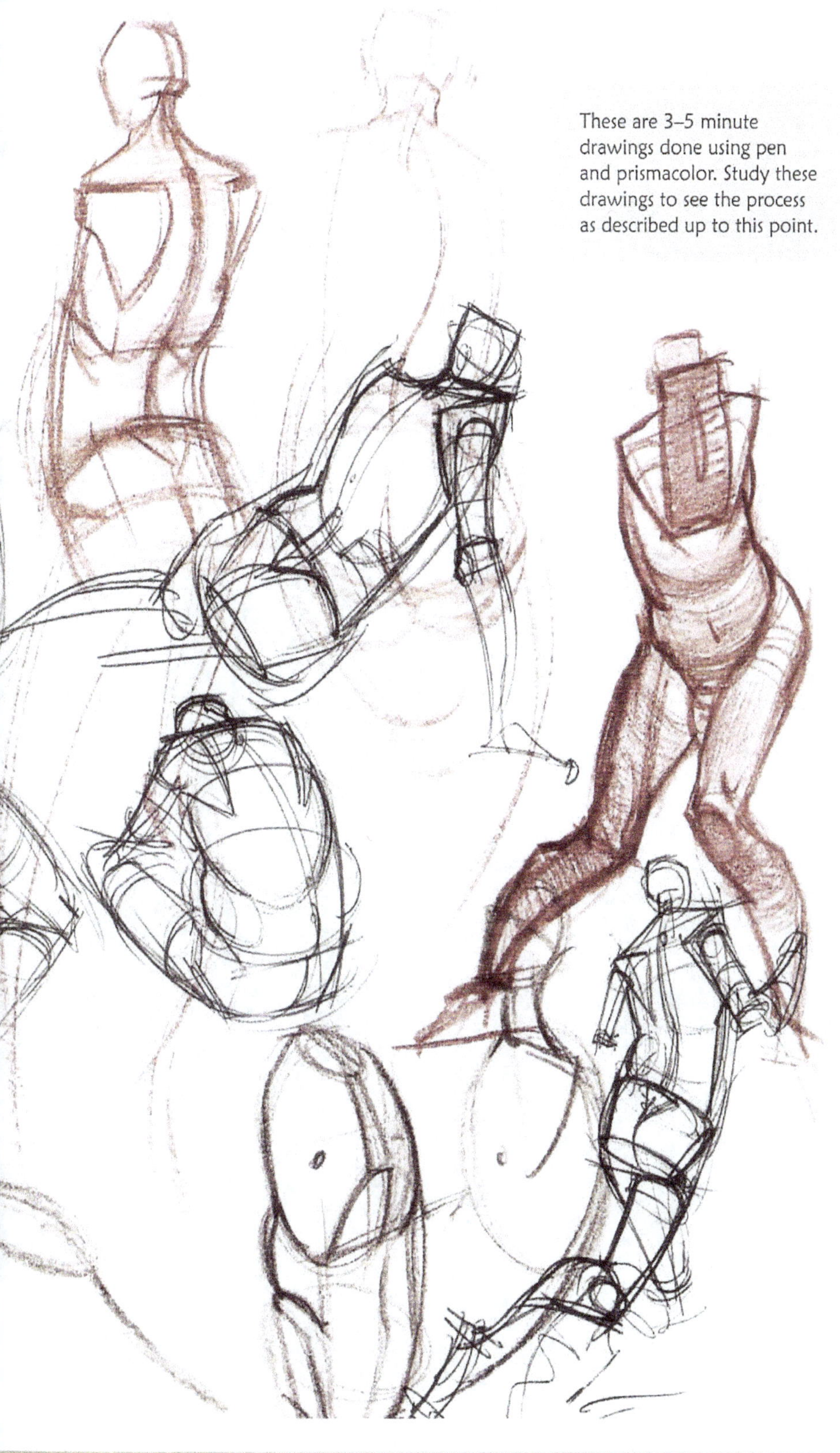

These are 3–5 minute drawings done using pen and prismacolor. Study these drawings to see the process as described up to this point.

ARMS AND LEGS

Having established the major masses of the rib cage and pelvis, the next step involves defining the arms and legs in their placement and perspectives.

One of the main goals when working through a construction-based study of a figure is to keep all the forms balanced as shapes and as volumes.

Remember, developing a construction drawing is not necessarily the desired finished product. Rather, this is one invaluable stage of knowledge (among many others) that has to be integrated into any image.

On the arm, represents the head of the humerus, and on the legs, the top of the great trochanter of the femur.

Represents the elbow (end of the humerus) and the radius and ulna.

Represents the knee and ankle.

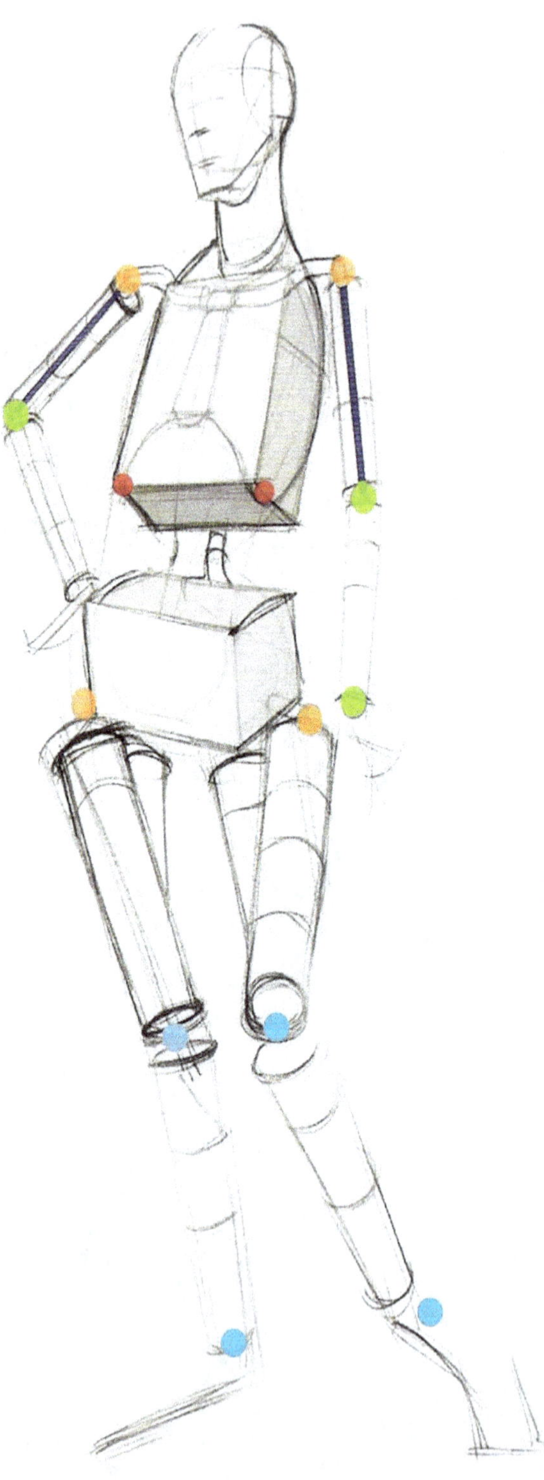

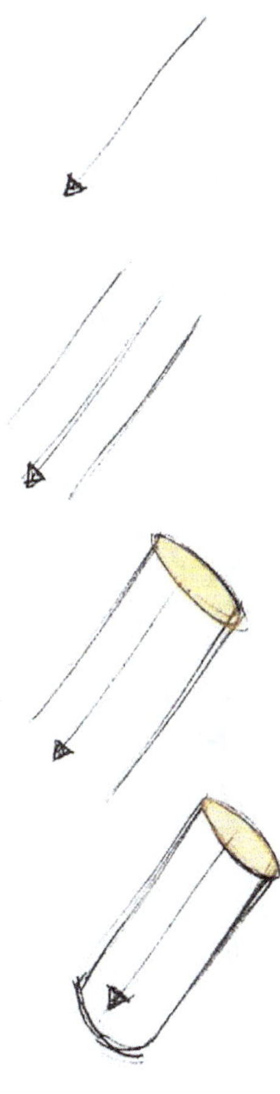

When developing the arms, one of the most difficult steps is showing the correct perspective. The four stacked drawings to the left show the process for developing the arm as a cylinder.

The first step is to use one line to place the direction and length of the arm.

In the second step, draw two lines parallel to the first to determine the width of the cylinder.

The third step is to give the cylinder a perspective based on the type of ellipse drawn.

The cylinder with the orange dot above it is an example of using the ellipse to show the view from above the object.

The cylinder with the green dot shows the view from underneath. When choosing an ellipse for the cylinder, alter the minor axis without affecting the width.

This illustration demonstrates how to turn a cylinder in space by changing the size of the ellipse. The fourth step (in the bottom right corner) completes the cylinder by putting a cap on the open end. This cap should be an exact copy of the arc of the ellipse from the other side.

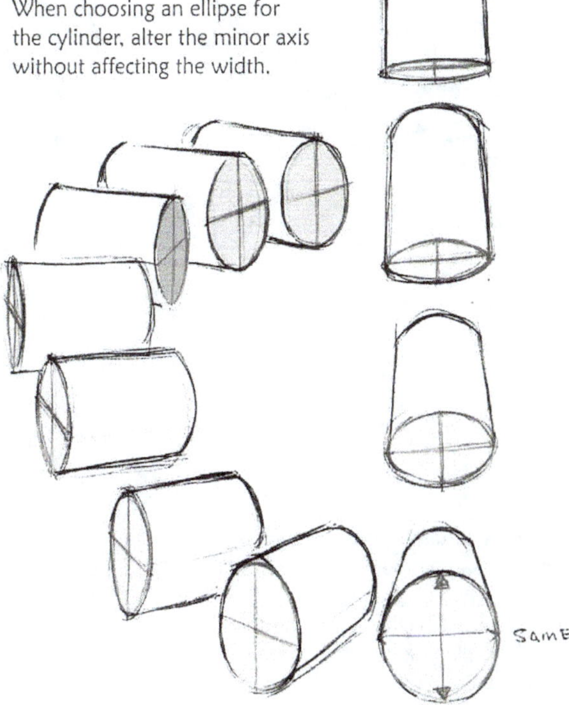

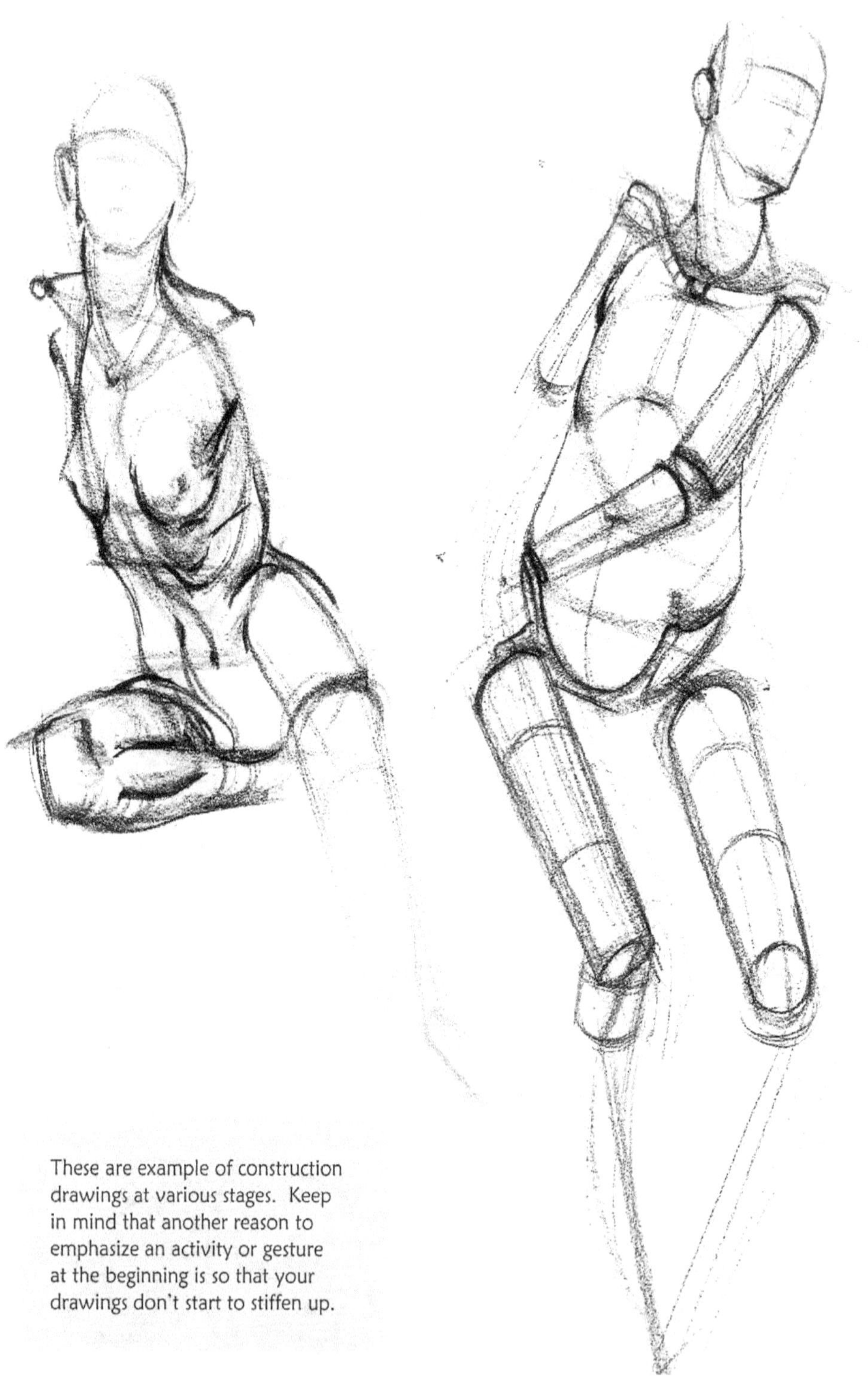

These are example of construction
drawings at various stages. Keep
in mind that another reason to
emphasize an activity or gesture
at the beginning is so that your
drawings don't start to stiffen up.

While including the skeletal structure
and the geometric volumes, the
drawings may run the risk of becoming
rigid. In every stage, you want to be
reinforcing the story or gesture. The
landmark stage should be a re-telling of
the gesture, describing the pose in space
and bringing more realism through the
skeleton.

Look for the simplest explanation of
how to connect parts. A "C" or "S" is
used at every transitional area on the
figure to secure a connection as well as
indicate the distribution of weight.

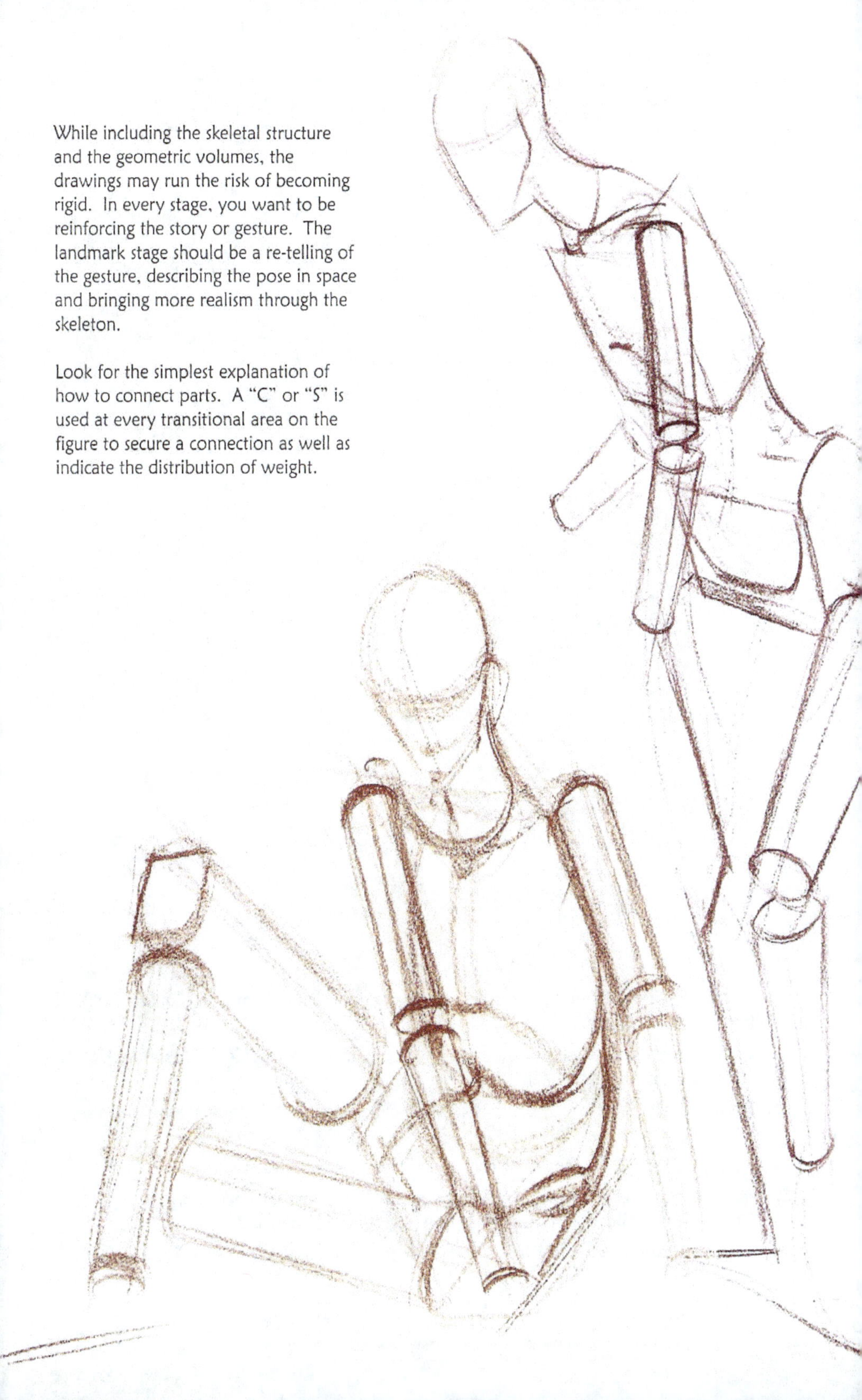

FORMS AND CONNECTIONS

These six exercises represent the basic training principles that allow you to render, invent, or conceive of a figure in space. At every point in the process of drawing, you should be involved in one of the six skills to a varying degree. The goal in this process is not to literally draw a nose, mouth, contour line, etc., but to be engaged with the underlying principles that develop a sense of illusion.

Do these exercises when daydreaming, at work, in class, or as warm-ups.

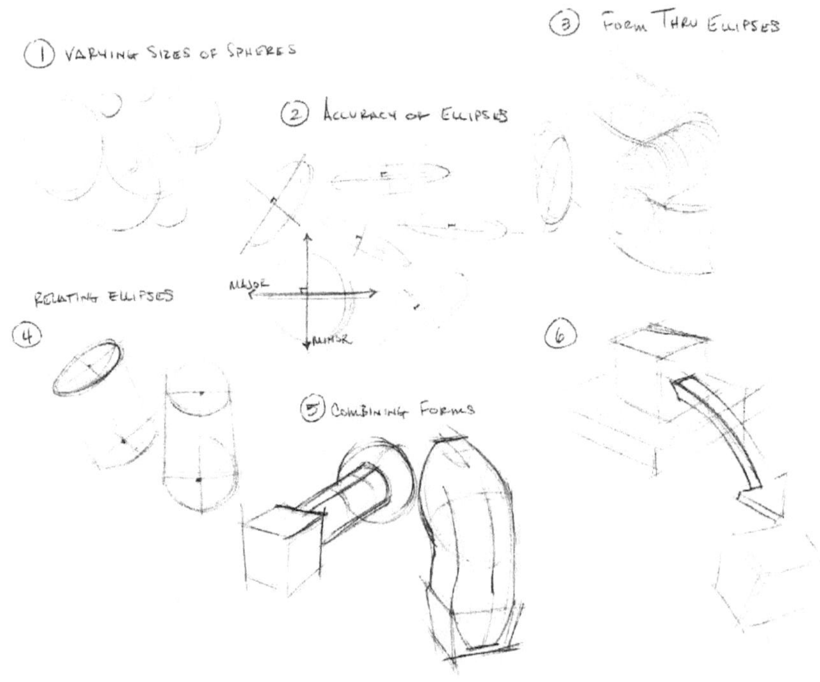

1. Practice drawing spheres using one line, in varying sizes. Draw from your shoulder, not from your wrist.

2. Draw different sizes of ellipses using one line (again, from the shoulder), double-checking its accuracy by adding in the major and minor axes.

3. Practice thinking spatially by drawing only wrapping lines. Imagine you are drawing a garden hose, snake, or slinky.

4. Build on Exercise 2 by pairing the ellipses, and developing cylinders.

5. Take the sphere, box, and cylinder as departure points for more complex, organic forms The entire figure builds out of spheres, cylinders, boxes, ellipses, and curves.

6. Imagine a box falling off of a ledge, and rotate the planes to show a sense of fall and turn.

Before moving on to the development of anatomy, it is paramount to understand how to keep a consistent feeling of form. This stage will emphasize the importance and technical principles needed to continue to emphasize the underlying shapes. The box, sphere, and cylinder still have to be reinforced with the use of line and connection.

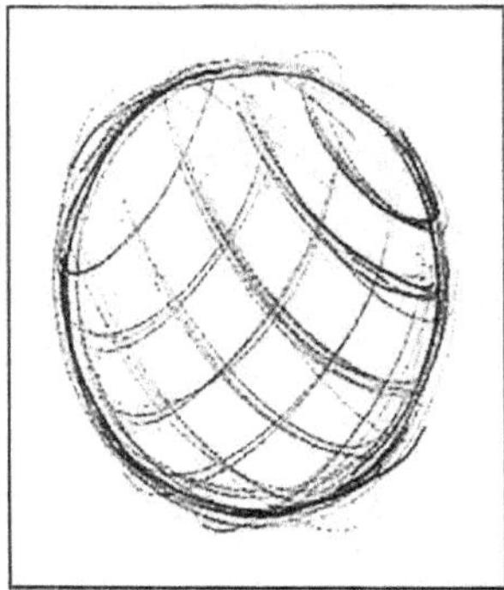

Whenever adding to or describing a form built from a box, cylinder, or sphere, it is important to never use line in a way that contradicts that form. Notice how the lines around the volumes to the left always work around the volume. These lines travel across the surface as if they had to literally walk across them. This is one of the best ways to convince your viewers that what they are seeing occupies space.

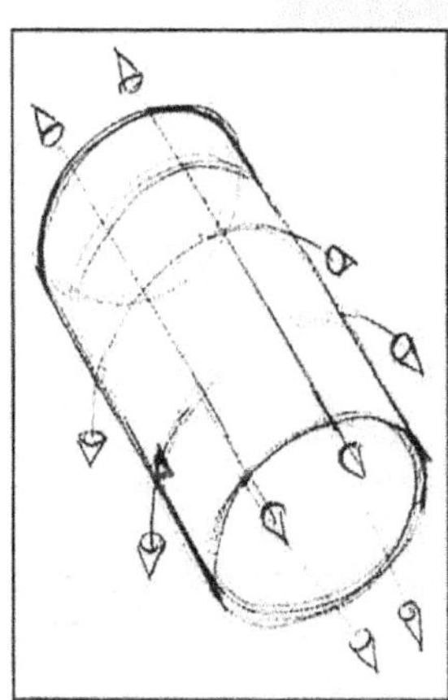

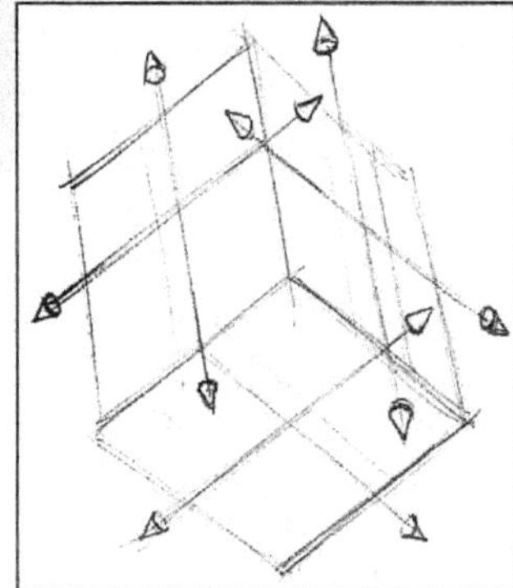

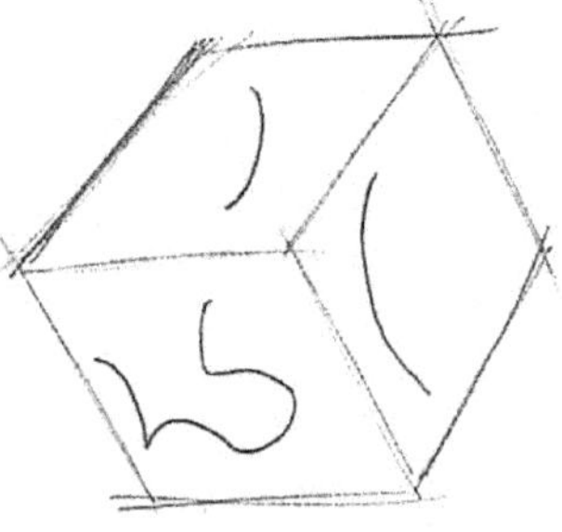

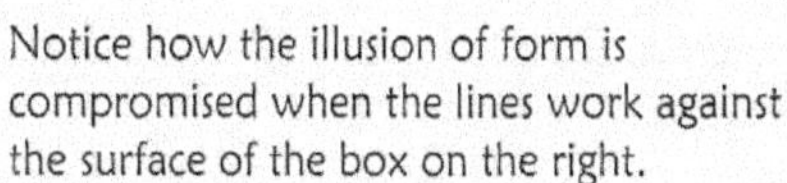

Notice how the illusion of form is compromised when the lines work against the surface of the box on the right.

The key to combining these perspective forms is to imagine one volume being pushed into the other while using the line work around the forms in order to integrate their surfaces.

In the two examples on this page, notice how the forms feel as though they are joined as one. A more organic form like the one of the cylinder and sphere could be used to represent the shoulder or fingers.

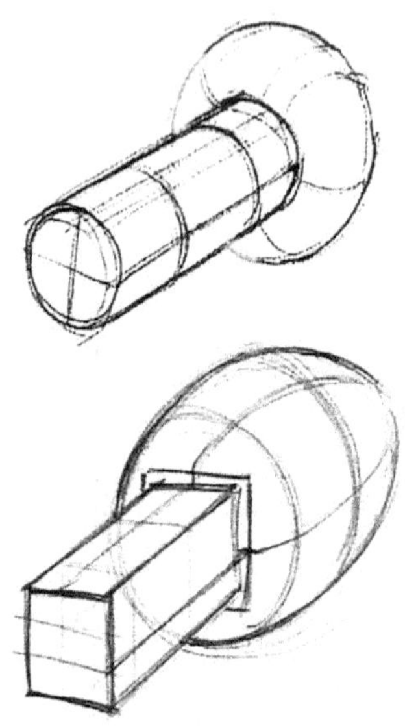

The form of the ovoid meeting the box follows the same laws and could easily be used to describe the upper leg ending in the knee. The development of organic forms is an incredibly valuable exercise and one only limited by your imagination.

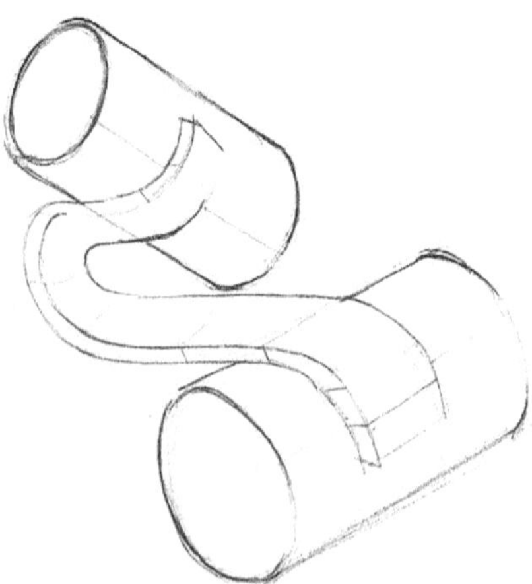

The example on this page shows a perspective situation similar to anatomy connecting disparate parts. With the cylinders moving in separate perspective directions, the goal of the elongated box is two attach on their surfaces in a way that describes the larger perspective.

Notice how the elongated box conforms to the larger perspective and overlaps onto and around the surface of the cylinders.

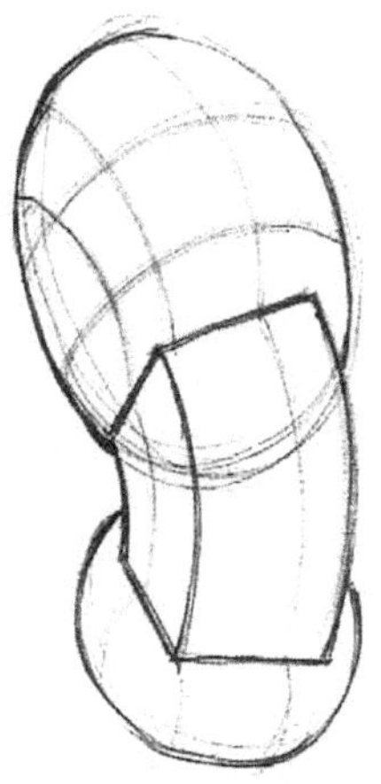

Study these examples
to see how multiple
objects have been used
to describe the organic
qualities of the rib cage,
midsection, and pelvis.

These examples illustrate
the more abstract
elements integral to the
perspectival showing
of anatomical shapes
resting on top of larger
perspective volumes.

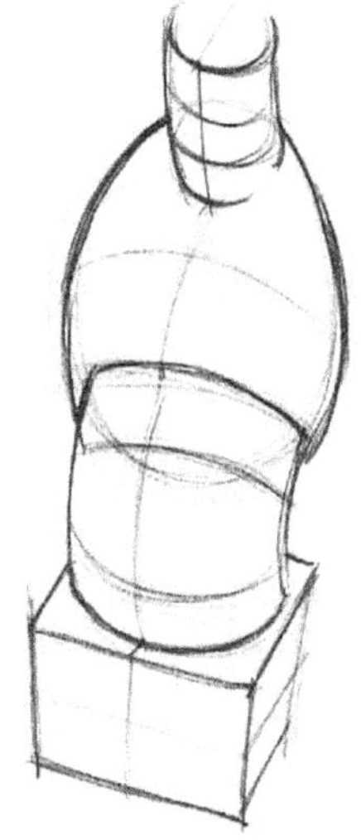

To summarize, there are three technical ways of
connecting and adding forms:

The first is the use of "T" overlaps to emphasize a form
moving in front or behind another form.

Second, adjusting the shape of a form to give it
continuity with the larger perspective of the form it
rests upon.

Lastly, using a intersecting or transitional form, which
transitions the perspective from one volume to another
by starting on one perspective and ending on the other.

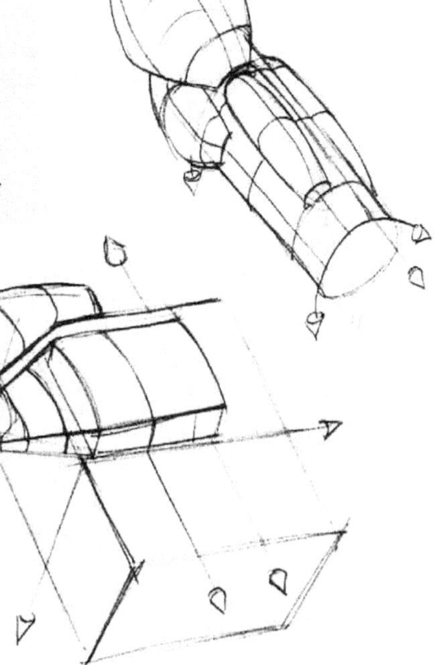

SUMMARY

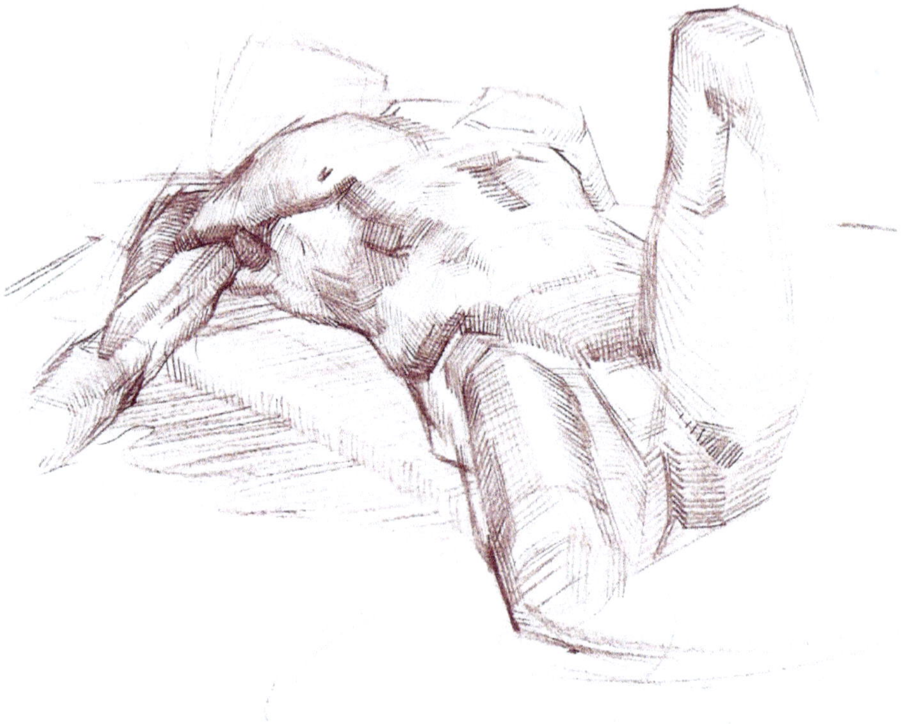

The preceding chapter can be summarized by thinking through these four steps:

1. VOLUMES

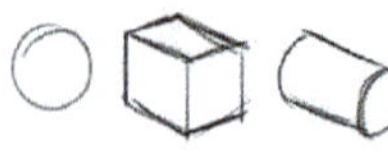

The entire figure will be constructed out of spheres, boxes, and cylinders — get comfortable with these volumes and become as familiar as possible with drawing them.

2. MAINTAINING THE VOLUME WITH LINE

If the sphere, box, and cylinder are your building blocks, you will never want to work against the effects they develop. In order to support these effects, become sensitive to how line enhances or destroys these surfaces (wrapping lines).

3. COMBINING VOLUMES

Combine the sphere, box, and cylinder together in order to develop organic forms that can more easily approximate the forms of the figure.

4. ADDING TO & CONNECTING VOLUMES

Adding and connecting volumes involves maintaining the first three steps while integrating the complex volume into others.

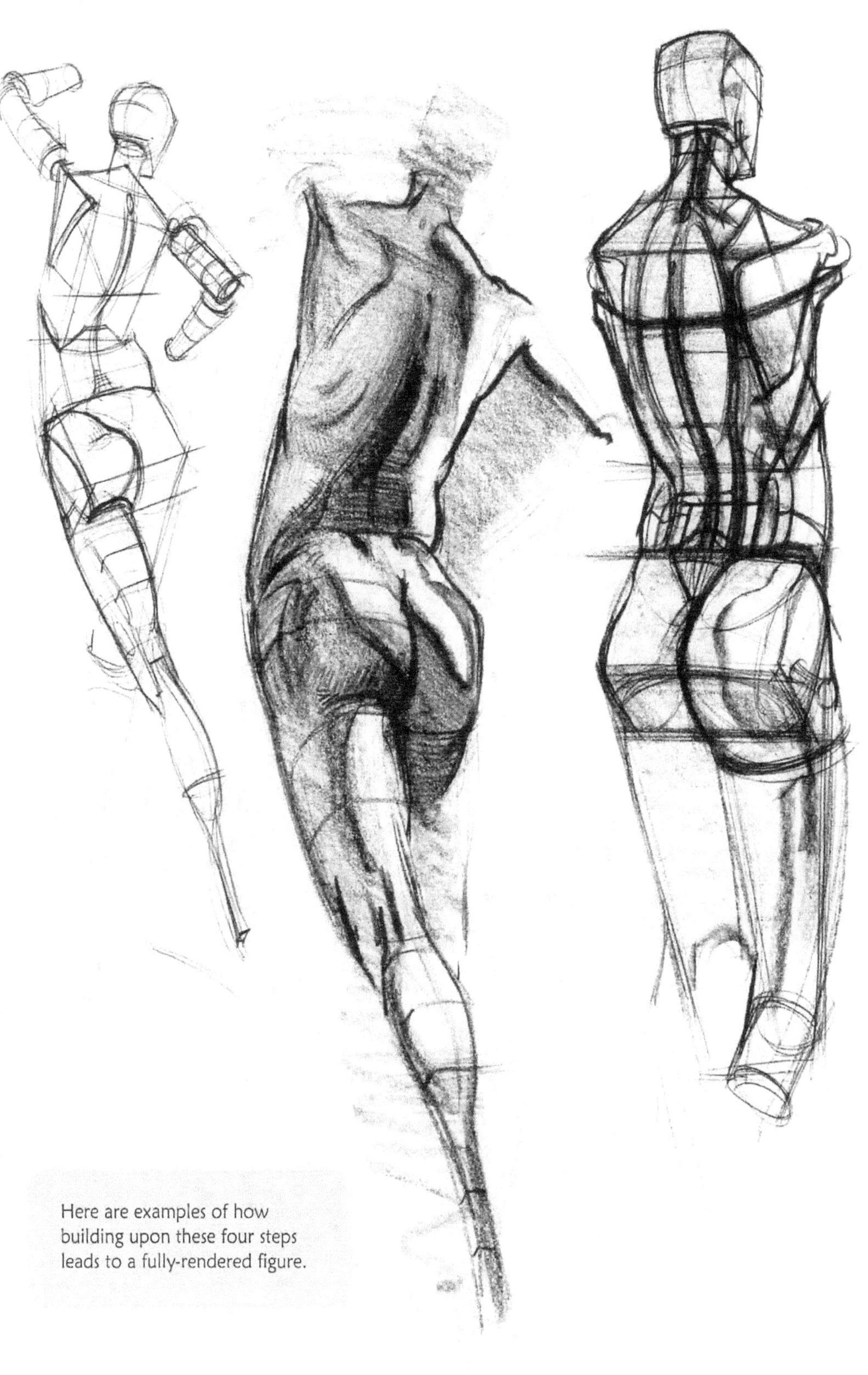

Here are examples of how
building upon these four steps
leads to a fully-rendered figure.

HEAD DRAWING

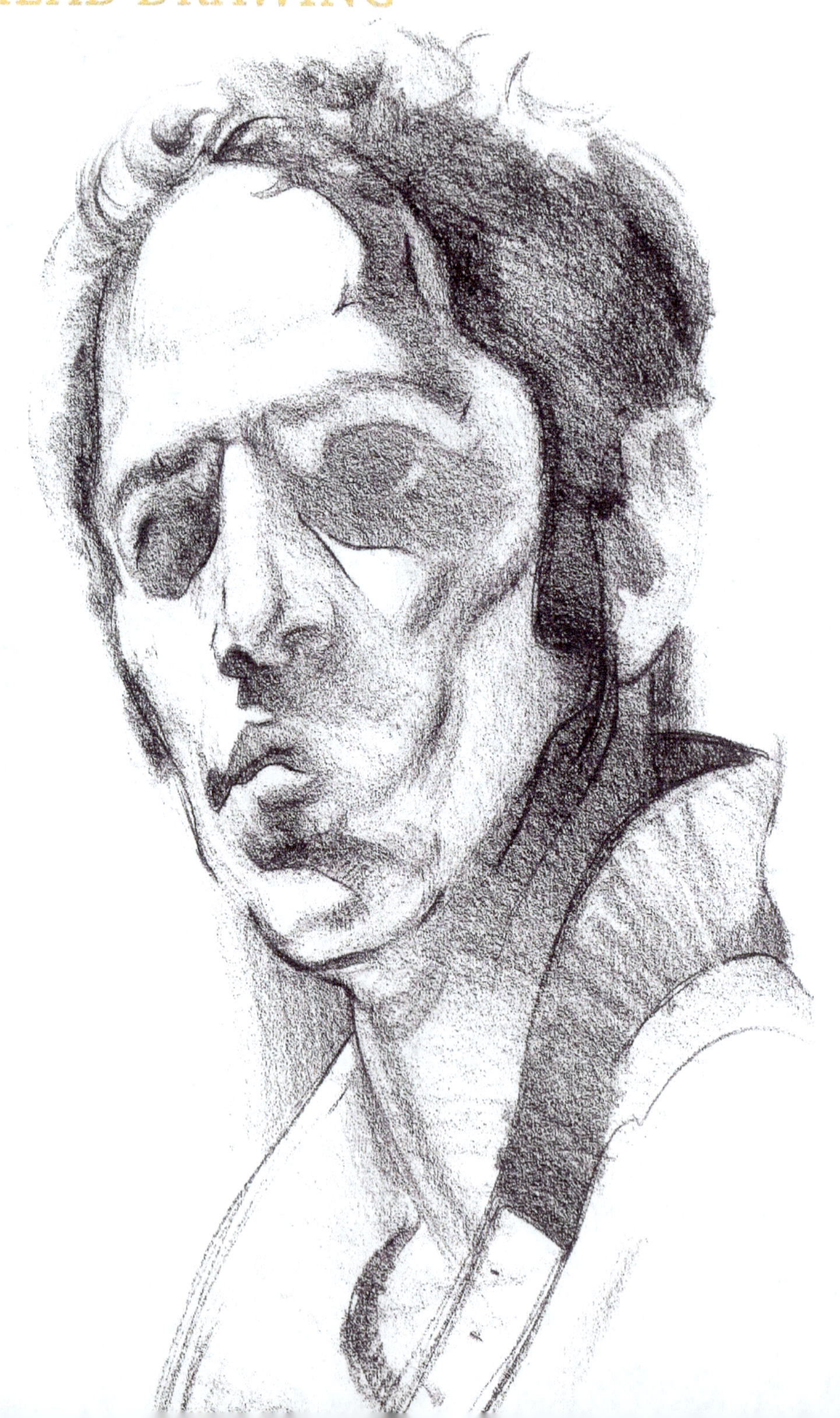

The process for drawing the head is based on development and form. The goal is to understand all the parts of the skull in order to create a believable, volumetric model that you can use to work from out of your head.

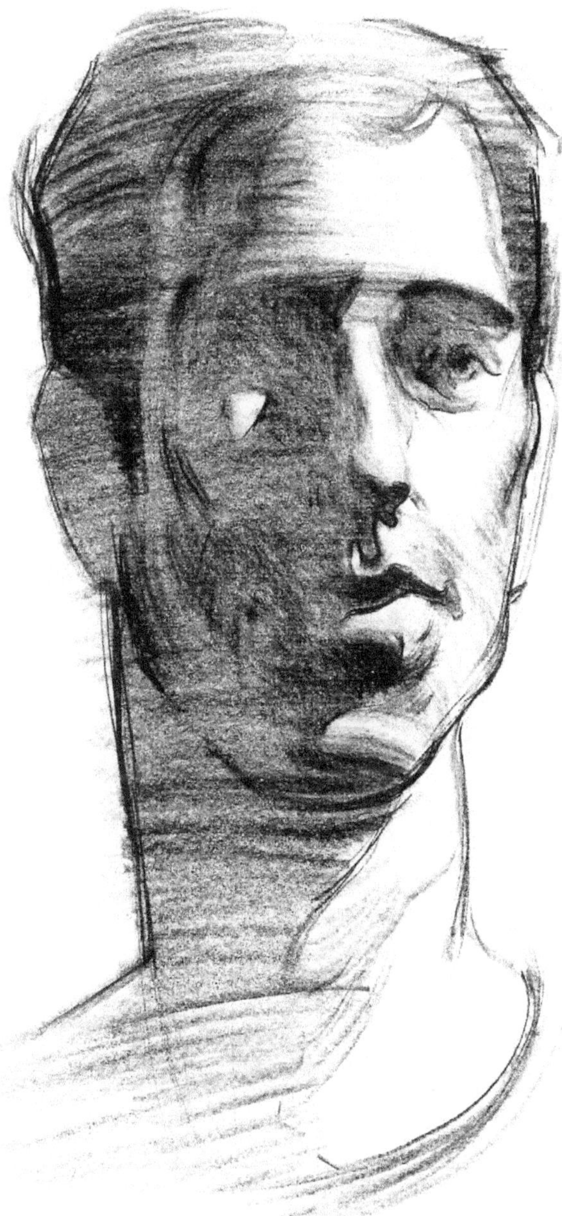

The parts of the skull are first broken down into the most basic forms, and then reassembled — similar to a sculptor progressively building up forms with clay. In this chapter, we will cover a generic understanding of the skull. All of the planes and corners, once understood, can then be manipulated exhaustively to present any character type.

Having learned this process as a foundation for a rendered drawing, head invention, or exercise in plane separation, you should use it for exaggeration or expressive intent. At every stage in this process, manipulations can be made to develop the skull of an animal, creature, character, or just different types of people (individuals, races, etc.).

The sphere represents the cranial mass, which is 2/3 of the entire skull.

When drawing the head, continue to use the same process outlined in the first two chapters. Begin with broad 2-D ideas of position, orientation, and placement, then build into perspective, and finally construct the details on top of those developed surfaces. This process is repeated throughout the remainder of the book to emphasize a consistent understanding of the fundamentals.

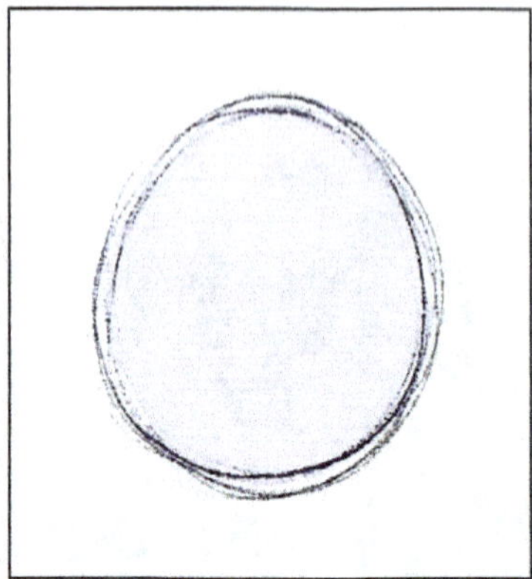

STEP 2: TILT

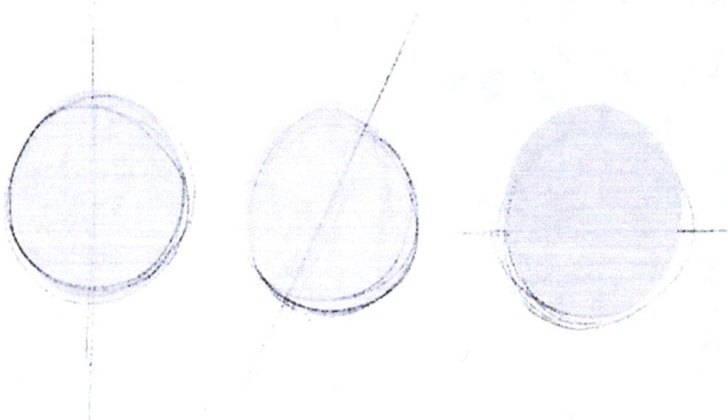

The second step involves giving the sphere a tilt. With a straight line drawn through the center of the sphere, the cranial mass is given a 2-D orientation. The first drawing on the left would show a head standing upright. The drawing in the middle shows a head starting to lean. The last drawing, with the horizontal line through it, shows a head that is lying down or flat. This step is important in developing the positioning of the head.

This step introduces the shape of the jaw back onto the cranial mass.

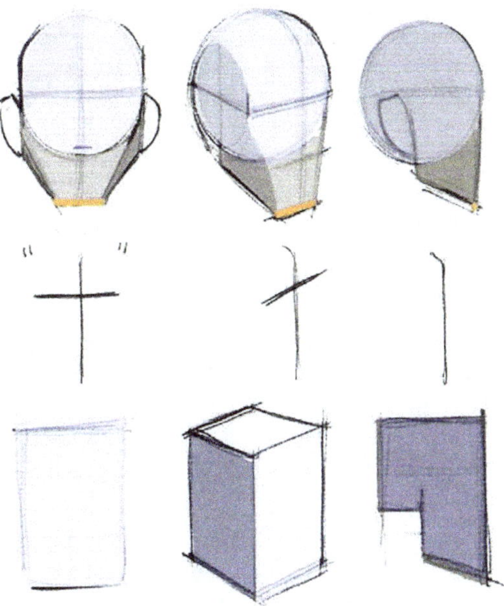

The shape of the jaw is formed by extending the lines of the cranial mass down to give the skull an overall egg shape.

The areas in these diagrams show the shape of the jaw from a front, three-quarter, and profile view. Observe how, with only shapes, the placement of the jaw starts to suggest a 3-D look or position in space.

The "t" of the face (shown most clearly in the top left illustration) further helps with positioning. The "t" is the line of symmetry in the face. When the face is seen straight on, the vertical line divides the face into two equal halves and the horizontal cross line represents the eye line.

Note that when the face changes positions, the "t" favors one side of the face. For example, as the head turns to the right, the center line of the "t" starts to favor the right side of the face. When the face is seen in profile, the "t" is lost.

The addition of the jaw will represent the remaining 1/3 of the skull, unless otherwise changed for exaggeration in character, animal, or creature.

Finding the "t" is extremely important — not only for correct placement of the front plane of the face, but also for a solid organization of the features.

At this stage, you should be focused on using the jaw, indicated with a straight, horizontal line for the chin, to show a complete turn. This involves an awareness of the symmetrical view, and what happens to the chin-line as the head turns right or left (shown in orange).

With the tilt, the "t," and chin, you can establish all possible leans of the head and implied dimensional turns with only shape.

Notice that because the head is a symmetrical form, we can use the same process for establishing volume that was used for the rib cage and pelvis.

This step introduces perspective into your drawing. Before this point, everything has been organized through the use of shape to arrange the placement of the larger forms. Perspective is now added to create the illusion of three dimensions.

Drawing a more complex form like the head requires a strong familiarity with the skills and exercises discussed at the beginning of the Form and Connections chapter. If your head drawings look a little off, try to pinpoint errors within the six skill areas so you can practice those specific skills.

The first step in deciding on a perspective is to simply determine whether you are underneath (A) or above (B) the head. Illustration A shows what a head looks like when seen from underneath. Note that the "t" has changed — it now tilts back across the face.

To the right, note that this perspective is simplified by the cylinder. When wrapping the "t" across the cranial mass, always draw through and around that sphere as if it were made of glass. The dotted lines on the drawing show how this appears. It is important that this line bisects the sphere at its equatorial center — think of this line traveling around the sphere like the line of the equator traverses the surface of the earth.

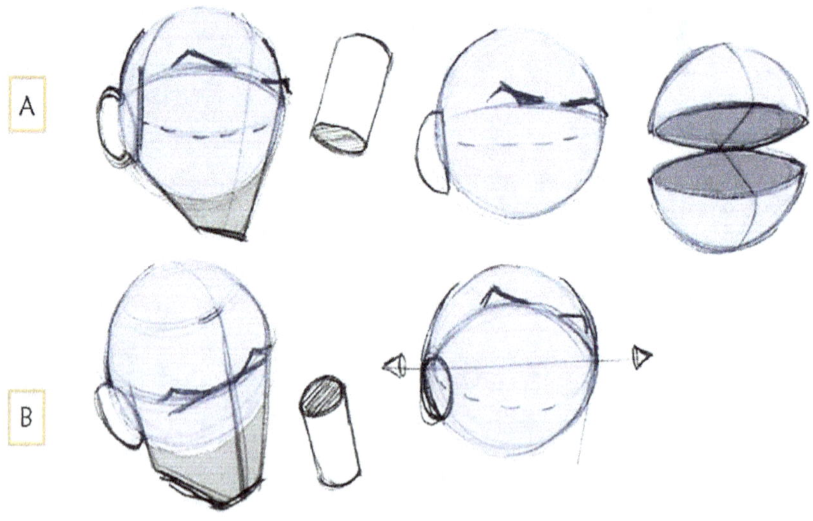

The two drawings in the center show how to easily place this "t" or eye line. By relating the eye line to the brow and top of the ear, you will more accurately describe the perspective angle.

Illustration B shows the head seen from above. The cylinder to the right of the head shows how all the line work is still geared toward describing a basic perspectival idea.

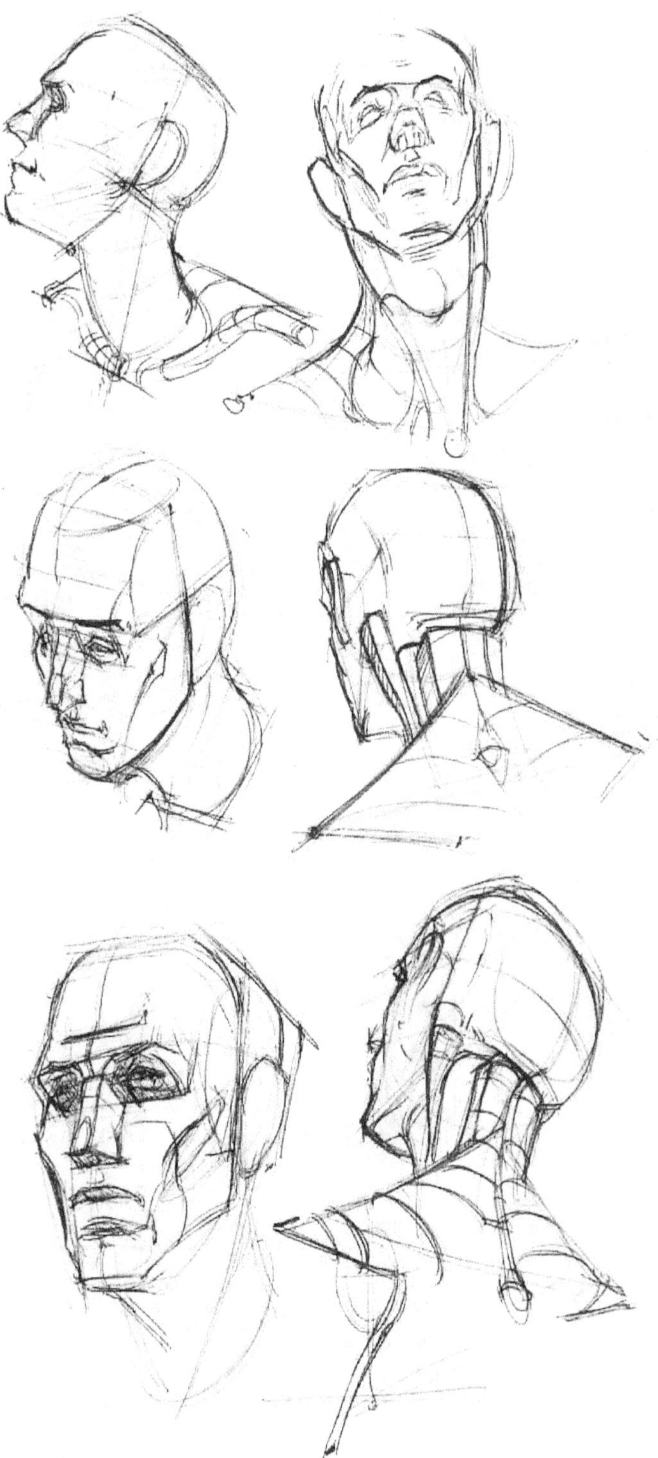

Here are some 5 minute sketches.

See if you can analyze them for the first four steps discussed so far. Start by finding the cranial mass, its tilt, develop the jaw, then establish the perspective leans through the eye line.

These four steps constitute the gesture and foundation of positioning the head in space.

STEP 5: PROPORTIONS

Having set up the major forms through the use of shape and perspective, the placement of the smaller forms can be established by setting proportions.

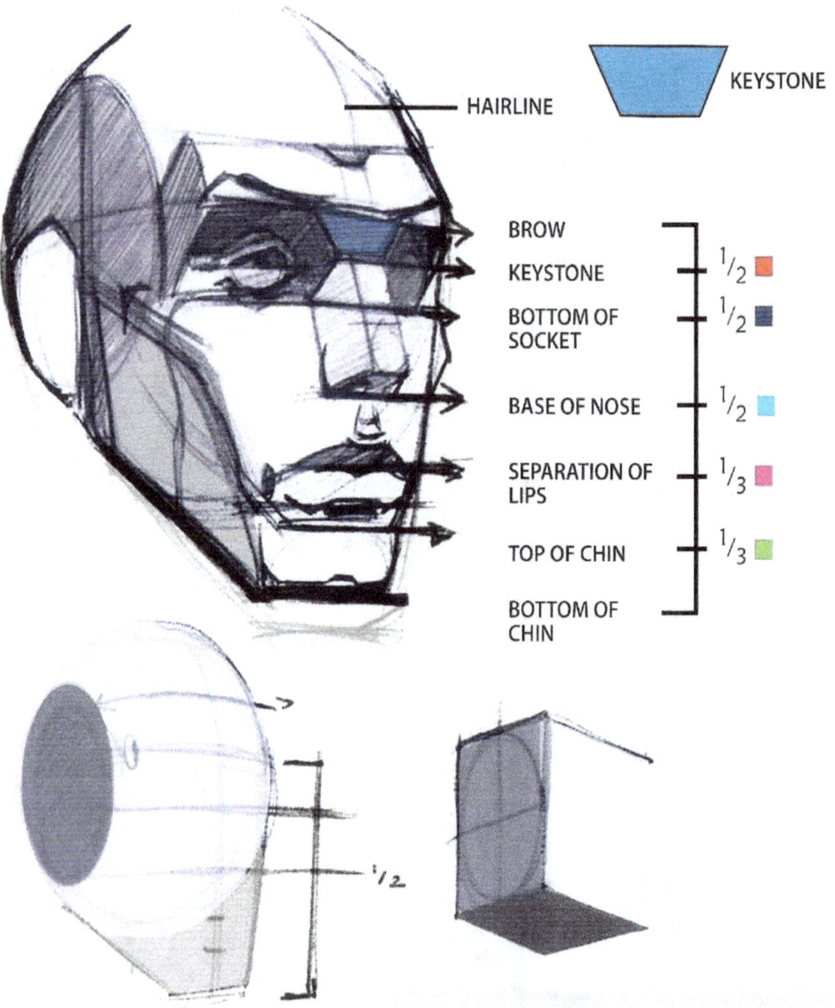

This method of finding proportions is based on identifying pronounced areas of bone on the skull (landmarks). These proportions are always consistent, despite the perspective placement (extreme views). Begin these measurements by finding the brow line and the bottom of the jaw.

The base of the nose is half-way between the brow and the bottom of the jaw.

From the base of the nose to the brow above, the face is continually broken into halves:

- The first mark is the bottom of the eye sockets, which is half-way between the base of the nose and the brow.

- Halfway between the bottom of the sockets and the brow is the center of the eyes and the bottom of the keystone shape (shown in blue). The shape in the upper right corner shows the basic appearance of the keystone shape when seen straight on. This keystone shape is the area of bone that separates the eyes. This is the most important area to closely observe in order to create a likeness.

The area between the base of the nose to the bottom of the jaw is divided into three equal parts. The two marks that separate those three equal parts determine:

- the separation of the lips

- the top of the chin

The last point of reference is the hairline. While not an area of bone, the hairline will help in transitioning to the next step. The hairline's placement will change depending on the particular character type — some will have one and others won't.

> **TIP** Keep in mind that once these proportions are learned, their manipulations and exaggerations give endless possibilities of character, type, animal, etc. Currently, you are building the foundation that will allow later inventions to be grounded in reality and solid draftsmanship.

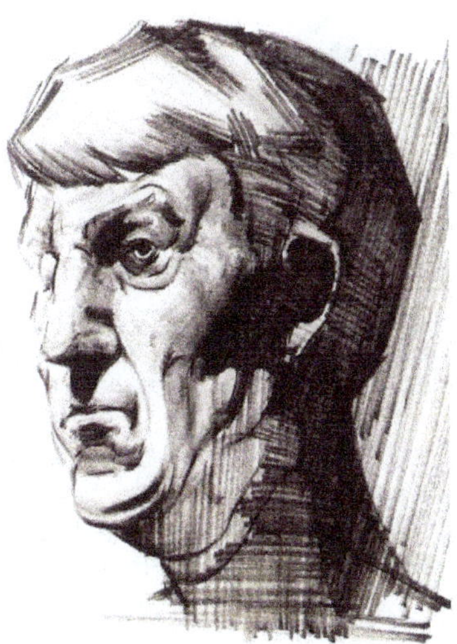

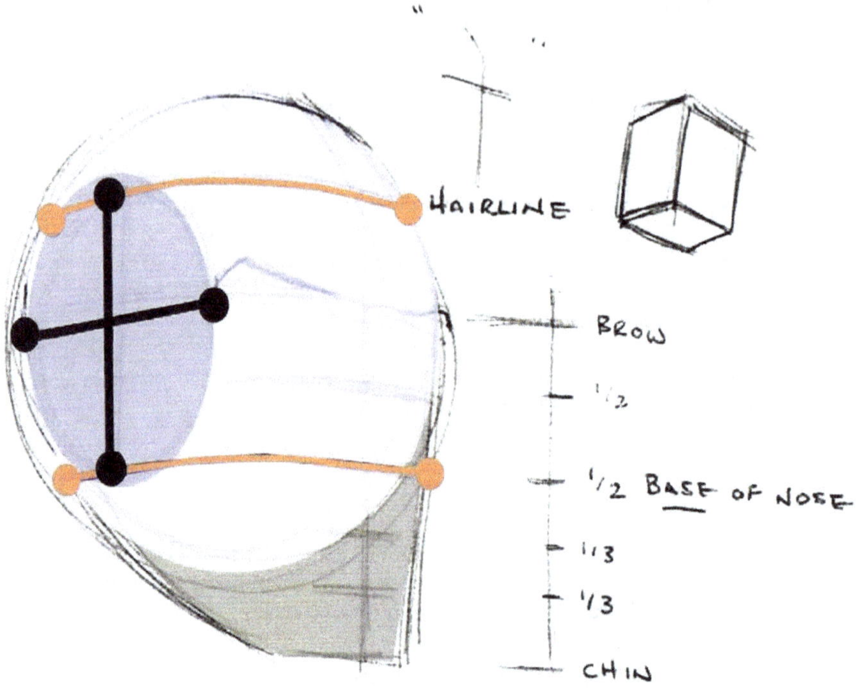

This step begins with finding the hairline and drawing it in perspective all the way around to the back of the skull.

Next, beginning from the base of the nose, draw another line back across the form to the back of the skull.

Between the back of the skull, the top of the line drawn from the hairline, the outside of the brow, and the line drawn from the base of the nose, draw an ellipse to represent the side plane of the skull. These four points are always used to find this side plane; however, the orientation and the size of the ellipse will change depending on the perspective.

This is the most important stage in giving the head a 3-D appearance. In a very simple way, the box (top right of the illustration) describes what is beginning to take place on the skull.

Between the four points mentioned above, draw two straight lines connecting them. One should be horizontal going from the back of the skull to the end of the brow, the other should be drawn from the line at the base of the nose to the top line at the hairline. During this step, keep the vertical line as close to matching the line of the tilt (step 2) as possible.

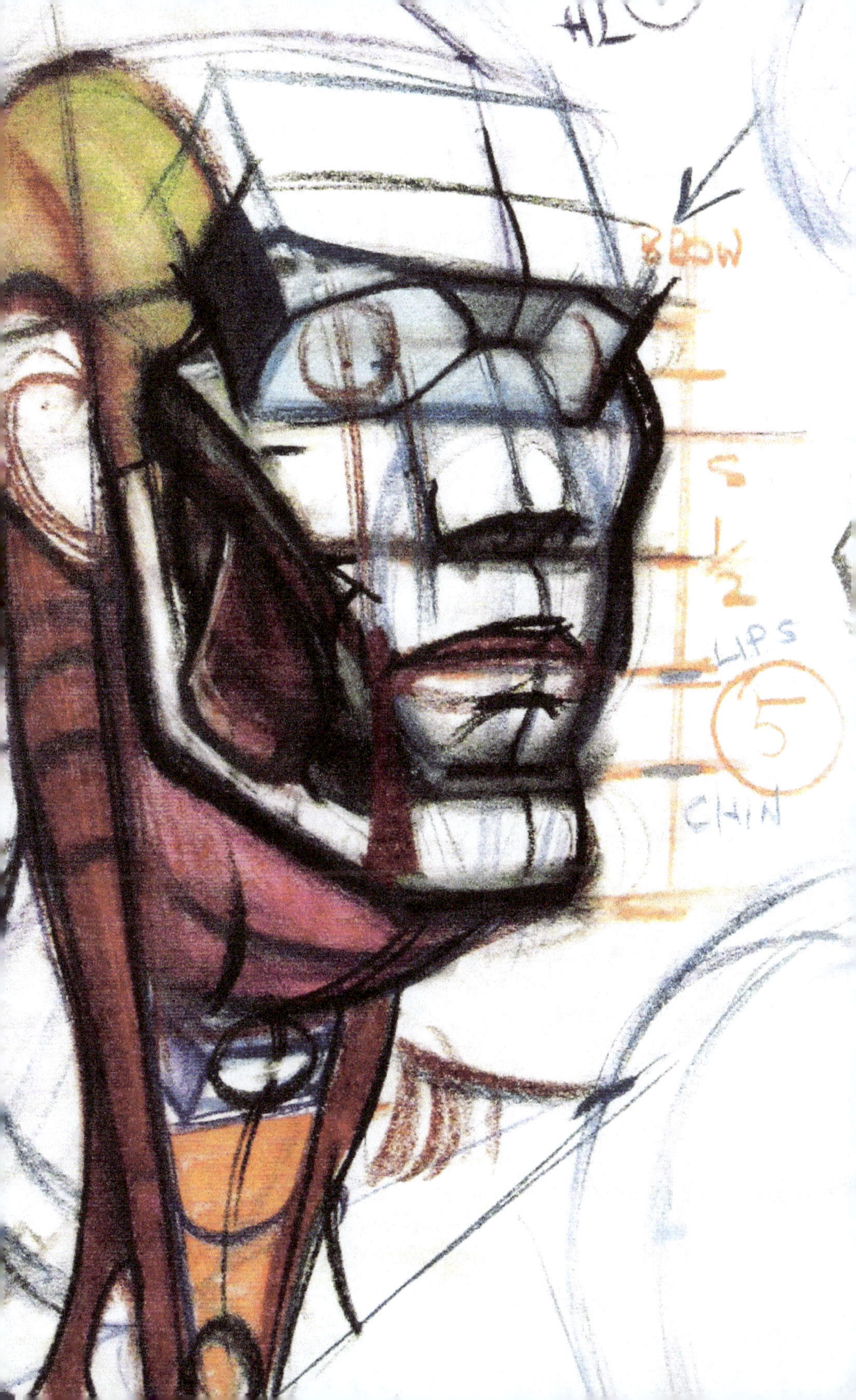

H.L
BROW
5
1/2
LIPS
5
CHIN

This step begins by placing the ear. The ear sits in the lower quarter, created by drawing the horizontal and vertical divisions in step 6. As shown in dark blue, the ear should be kept as a simple "C" shape that is no taller than the base of the nose to the line of the brow.

The second part of this step is to find the line of the cheek. As shown in dark blue, the line of the cheekbone begins at the top of the ear and continues as a "C" curve down to the corner of the jaw. The inclusion of the cheekbone adds another corner, showing the front plane of the face, and now side plane for the jaw.

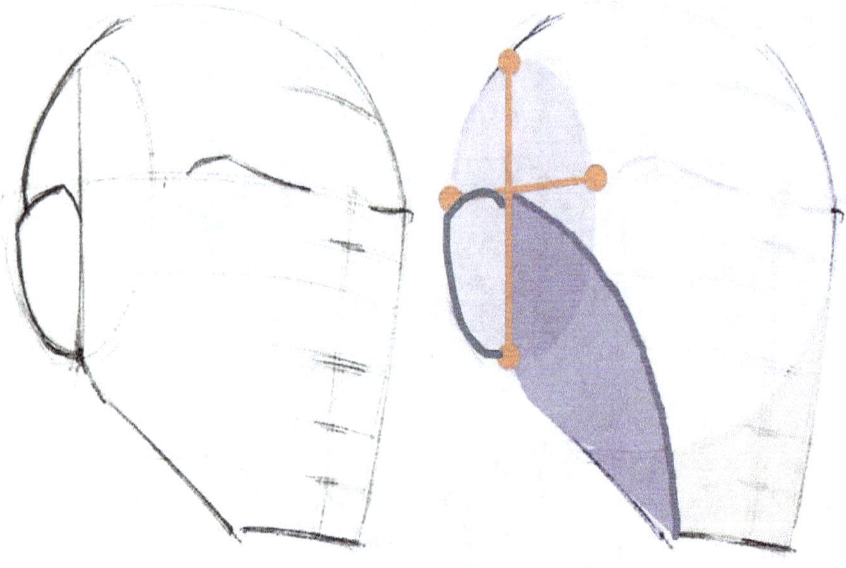

Note that as this process builds, some of the early line work is no longer emphasized. For example, the line work showing the beginning shapes is no longer needed because the forms are becoming more specific and exact.

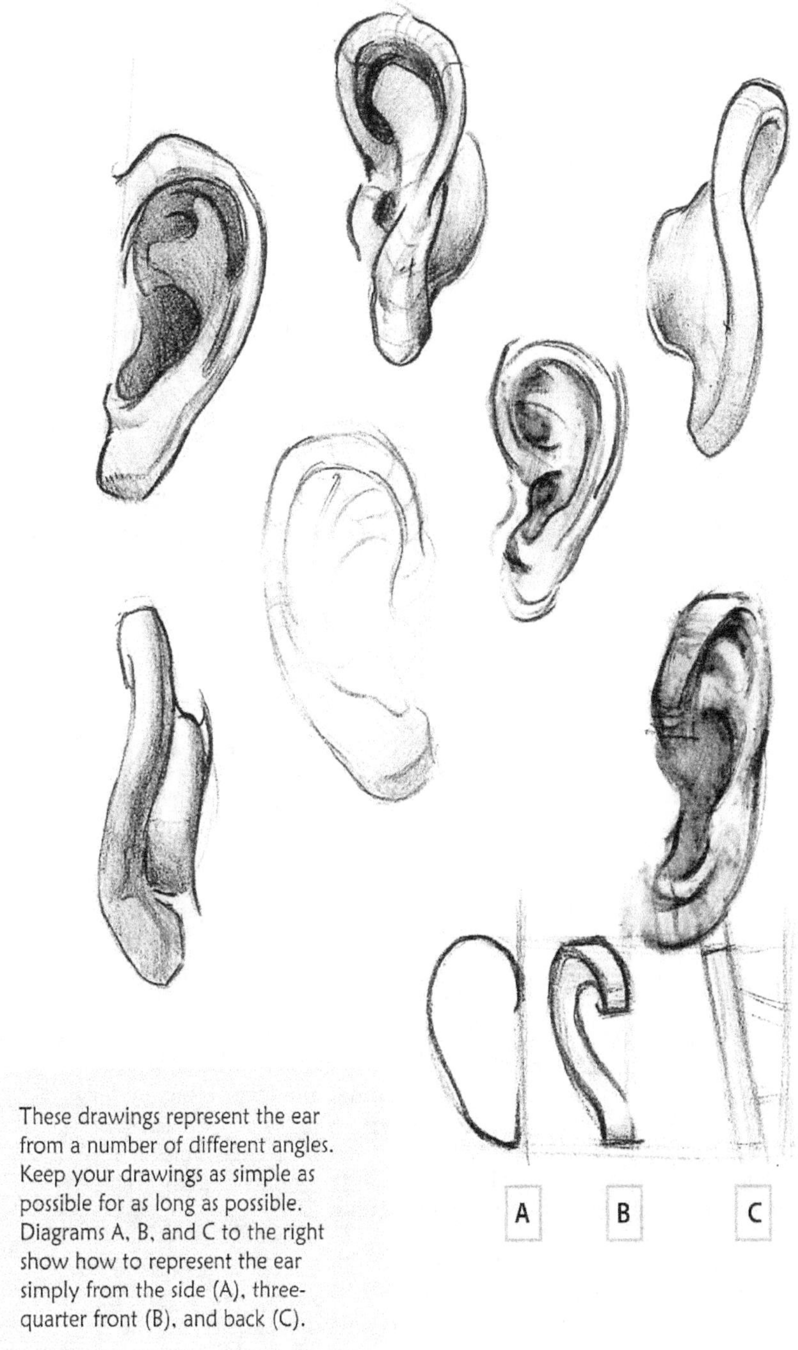

These drawings represent the ear from a number of different angles. Keep your drawings as simple as possible for as long as possible. Diagrams A, B, and C to the right show how to represent the ear simply from the side (A), three-quarter front (B), and back (C).

This step begins by finding the recessive plane for the eye sockets (shown with orange dots). It is important to show that the plane for the sockets pushes into the skull at an angle. Also note that all four dots are connected to show one plane that begins at the brow line and ends at the proportion line for the bottom of the sockets. This plane does not go past the cut-out for the side plane of the head.

Having established the plane for the sockets, the structure of the nose can be built. First look for the relationship between the tip and base of the nose. In the example at left, the tip of the nose (blue line) is drawn above the base of the nose (black line). This is an important stage to reinforce your perspective. If you were to draw the tip of the nose below the base, it would suggest that the viewer is on top of the head looking down, rather than underneath the head looking up.

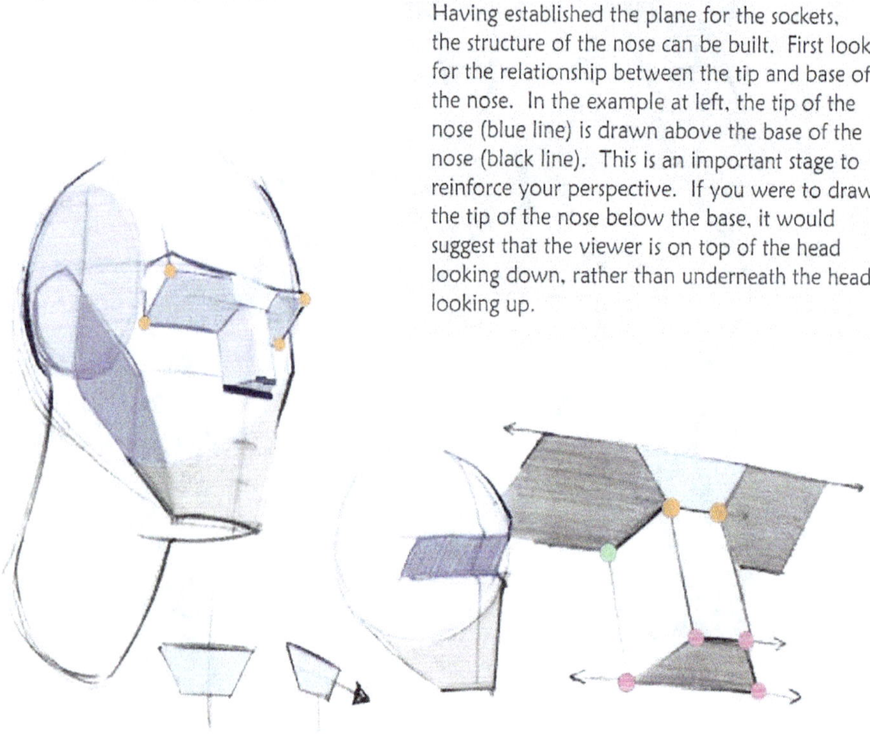

From the tip of the nose, at either end, two lines are drawn down to connect it to the base. You should now have one plane, which represents the bottom plane of the nose. The next step (above right) is to draw two straight lines from the tip of the nose (pink dots) to the ends of the keystone shape above (orange dots). This form gives you the front plane for the bridge of the nose.

To complete this structure, drop a line from the point at the keystone (orange dots) to the line for the bottom of the sockets (green dot) at about the same angle as was used for the side of the base of the nose. After connecting this again to the base of the nose (connect green dot to pink), you have the side plane of the nose drawn to proportion, accurately placed on a believable perspective-based form.

This page shows a variety of different shapes and views for the nose. Notice that the red lines allow you to see the importance of the box for use in establishing the placement of the nose on the face and its perspective. The varying angles and views of the nose are totally dependent on how well you understand the placing a box in space.

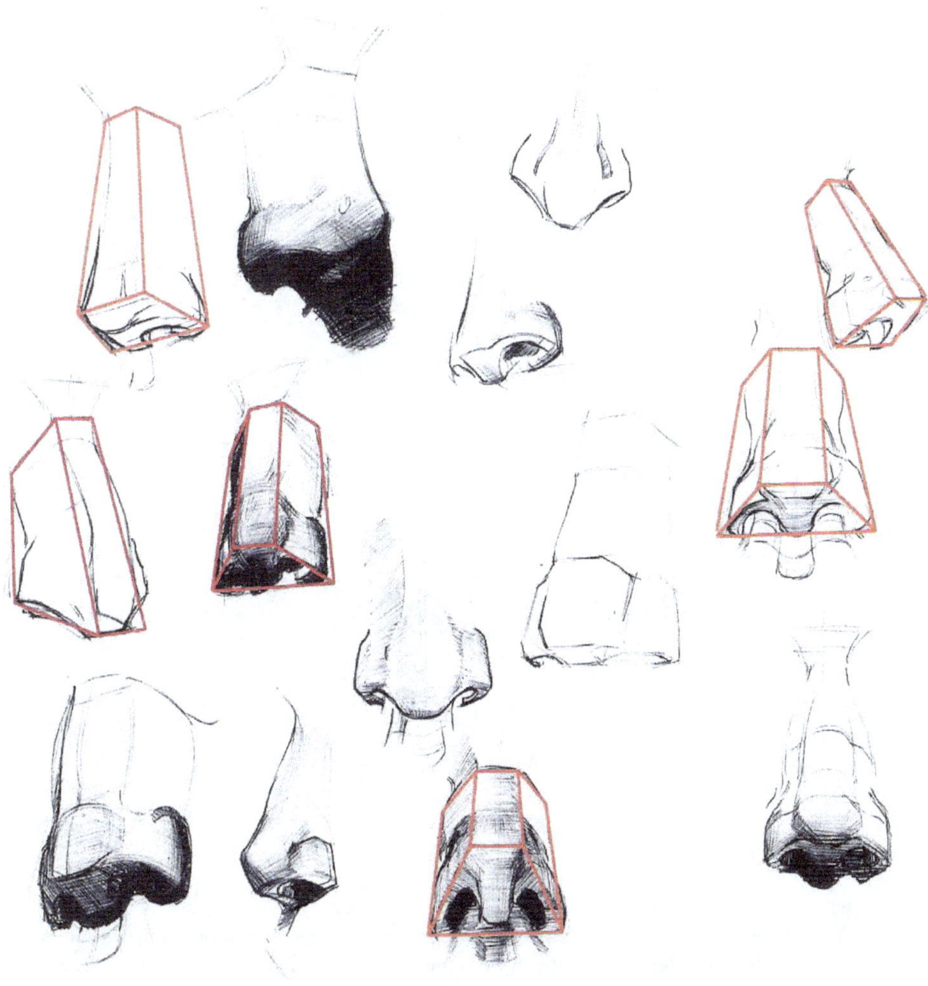

Having the developed the box, notice that different types of noses come from exaggerating any of the straight lines to become a variety of curves. Additionally, the underside of the nose in these examples has been broken into the septum and nostrils.

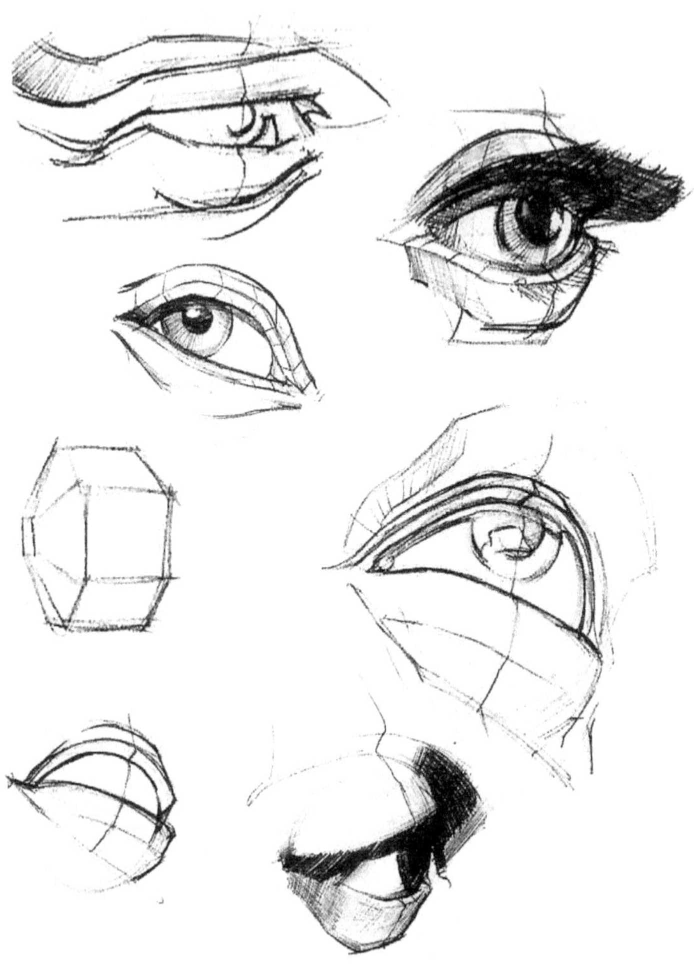

These pages contain example drawings of the form, placement, and planes of the eyes. When drawing the eye, always begin by describing the sphere of the eye and where it sits within the socket. It is most important to give the eye its context before going right in for the highlight. When describing the eyelids, be sure to think of them as wrapping lines. The lids should feel as if they travel across and around the underlying form of the eye.

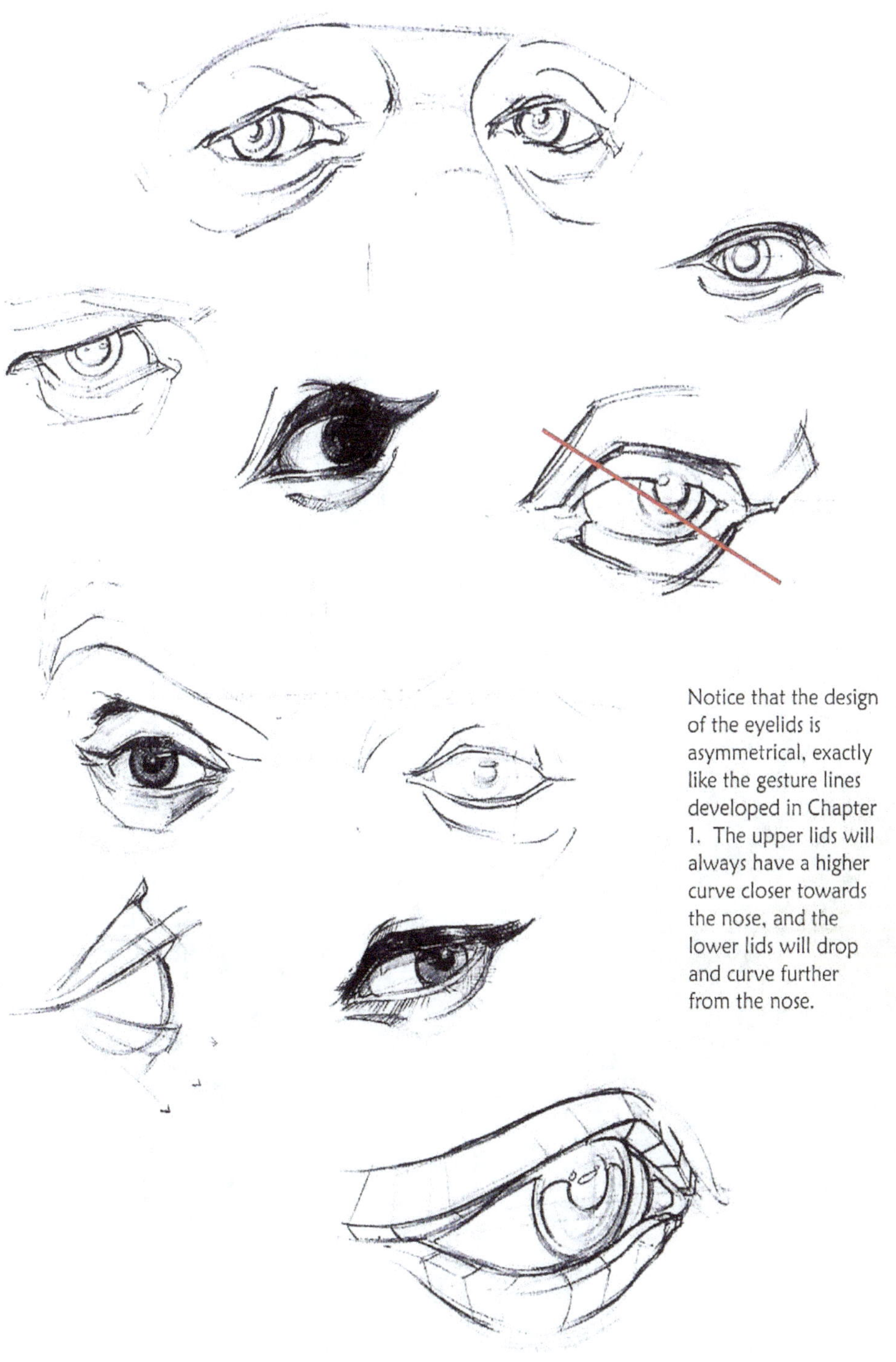

Notice that the design of the eyelids is asymmetrical, exactly like the gesture lines developed in Chapter 1. The upper lids will always have a higher curve closer towards the nose, and the lower lids will drop and curve further from the nose.

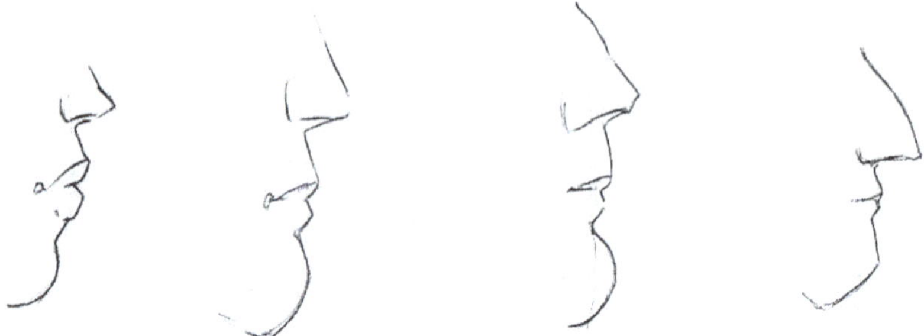

This step involves developing the area of the tooth cylinder or denture sphere. This area of bone, which includes the teeth, pushes out and away from the face. This is the main characteristic of the lower portion of the face to develop before drawing the lips.

The denture sphere is an oval. This oval is drawn from the top of the chin up underneath the area of the nose (this is shown in the example on the facing page in the lower left).

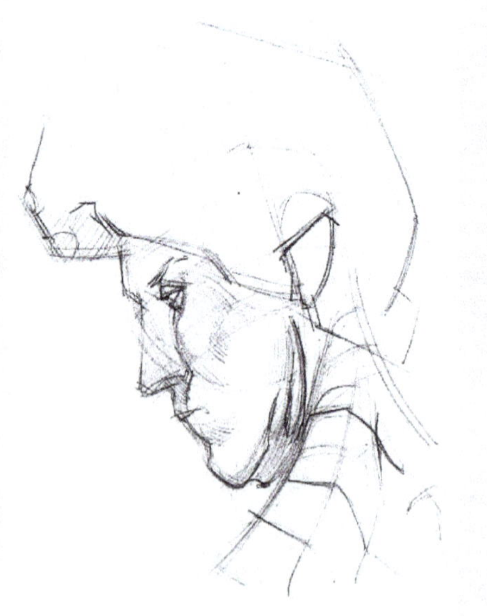

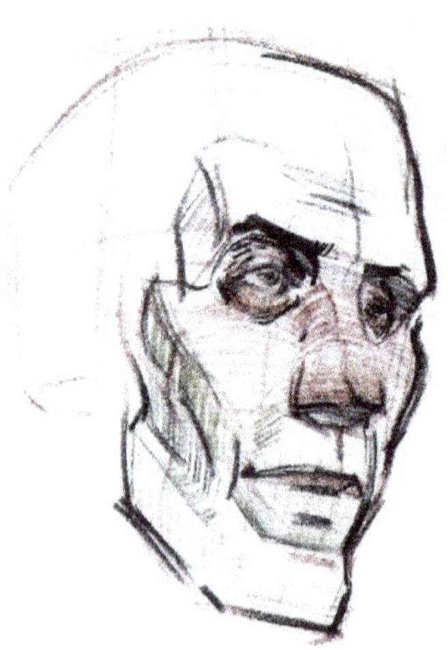

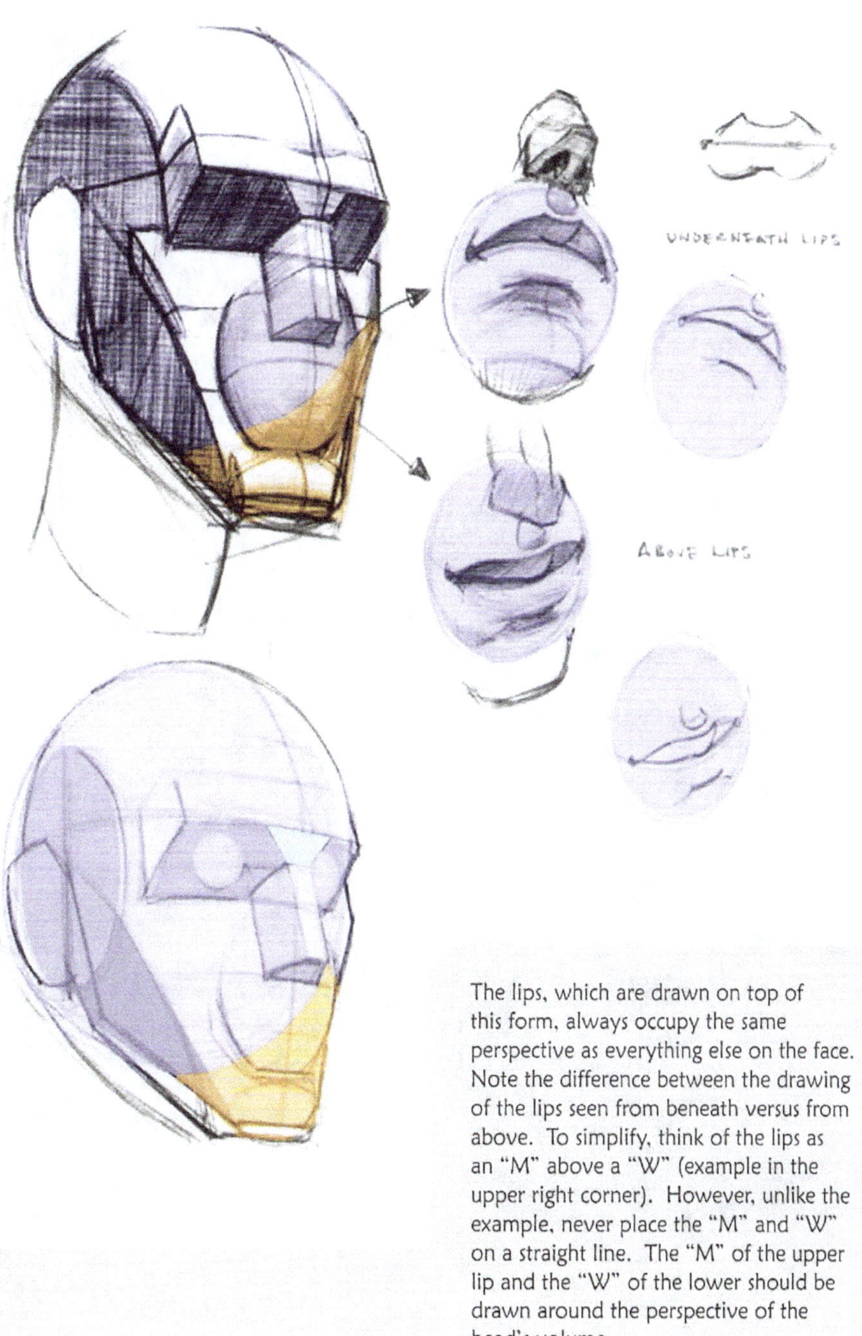

The lips, which are drawn on top of this form, always occupy the same perspective as everything else on the face. Note the difference between the drawing of the lips seen from beneath versus from above. To simplify, think of the lips as an "M" above a "W" (example in the upper right corner). However, unlike the example, never place the "M" and "W" on a straight line. The "M" of the upper lip and the "W" of the lower should be drawn around the perspective of the head's volume.

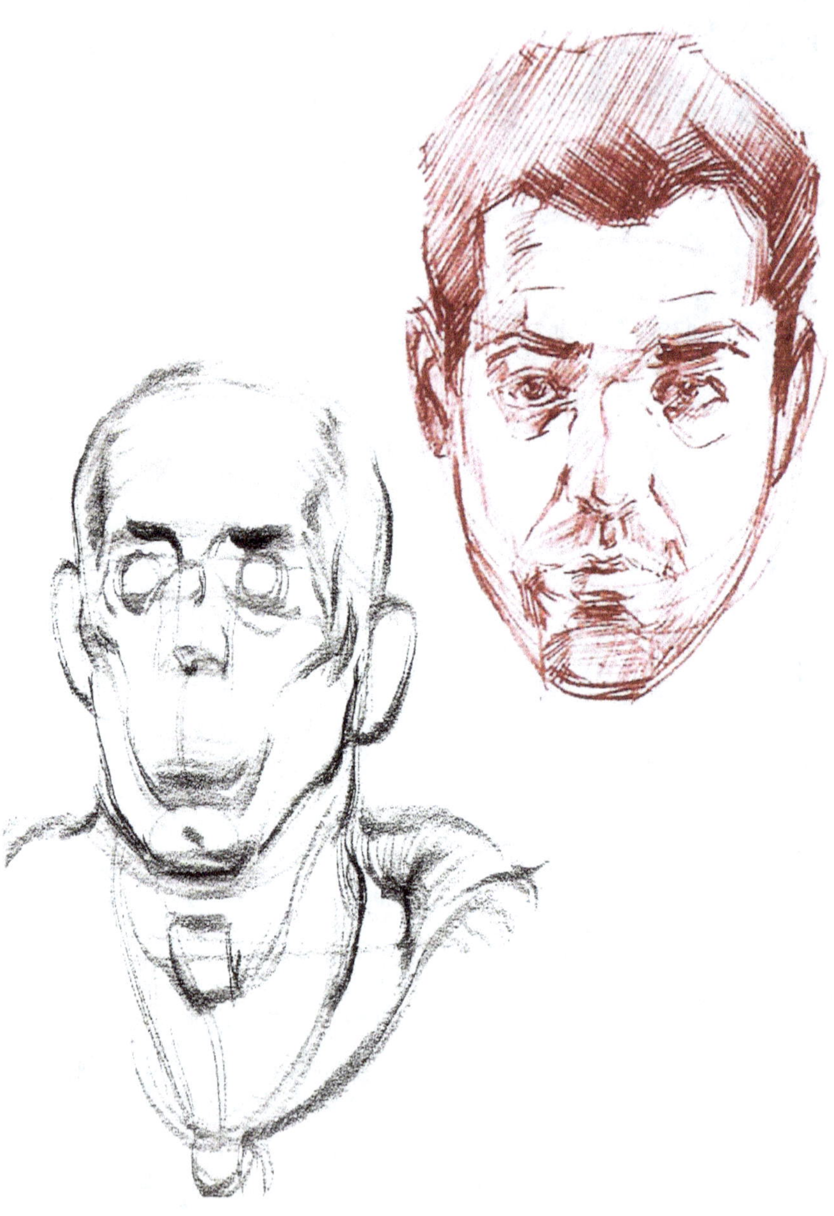

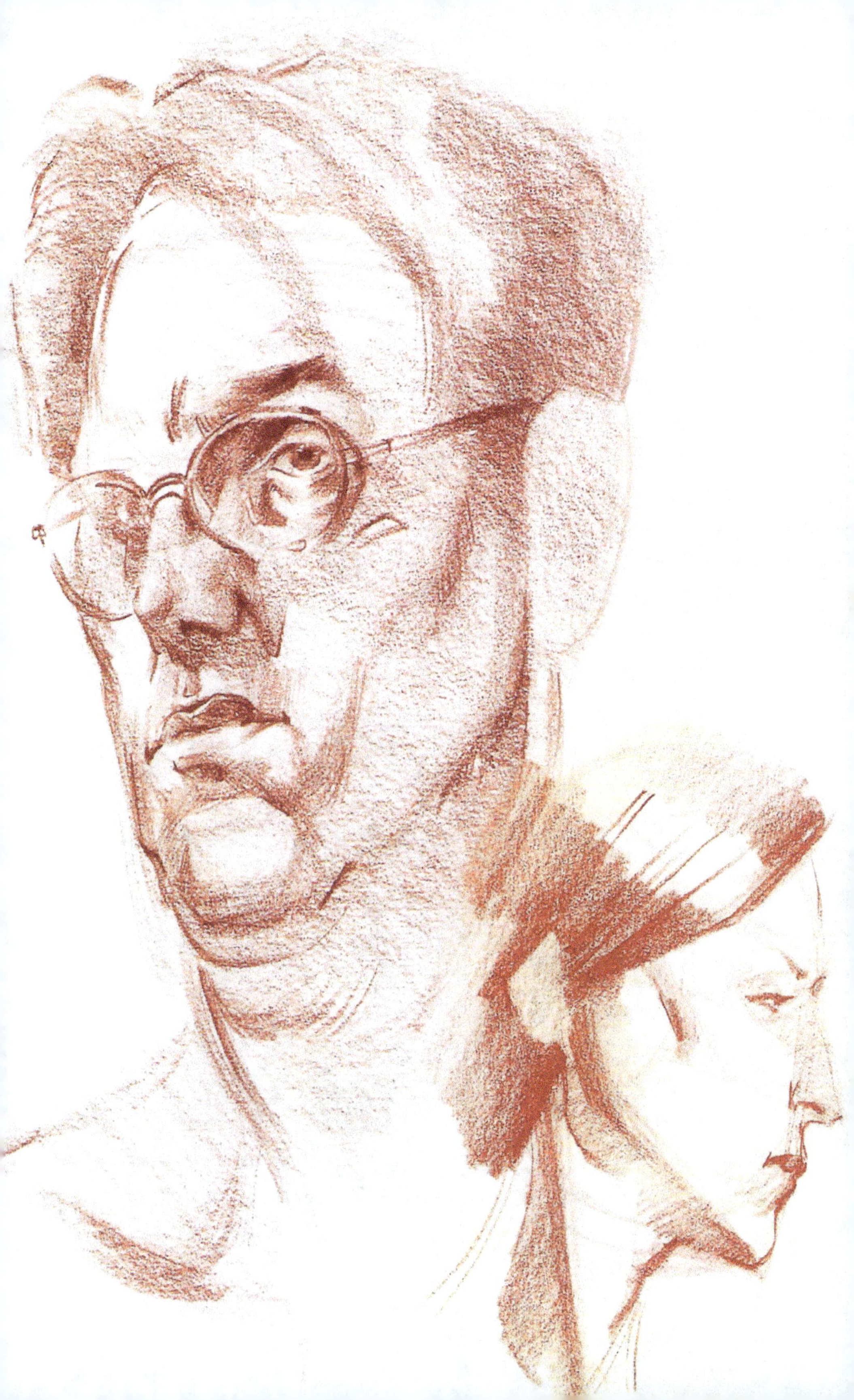

COMPLETED LINE DRAWING

This is an example of how your completed line drawing should look. Before moving on to any finishing work on the character or head, make sure there is a fundamental development of the basic forms. These forms are what give your drawings the feeling of believability, naturalism, and volume.

The development of the features should, at this beginning stage, be a tertiary concern.

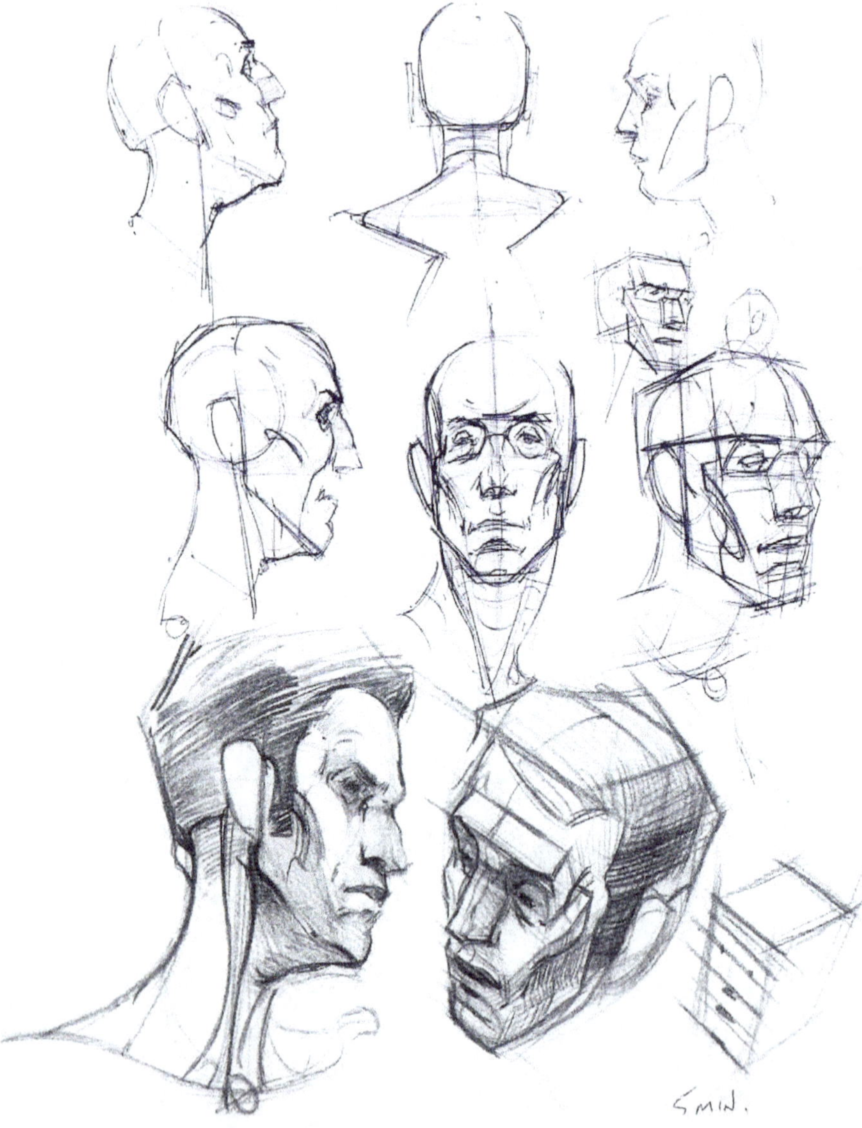

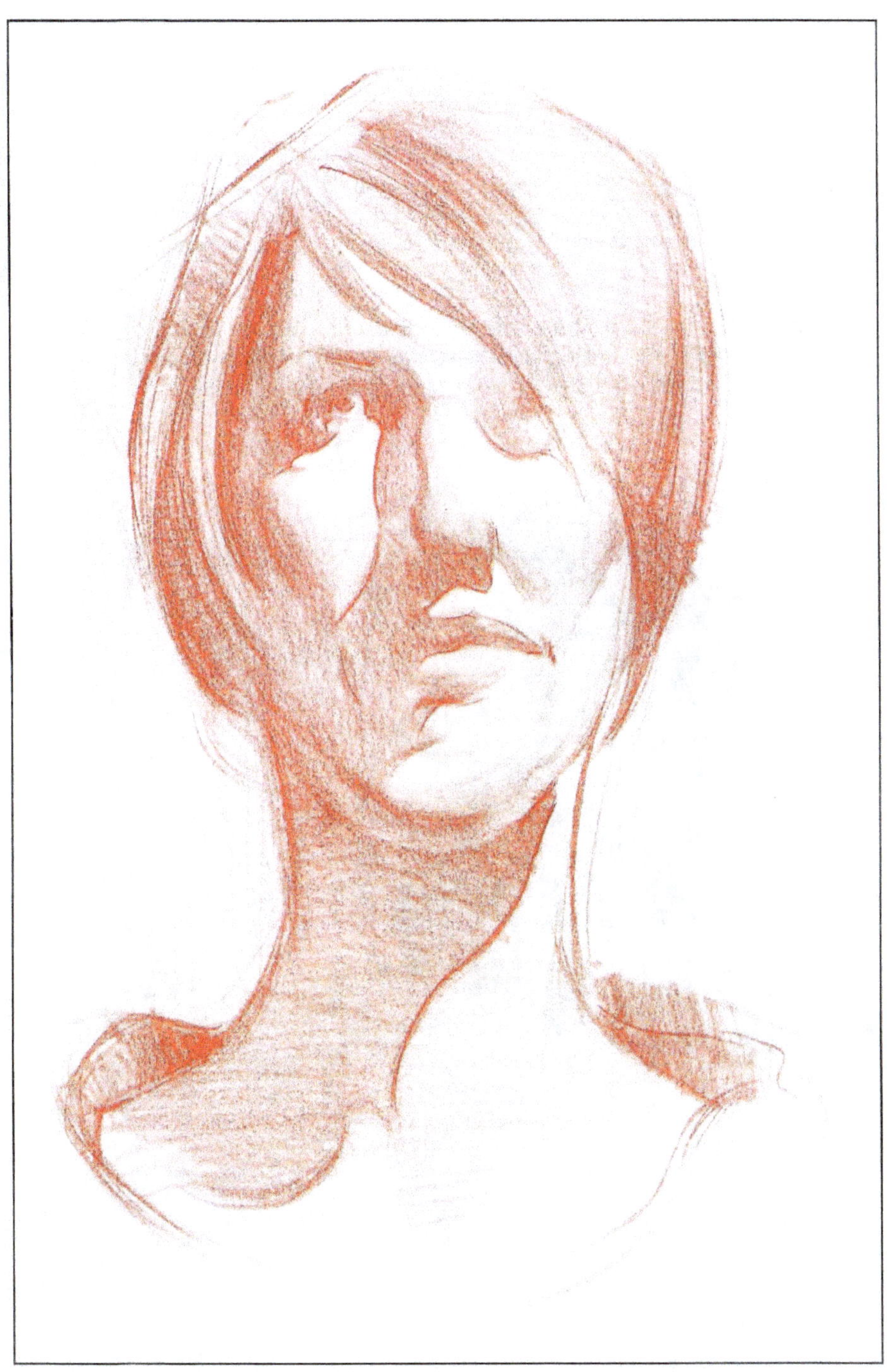

Developing the profile view can be handled with a slightly different set of tools. The first steps, however, should be the same. It is still most important to establish the shape, tilt, and perspective. In doing this, your drawing should look like example B (straight on), C (view from beneath), or E (view from above). Depending on your intention, it may be easier to demonstrate the feeling of volume by cheating or exaggerating a perspective viewpoint (in the case C or E).

After finishing this step, and using the same proportions, the profile for an individual can be introduced as a design. Try likening your characters profile by thinking of how much the forms of the face project out from the front line (F) or push into it.

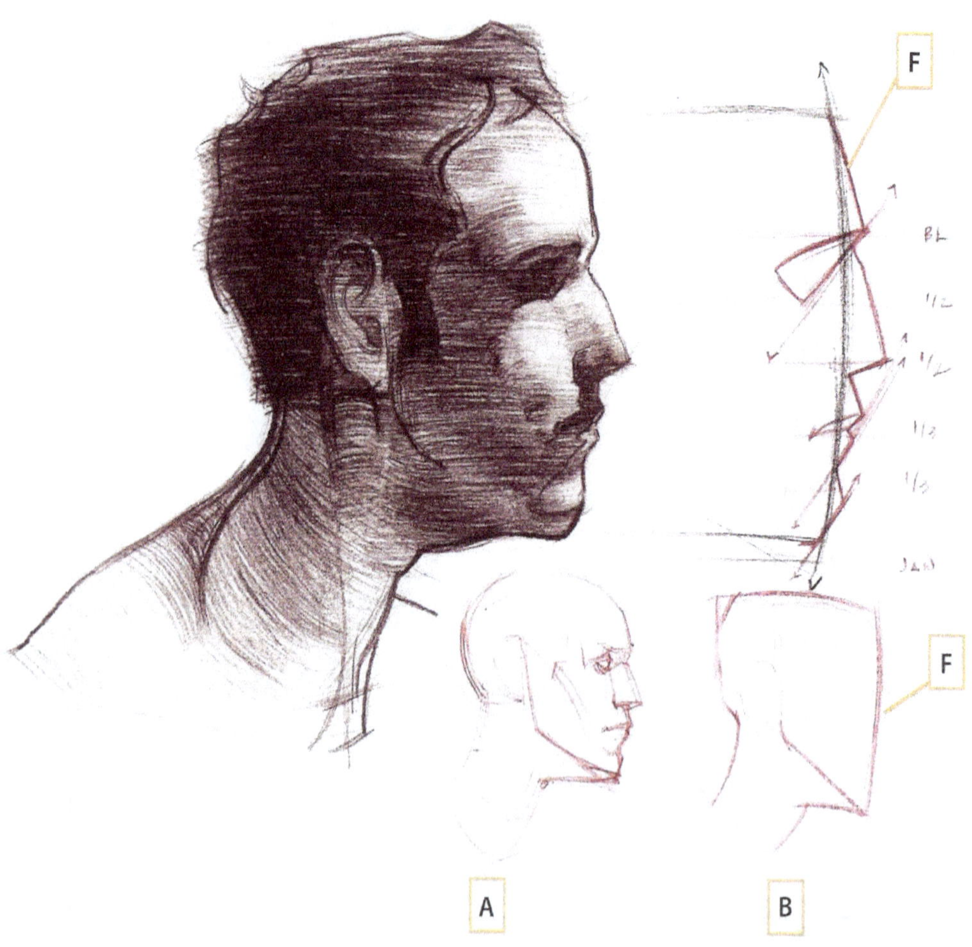

C

D

E

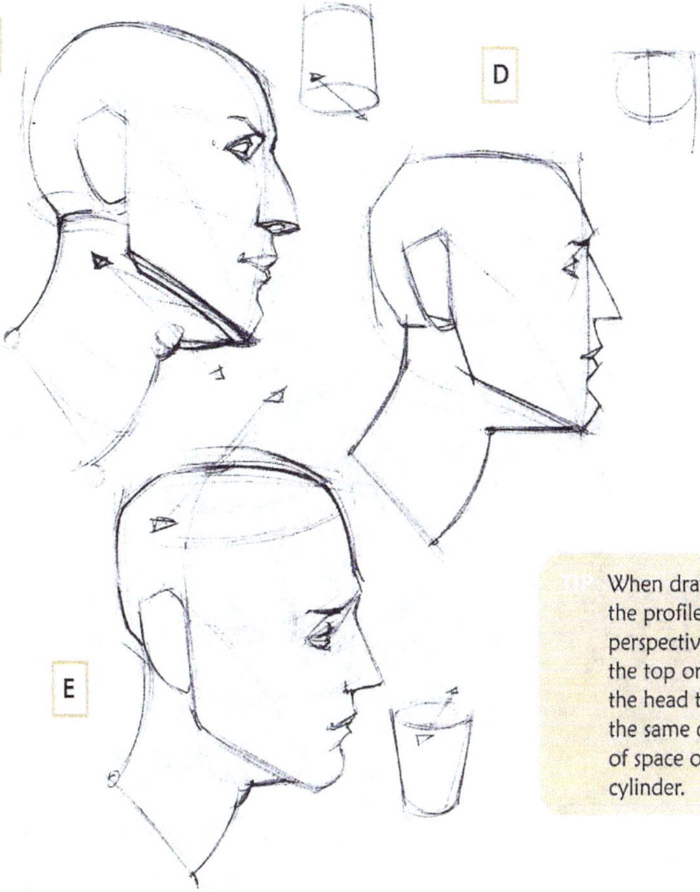

A feeling of naturalism in the profile is achieved with an idea from the Gesture Chapter. Notice that the forms of the face are in a balancing act, alternating between a form projecting out and a form receding in. For example, the forehead pushes out, the eye sockets recede, the nose pushes forward, etc. Keep this pattern in mind when designing your own characters or drawing from life.

G

H

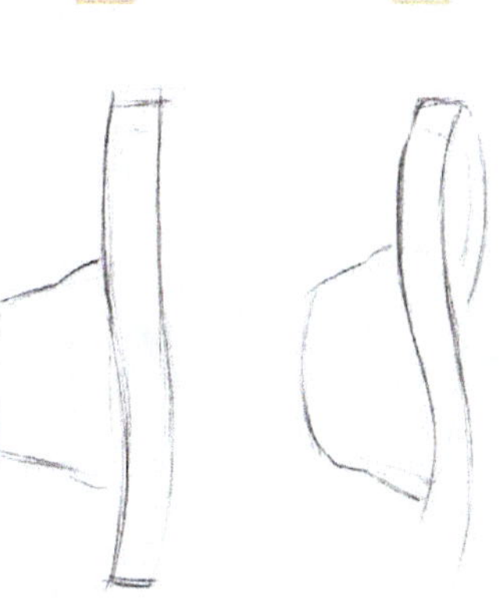

Notice the shape for the back of the ear is highly simplified. Example G shows the back view as essentially a cup or cylinder shape with an "S" on top, while the three-quarter view (H) describes the form turning by introducing a corner.

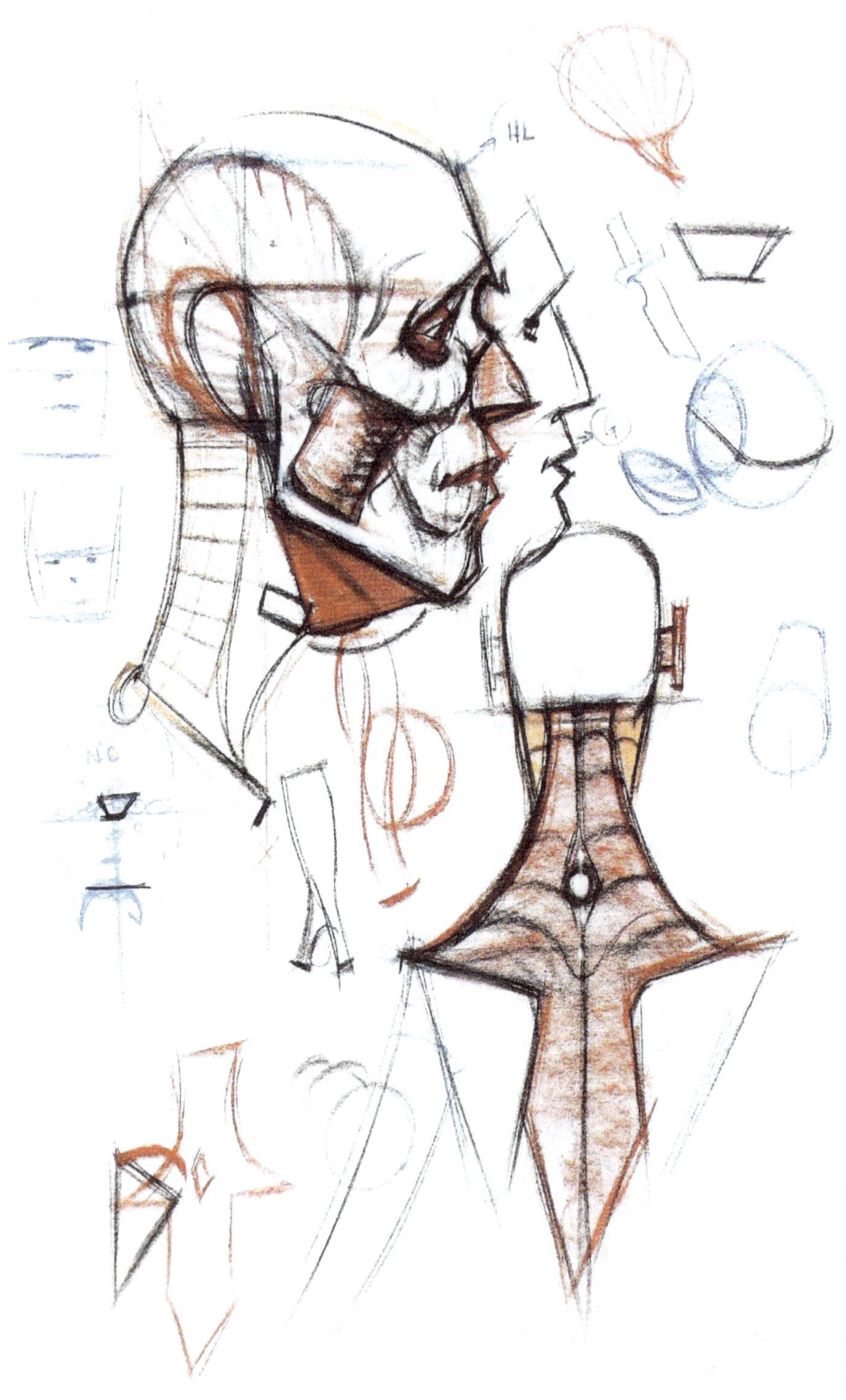

Similar to the profile, the back of the head also offers a unique set of problems. However, you will always begin with the first four major steps. Having set the placement up, the design for the back of the head is one predominately structured on the "T" overlap. The forms you are looking to overlap in this position (the anatomical forms will be discussed in the following chapter) are the Trapezius (1), the form of the neck (2), the sternocleidomastoid (3), the cranial mass (4), and ear (5). Simplify your drawing of the back of the head into an organization of these major forms.

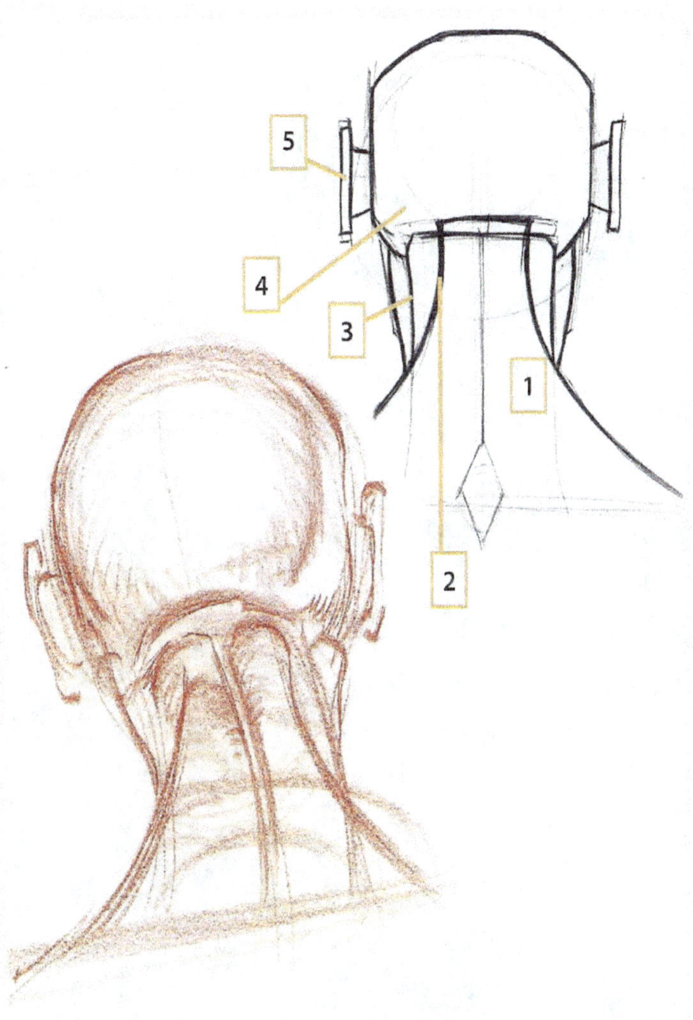

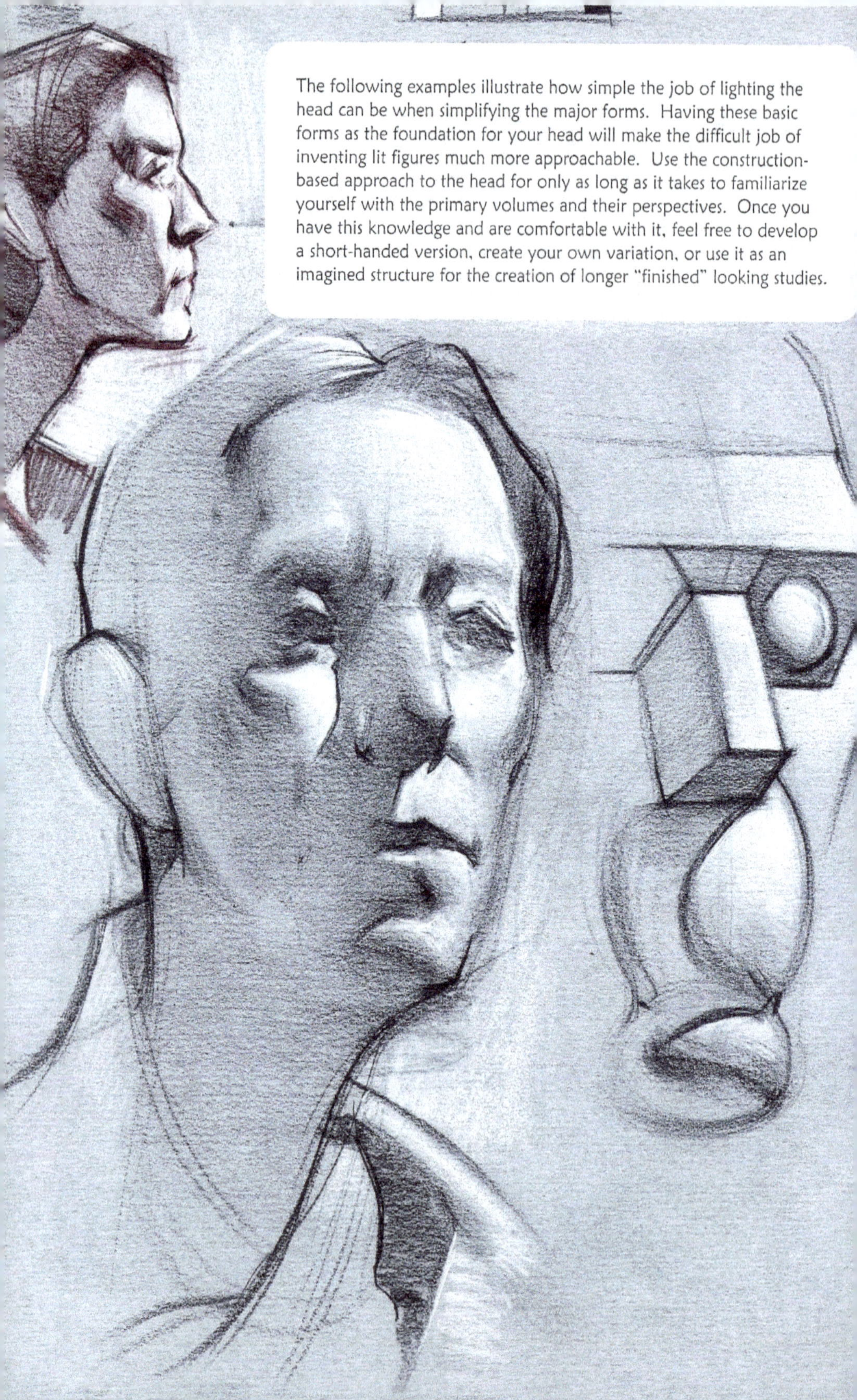

The following examples illustrate how simple the job of lighting the head can be when simplifying the major forms. Having these basic forms as the foundation for your head will make the difficult job of inventing lit figures much more approachable. Use the construction-based approach to the head for only as long as it takes to familiarize yourself with the primary volumes and their perspectives. Once you have this knowledge and are comfortable with it, feel free to develop a short-handed version, create your own variation, or use it as an imagined structure for the creation of longer "finished" looking studies.

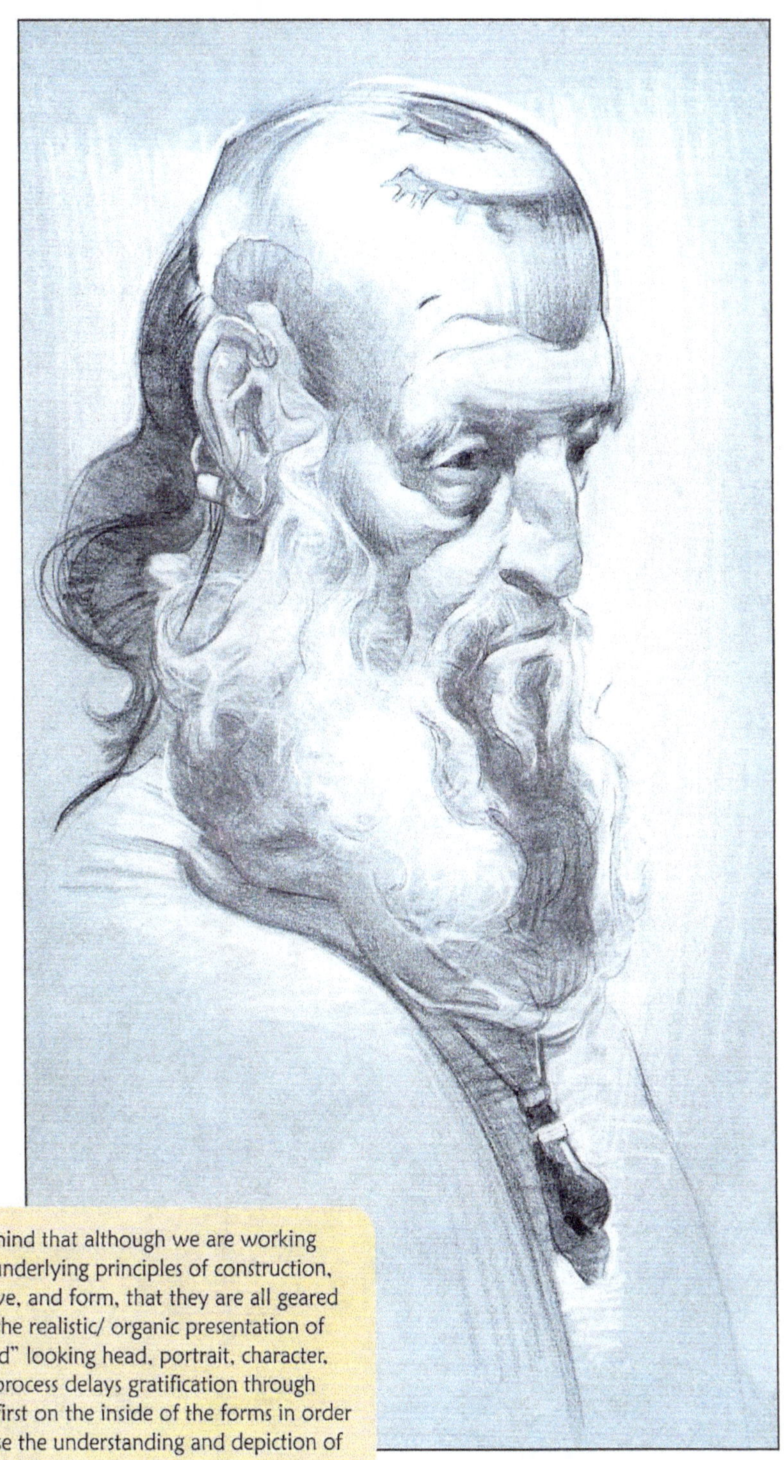

Keep in mind that although we are working
through underlying principles of construction,
perspective, and form, that they are all geared
towards the realistic/ organic presentation of
a "finished" looking head, portrait, character,
etc. The process delays gratification through
focusing first on the inside of the forms in order
to increase the understanding and depiction of
the outside.

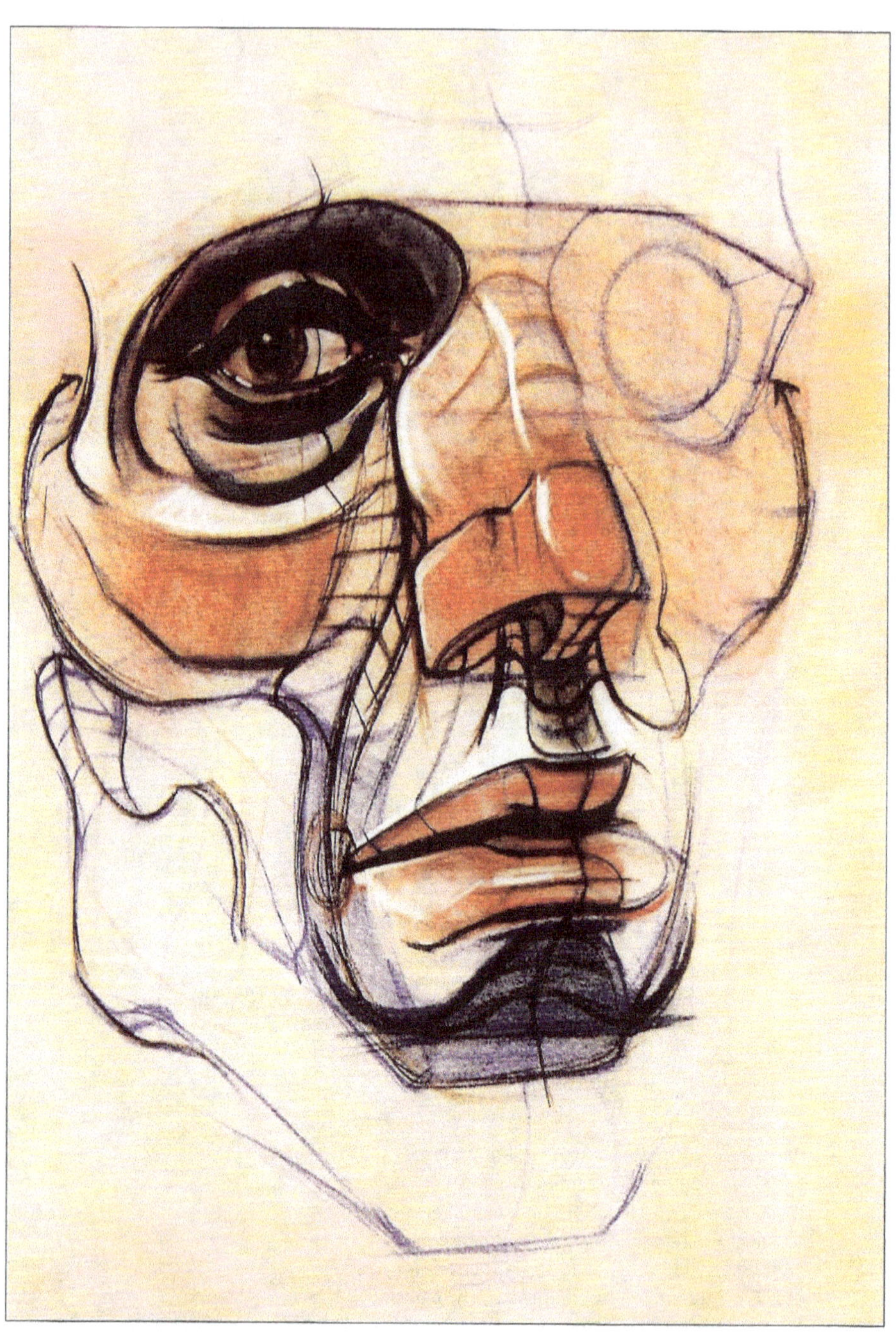

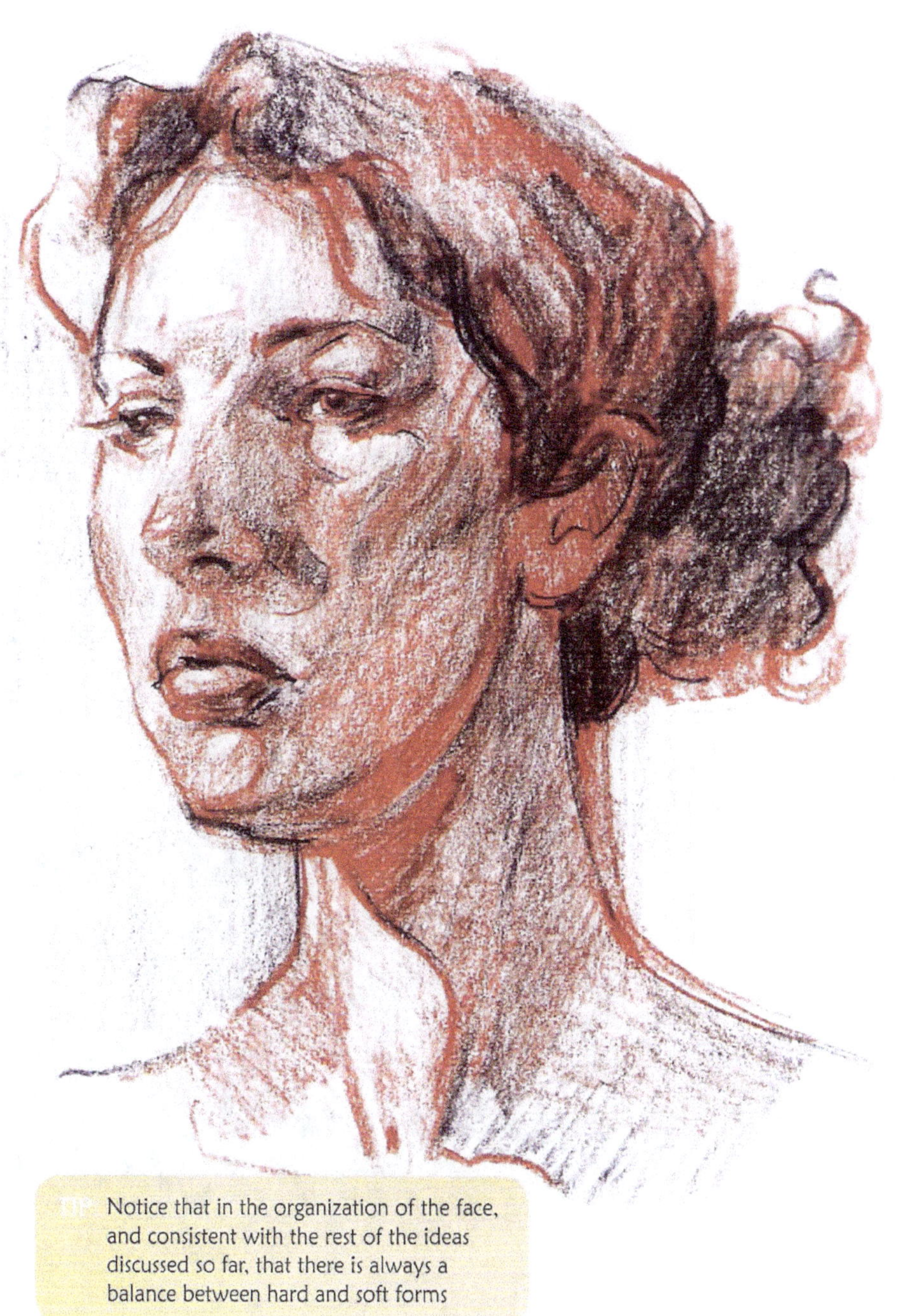

Notice that in the organization of the face,
and consistent with the rest of the ideas
discussed so far, that there is always a
balance between hard and soft forms

ANATOMY

The most important step in developing a drawing that includes anatomy is to first work through a process. Working through gesture, shapes, landmarks, and then volume gives your drawings a sense of solidity that the anatomy will need to respond to.

The muscles pointed out in this section are only muscles seen or affecting the surface of the forms.

This chapter is not meant to act as a reference manual for anatomy. This is a highly simplified approach to anatomy meant to serve as a foundation of a working process.   In understanding the muscles, the same approach to the figure is implemented: first, gesture and shape, then the development of volume and perspective.

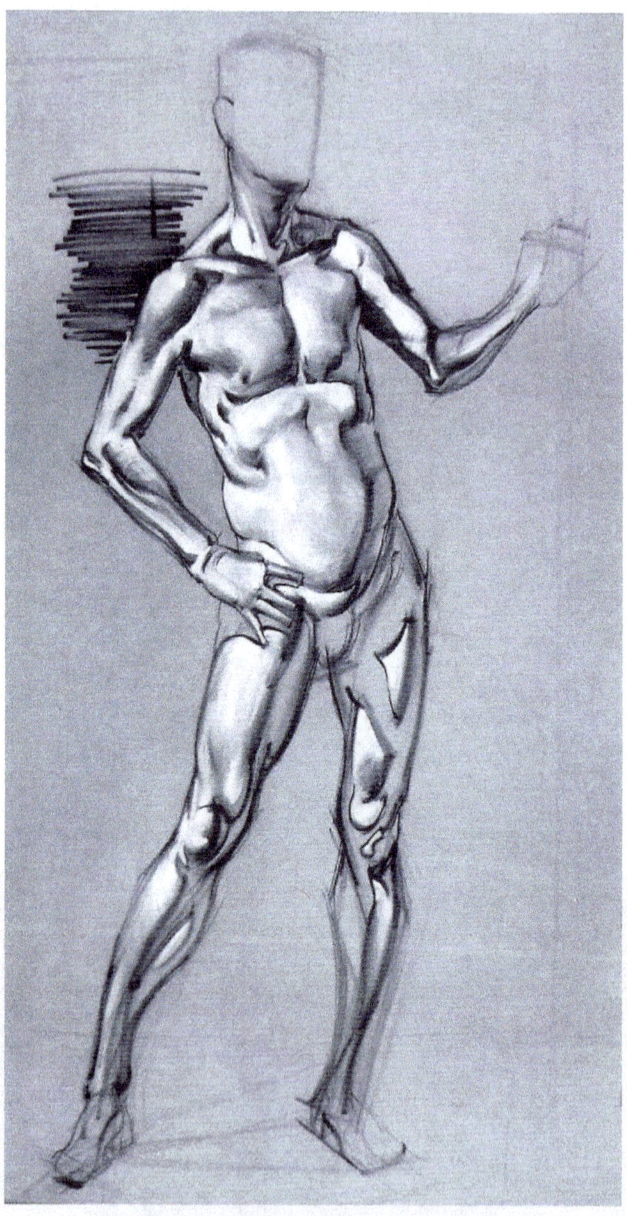

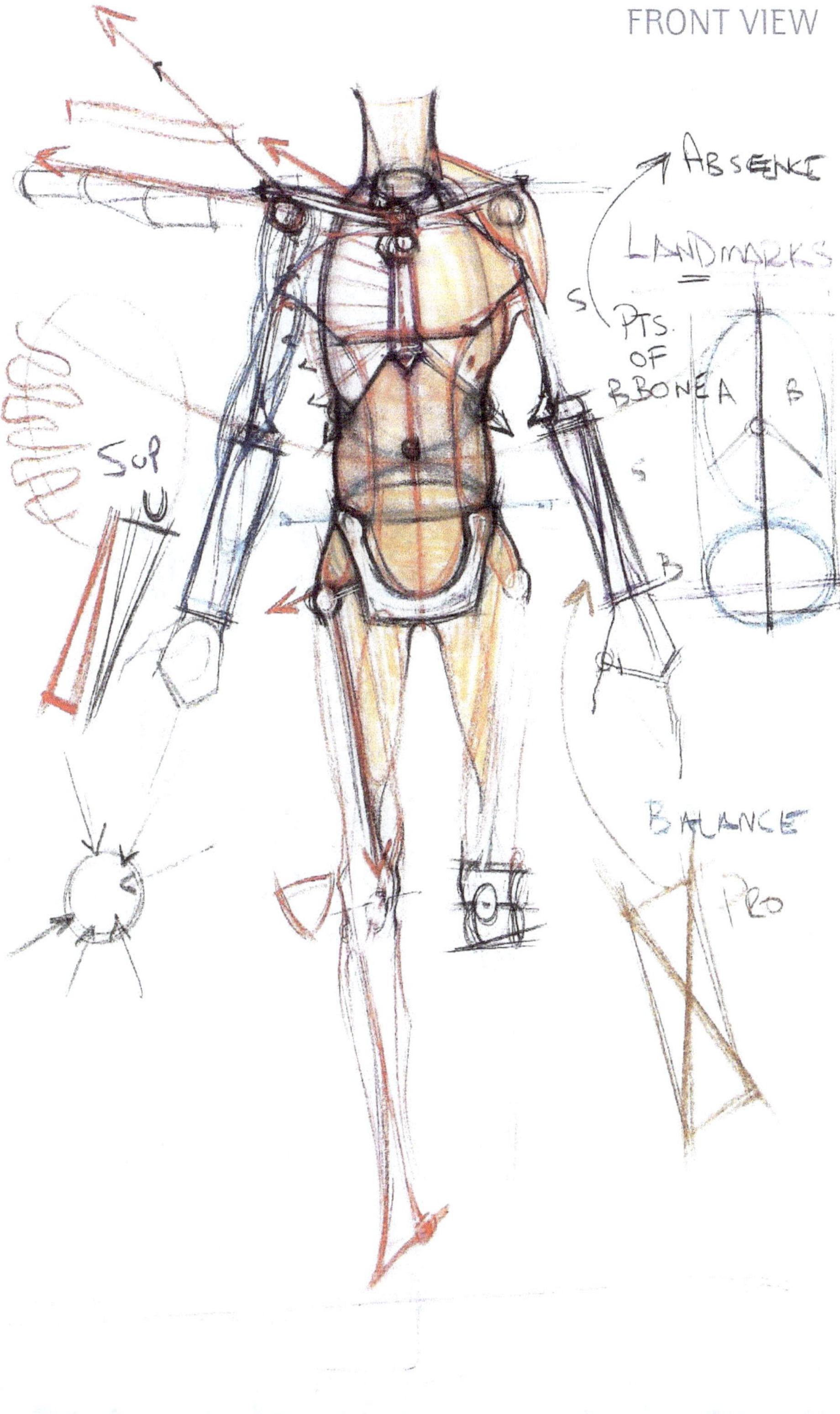

ABSENCE
LANDMARKS =
PTS. OF B.BONE
A
B
S
SUP
BALANCE
PRO

The color-coded drawings on these pages show the placement of the different muscle groups according to a simplified idea of their shapes. Study them for an understanding of how to use "T" overlaps. Because there is so much going on, it is extremely important to use the "T" overlaps to be clear about relationships and placement.

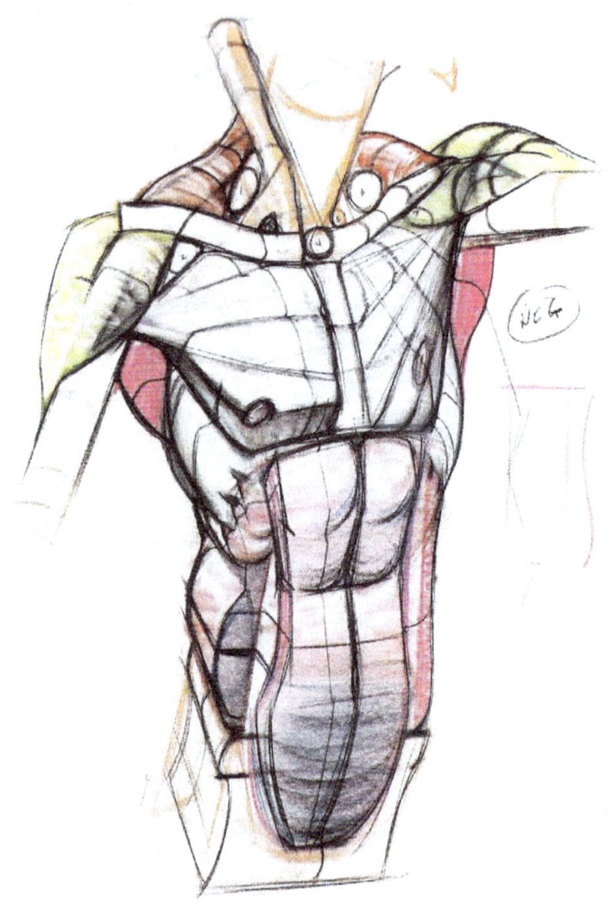

PROCESS

Following this step, every muscle is looked at in isolation, as a shape. First, you should be concerned with a muscle's shape and its placement; second, how that shape changes in relationship to the gesture or an action; and third, how to use the shape to keep a constant sense of volume and form.

ANATOMY PROCESS:

1. SHAPE
2. PLACEMENT
3. GESTURE
4. PERSPECTIVE

BACK VIEW

It is important to keep in mind the overall process at this point. Remember the progression:

> Gesture
> Shapes
> Landmarks/ Volume
> Anatomy
> Value

Think of these as all being related. Any time you move to another step, that step still needs to address all that came before it. For example, when we start to develop anatomy, we have to address the three steps that came before it in order to stay true to the original intention or gesture. Additionally, remember that drawing the pecs or deltoid is the description of a smaller or micro-gesture that is a smaller argument in support of the larger idea (the initial gesture).

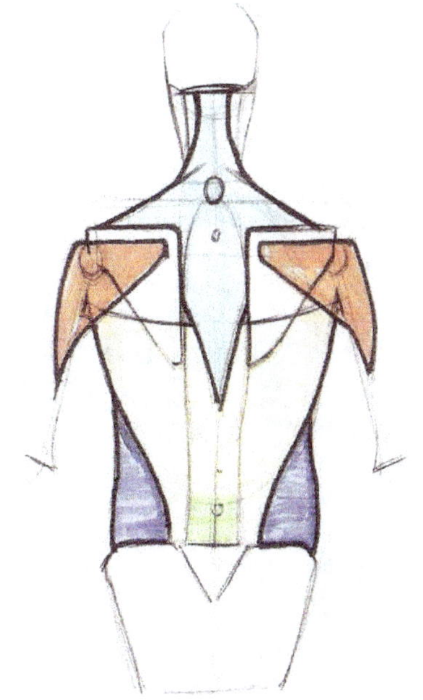

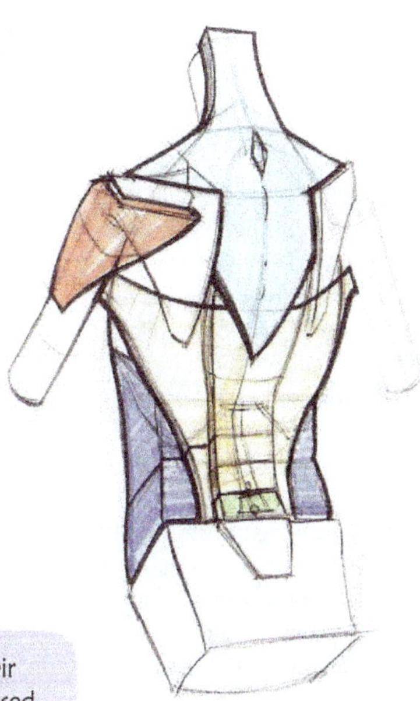

Consider the simplified shapes and their placements a map. This can be later used to help in identification or to create figures, animals, or other creatures.

ANATOMY AND MOTION

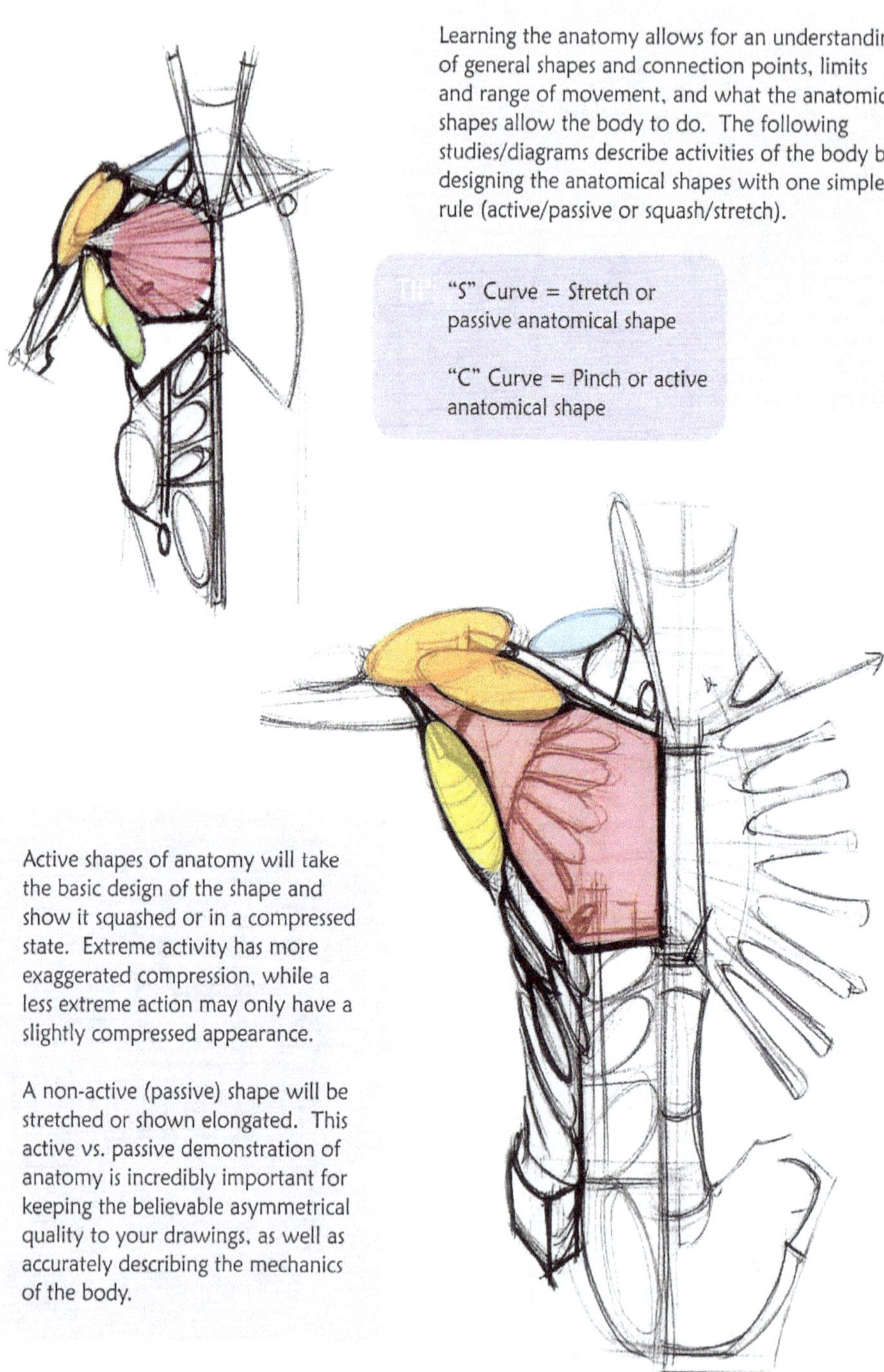

Learning the anatomy allows for an understanding of general shapes and connection points, limits and range of movement, and what the anatomical shapes allow the body to do. The following studies/diagrams describe activities of the body by designing the anatomical shapes with one simple rule (active/passive or squash/stretch).

TIP "S" Curve = Stretch or passive anatomical shape

"C" Curve = Pinch or active anatomical shape

Active shapes of anatomy will take the basic design of the shape and show it squashed or in a compressed state. Extreme activity has more exaggerated compression, while a less extreme action may only have a slightly compressed appearance.

A non-active (passive) shape will be stretched or shown elongated. This active vs. passive demonstration of anatomy is incredibly important for keeping the believable asymmetrical quality to your drawings, as well as accurately describing the mechanics of the body.

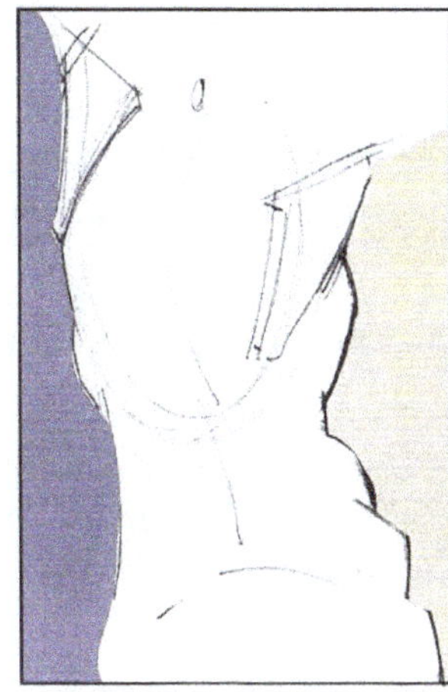

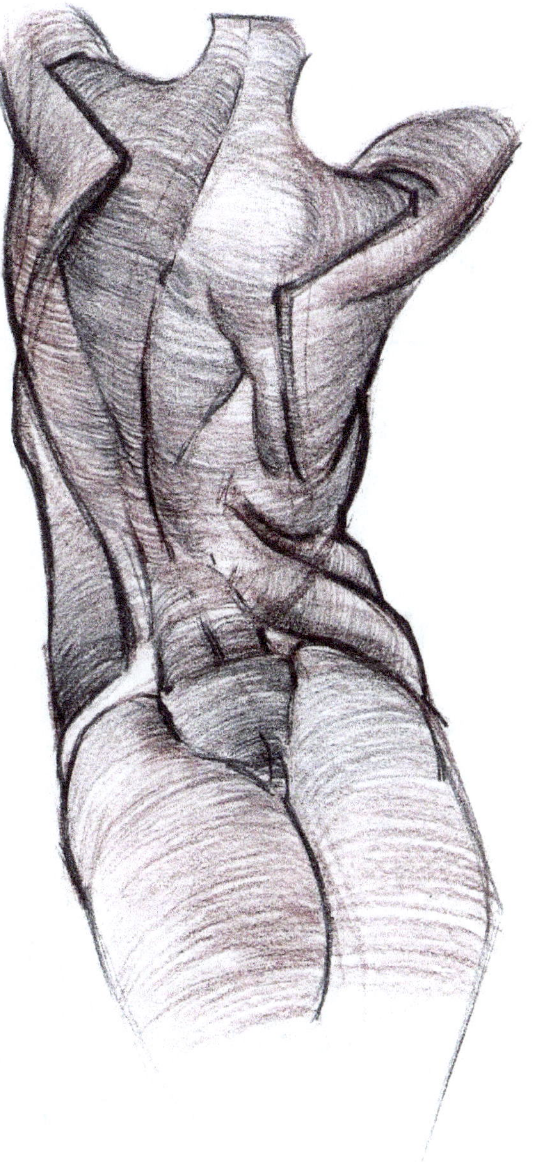

Notice that because the passive side of the figure is designed with all "S" curves, the negative shape is very fluid and rhythmic.

On the pinched side of the figure, the C-curve or straight has been exaggerated into a more jagged shape to suggest weight or flexion.

Simplifying your use of line will build a specific relationship to the negative shapes surrounding your figure. In the diagram above, notice the difference in the negative space created by specific ideas of active/passive anatomical gestures (rhythmic vs. angular). Remember, you are, through the process of drawing the figure, also responsible for the design of the surrounding areas.

STERNOCLEIDOMASTOID – GESTURE

The sternomastoid muscles work to pull the head and neck forward as well as rotate the head and face laterally. These muscles begin on the interior surface of the manubrium and clavicle and insert into the skull behind the ear.

STERNOCLEIDOMASTOID – SHAPE

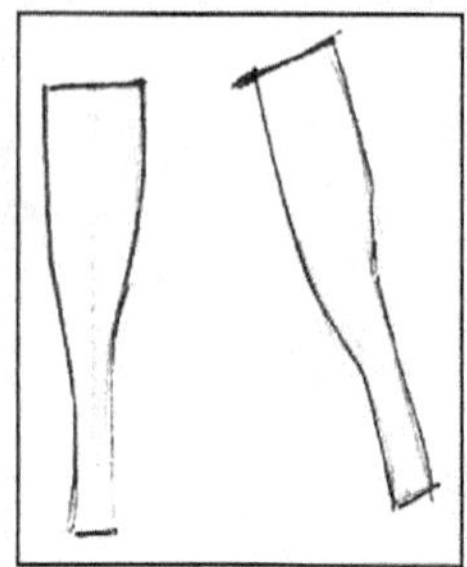

The sternomastoid can easily be remembered as a shape that resembles a baseball bat. When placing this shape, remember that it is aligned in a diagonal from the manubrium to the base of the skull.

Additionally, the shape of the sternomastoid should not be drawn symmetrically. One side of the shape is always higher, mimicking the design principles in our gesture.

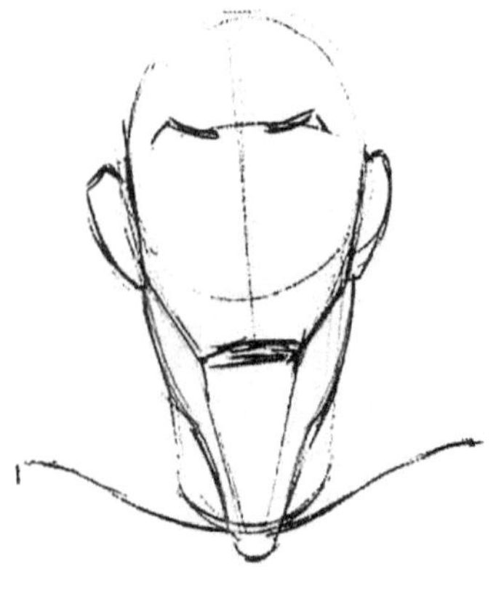

When the head turns, the involvement of these muscles can be indicated by elongating their shapes (indicating a passive state or stretch) or by contracting and shortening the form (which will indicate an active state or the muscle being used).

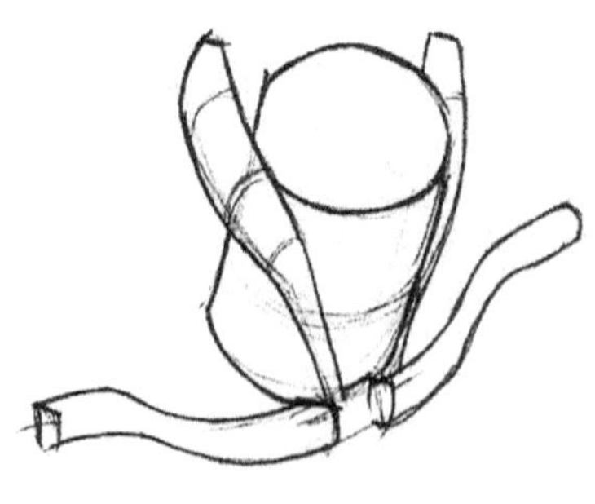

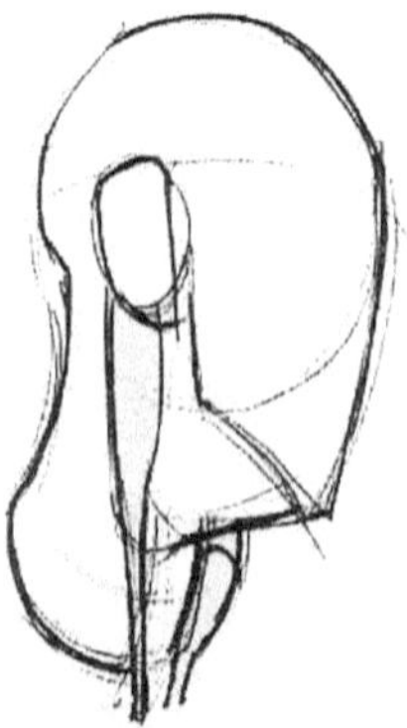

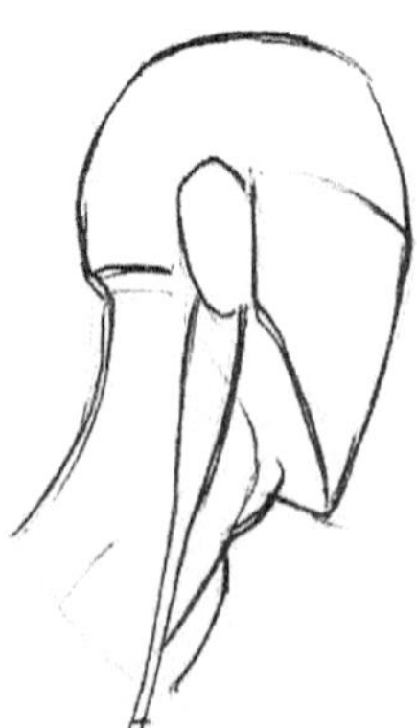

STERNOCLEIDOMASTOID – VOLUME

The volume of the sternomastoid should be shown wrapping around the cylinder of the neck while moving back spatially to suggest the distance from the manubrium back to the base of the skull.

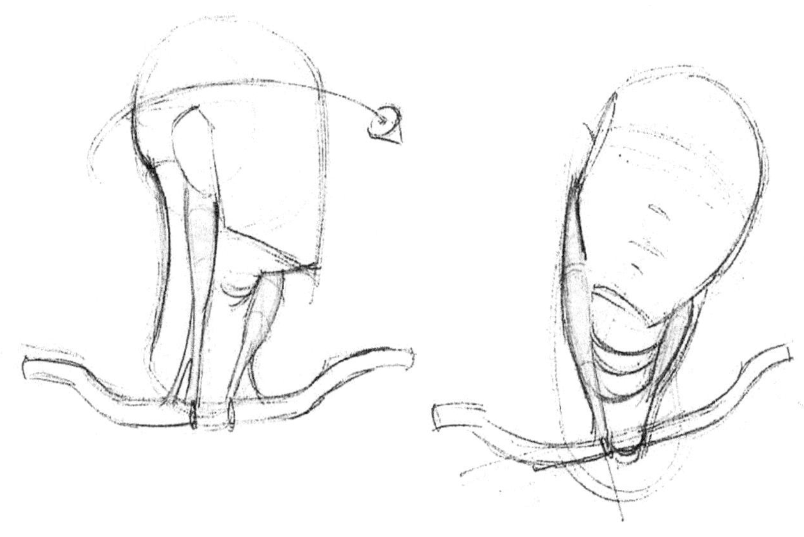

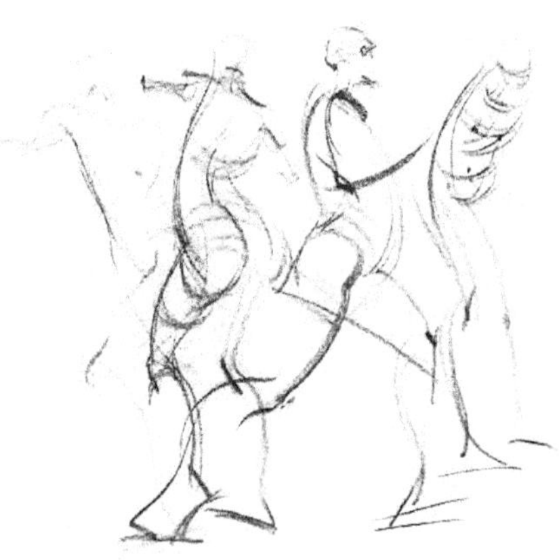

PECTORALIS MAJOR – GESTURE

The pectoralis muscle pulls the arm forward across the chest and rotates the arms medially. Remember the description of the muscle's action is what you want to look for when deciding the "C" or "S" curve, which again is the basic gesture to the muscle. It begins on the medial half of the clavicle, along the length of the sternum, and across the cartilages of the first six or seven ribs, inserting into the bicipital groove on the front of the humerus.

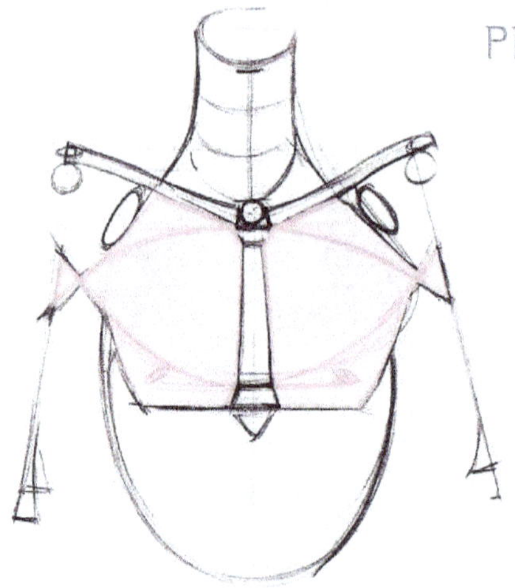

PECTORALIS MAJOR – SHAPE

The shape of the pectoralis resembles a fan, with the clavicular, sternocostal, and abdominal sections overlapping — or even more simply like a gold fish with its head missing. The flat portion of the head missing head sits along the sternum, while the tail can be seen wrapping and pulling to the front of the humerus.

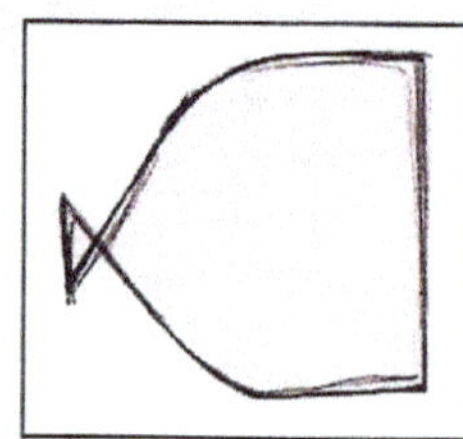

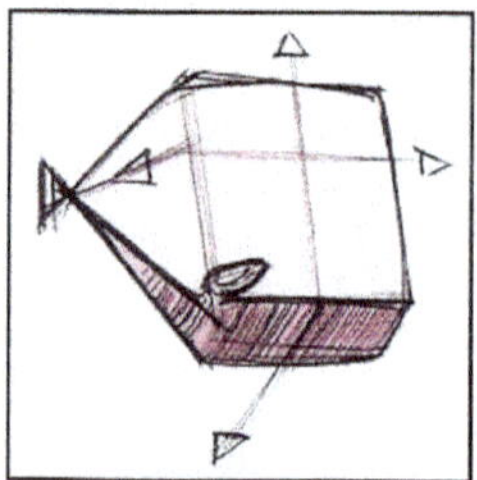

The pectoralis can simply be thought of as a small box or rectangle sitting on top of the rib cage. The width of the pectoralis should be shown towards the bottom of the form nearest the nipple. The volume will be displaced depending on the action. For example, if the arm is raised, the volume is spread more evenly and will have less of a noticeable corner. If the pectoralis is flexing or contracted, the volume becomes peaked and will have a more noticeable width.

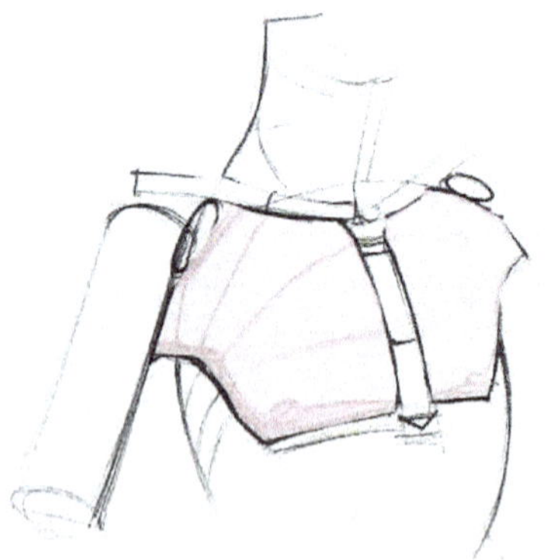

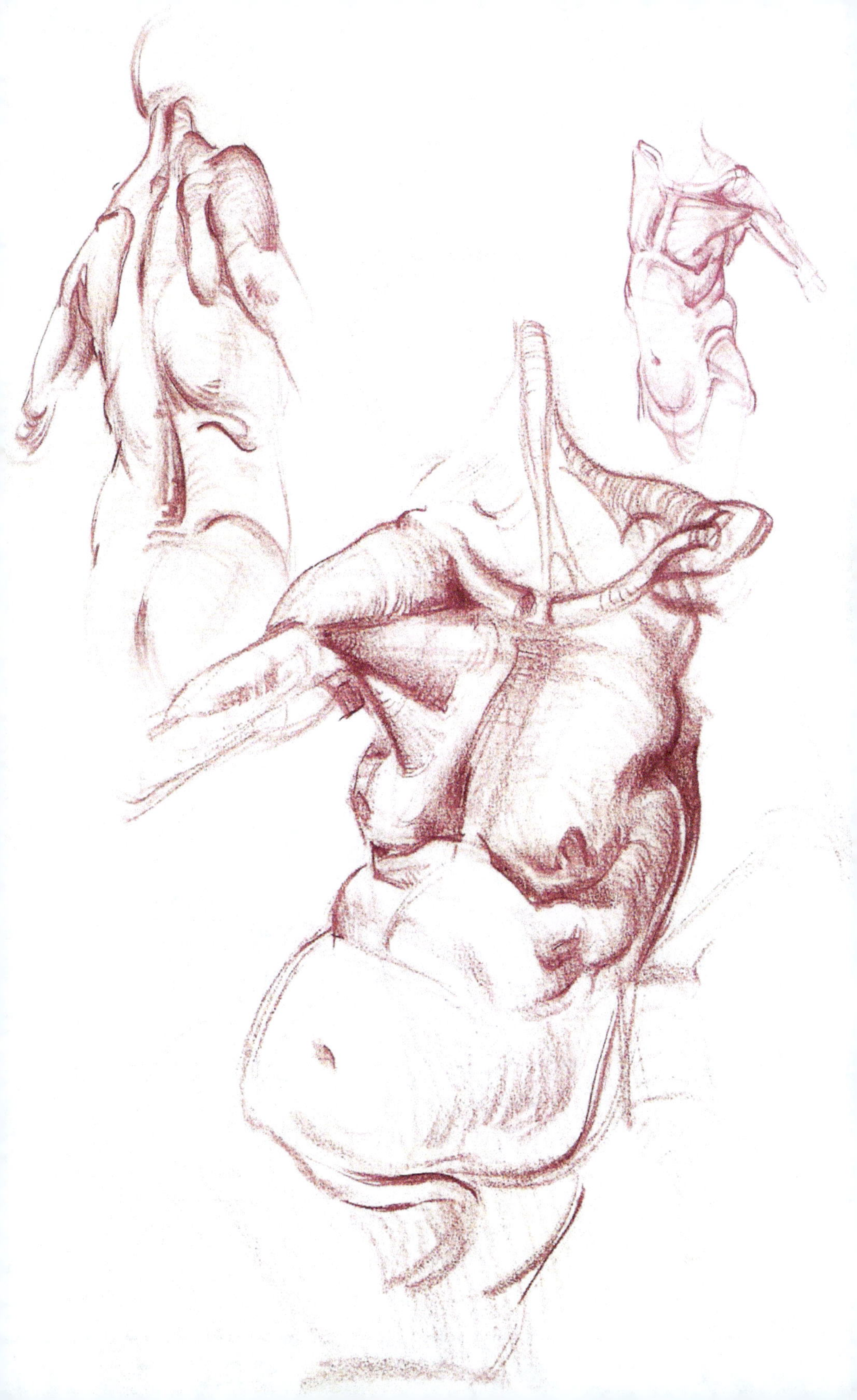

PECTORALIS MAJOR – VOLUME

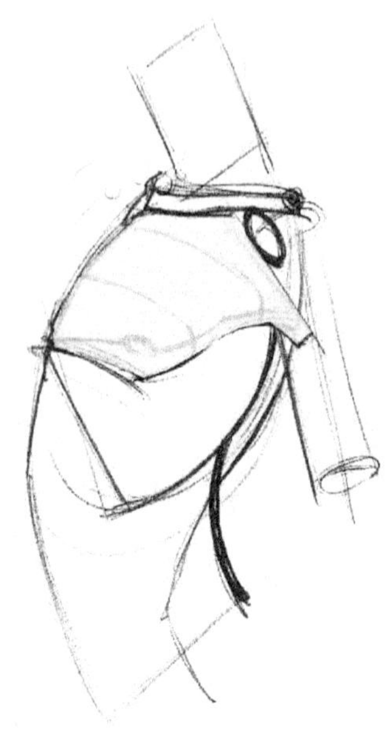

The shape will change when the arm raises by showing the tail of the goldfish unwrapping and becoming elongated.

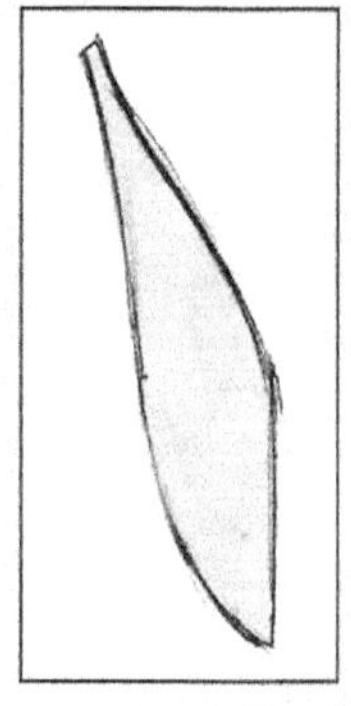

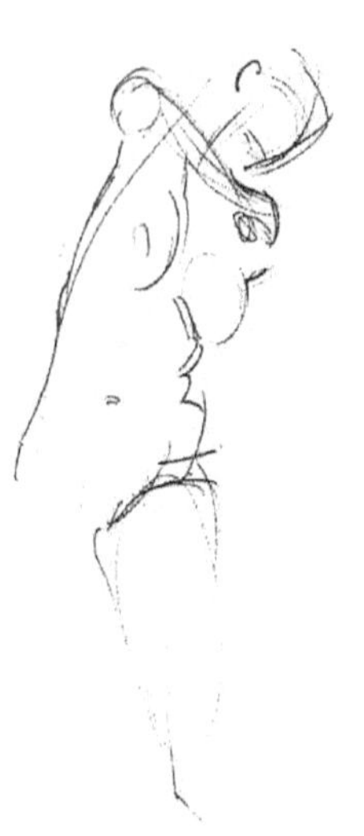

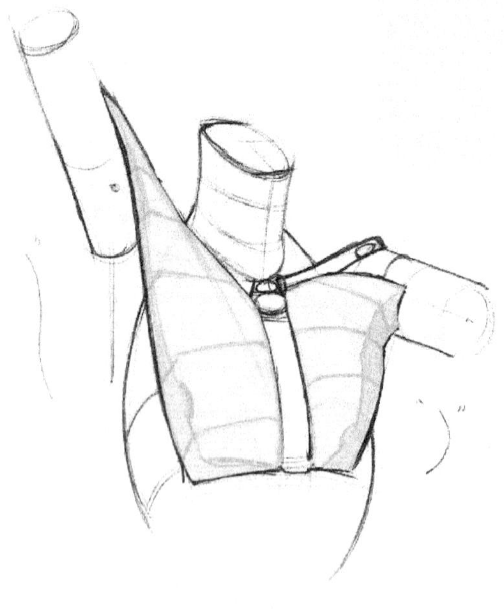

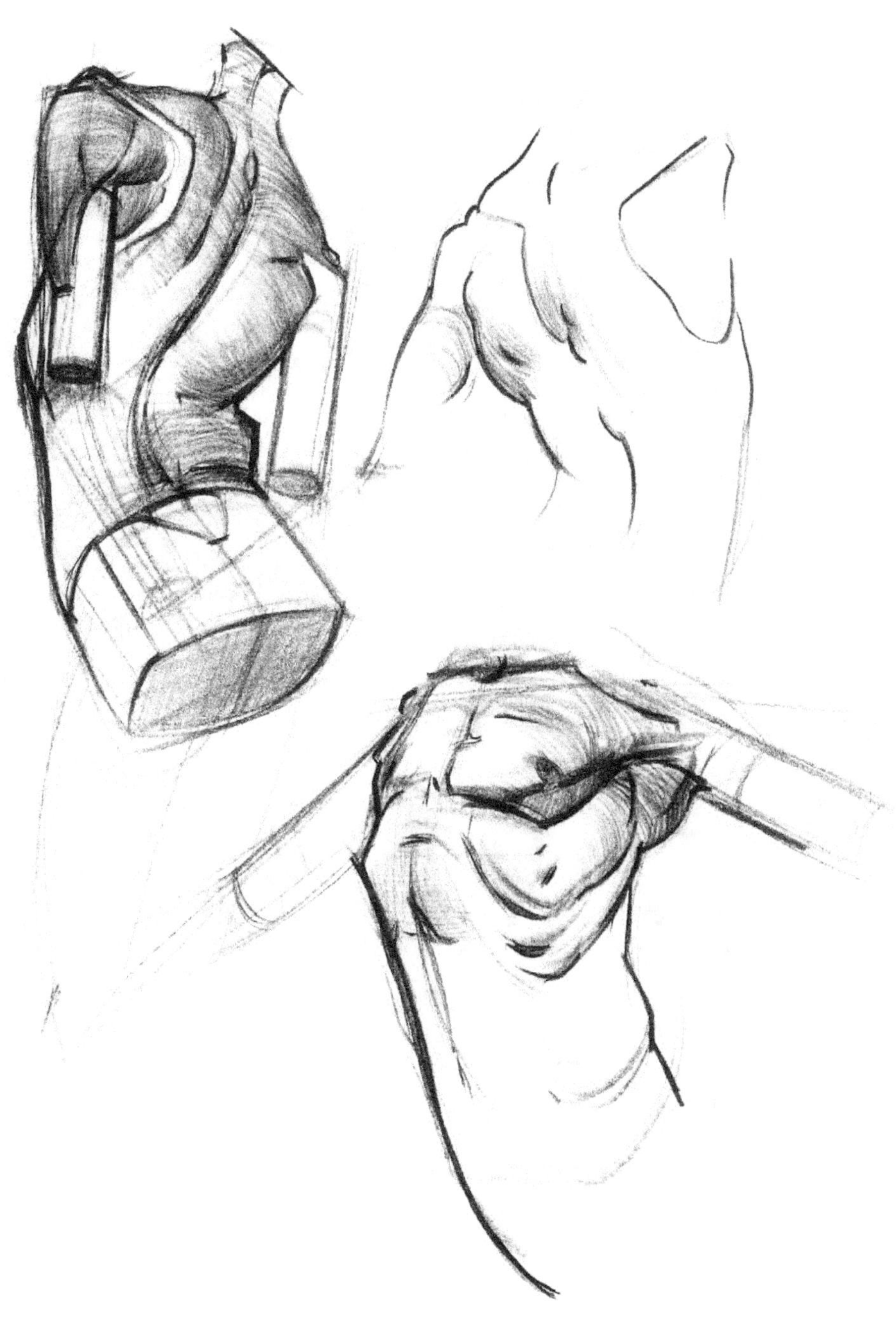

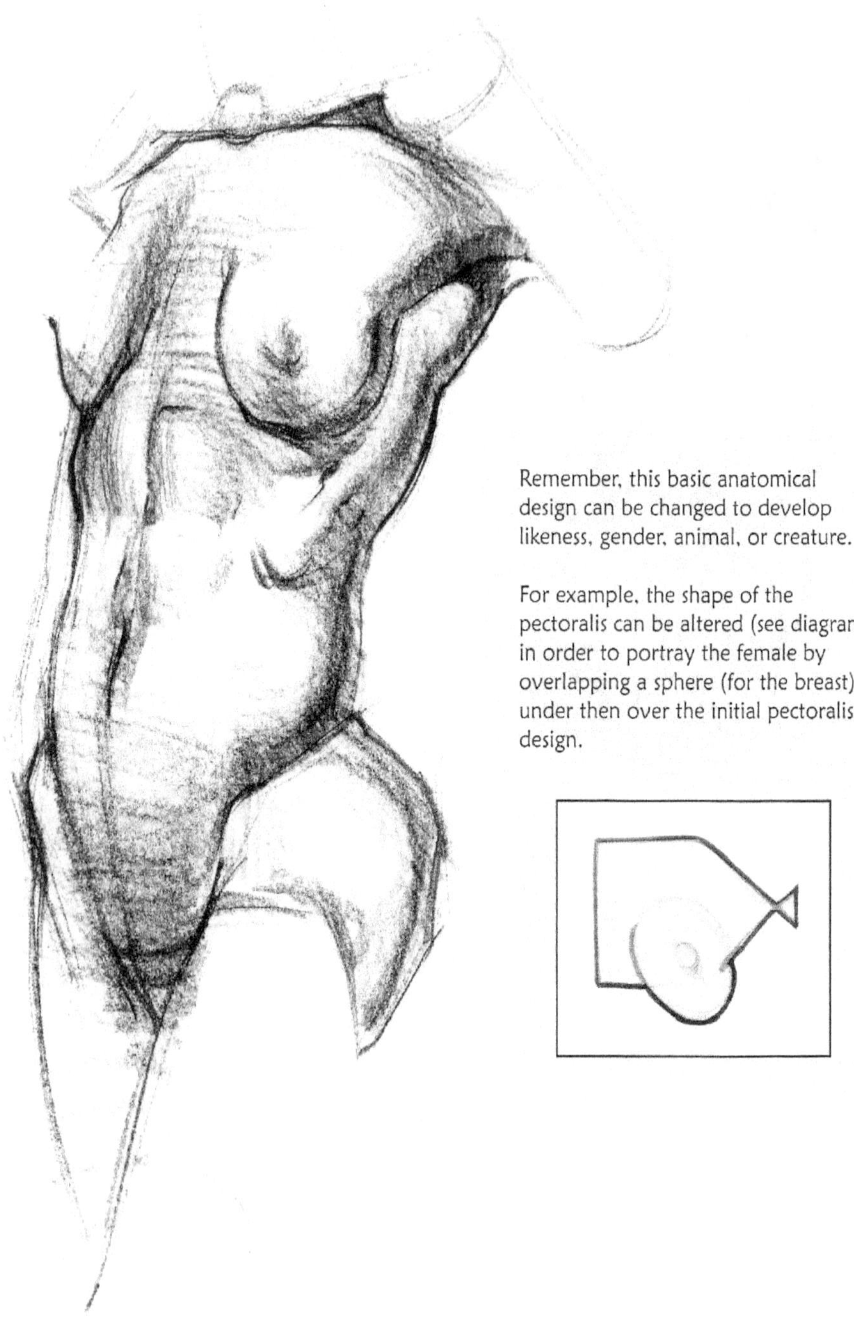

Remember, this basic anatomical design can be changed to develop likeness, gender, animal, or creature.

For example, the shape of the pectoralis can be altered (see diagram) in order to portray the female by overlapping a sphere (for the breast) under then over the initial pectoralis design.

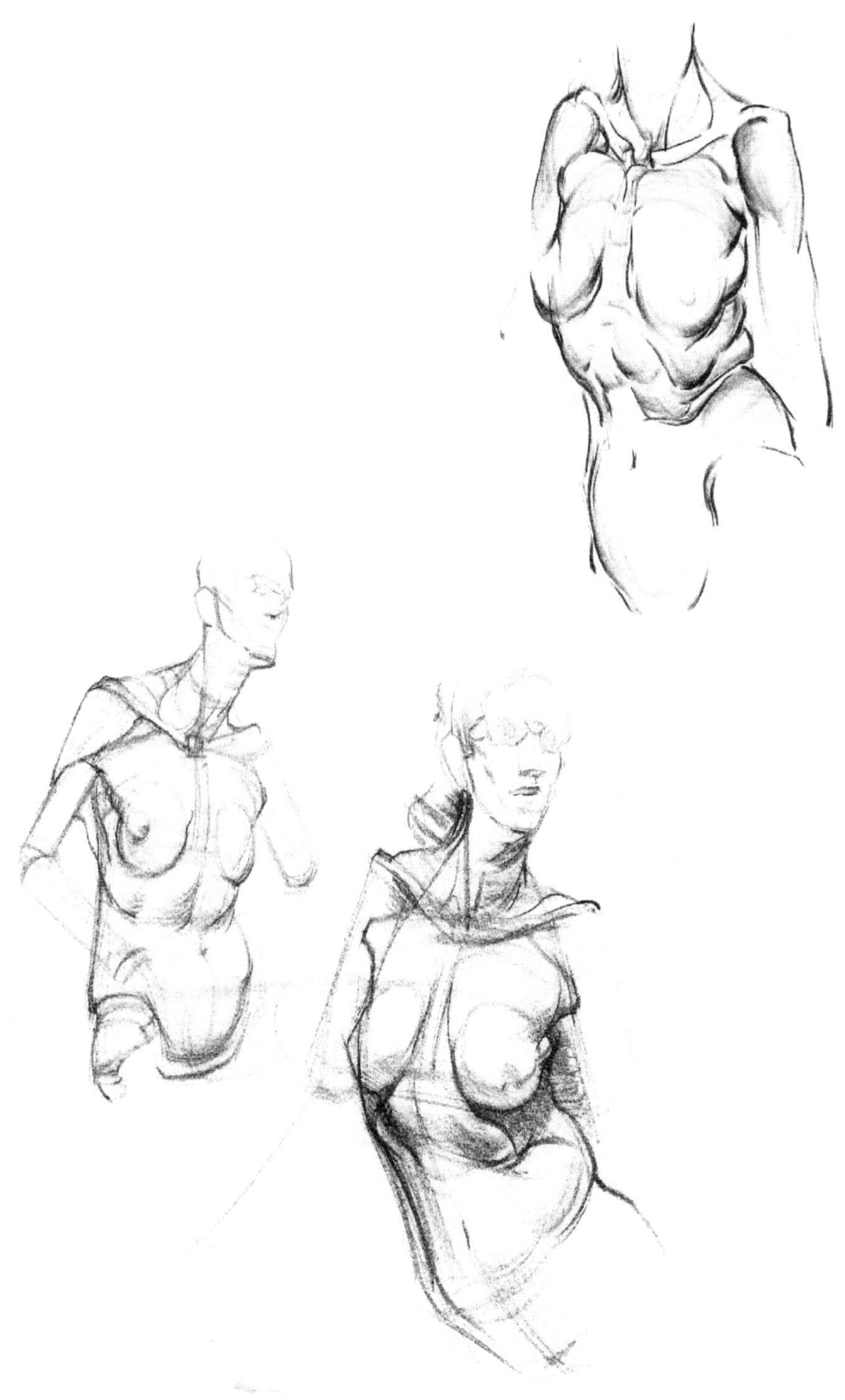

TRAPEZIUS – GESTURE

The trapezius rotates, lifts, and pulls the scapula. Its origin is at the base of the skull and down to the twelve thoracic vertebrae. The trapezius inserts along the lateral third of the clavicle, the upper edge of the spine of the scapula, and ending at the tubercle of the spine.

TRAPEZIUS – SHAPE

The trapezius can be more easily used and remembered by simplifying it into a basic shape. The complex shape of the trapezius can be thought of as an upside-down dagger.

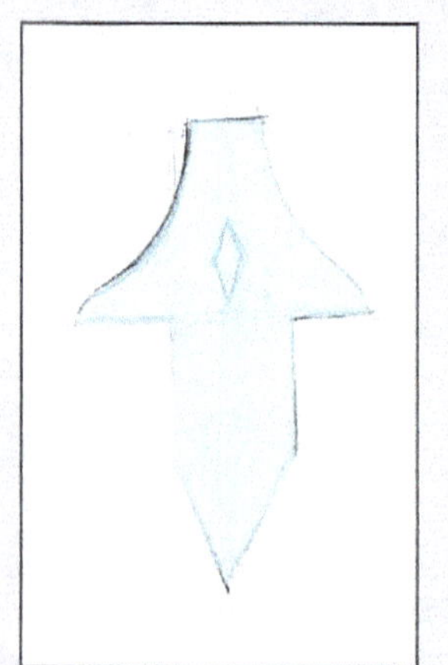

The top or handle of the dagger is the portion of the muscle inserting into the base of the skull. As the wings of the trapezius swing down to the top of the scapula, the handles of the dagger can be seen. Where the muscle pulls in a point to the spine, the blade of the dagger can be seen.

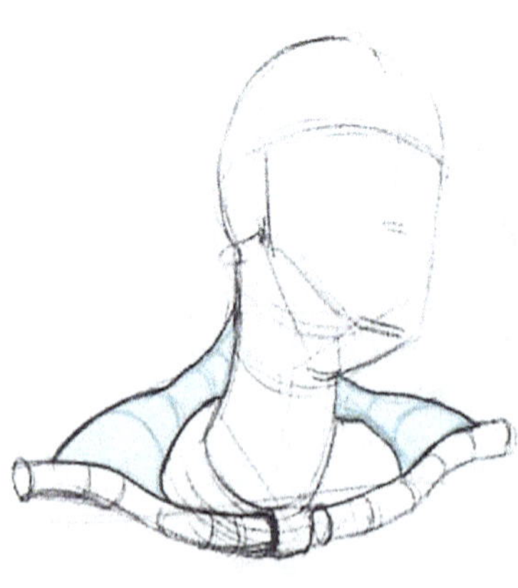

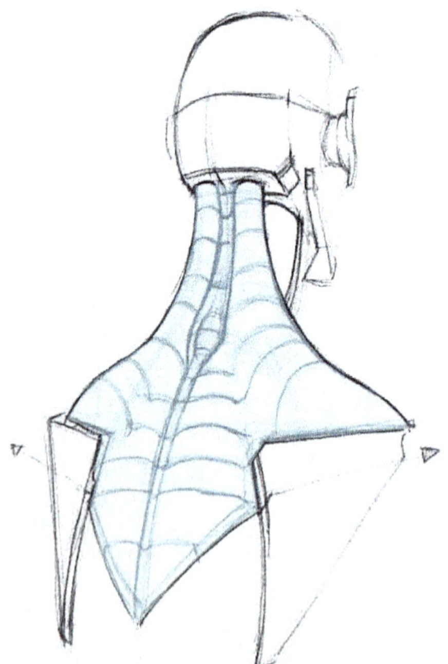

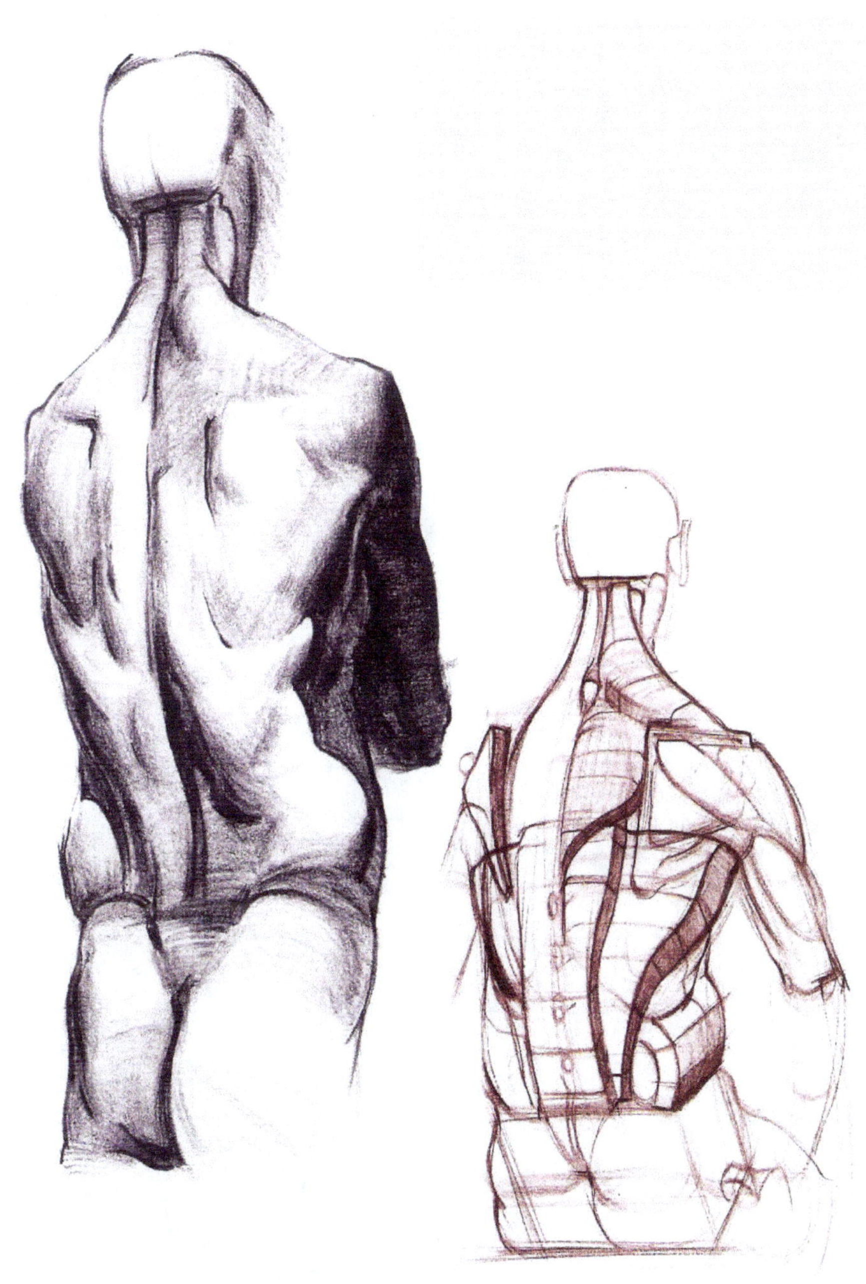

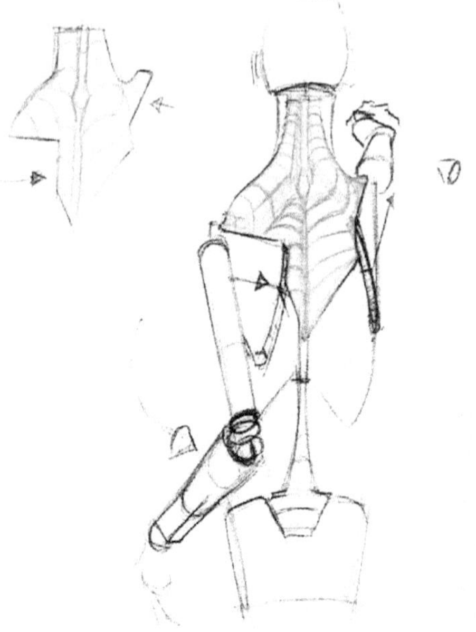

The shape of the trapezius can also change to suggest a movement or activity. Notice how the shape of the trapezius pinches when the scapula moves closer to the spine. Conversely, when the scapula pulls forward with the movement of the arm, the trapezius is stretched and is shown as thinner, pulling across the ribs beneath.

Additionally, the shape of the trapezius (along with all the overall shape of every muscle) can be altered to suggest gender, type, or character.

TRAPEZIUS – VOLUME

The volume of the trapezius should be shown moving across the major forms it rests on — the head, neck, scapulae, and rib cage. Additionally, its volume will be affected when the muscle's shape is stretched or pinched.

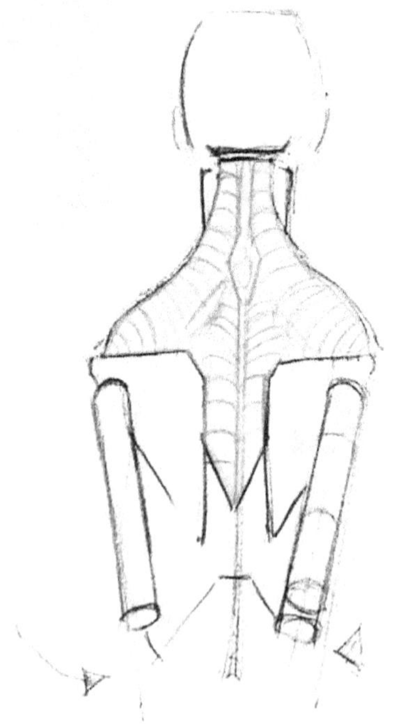

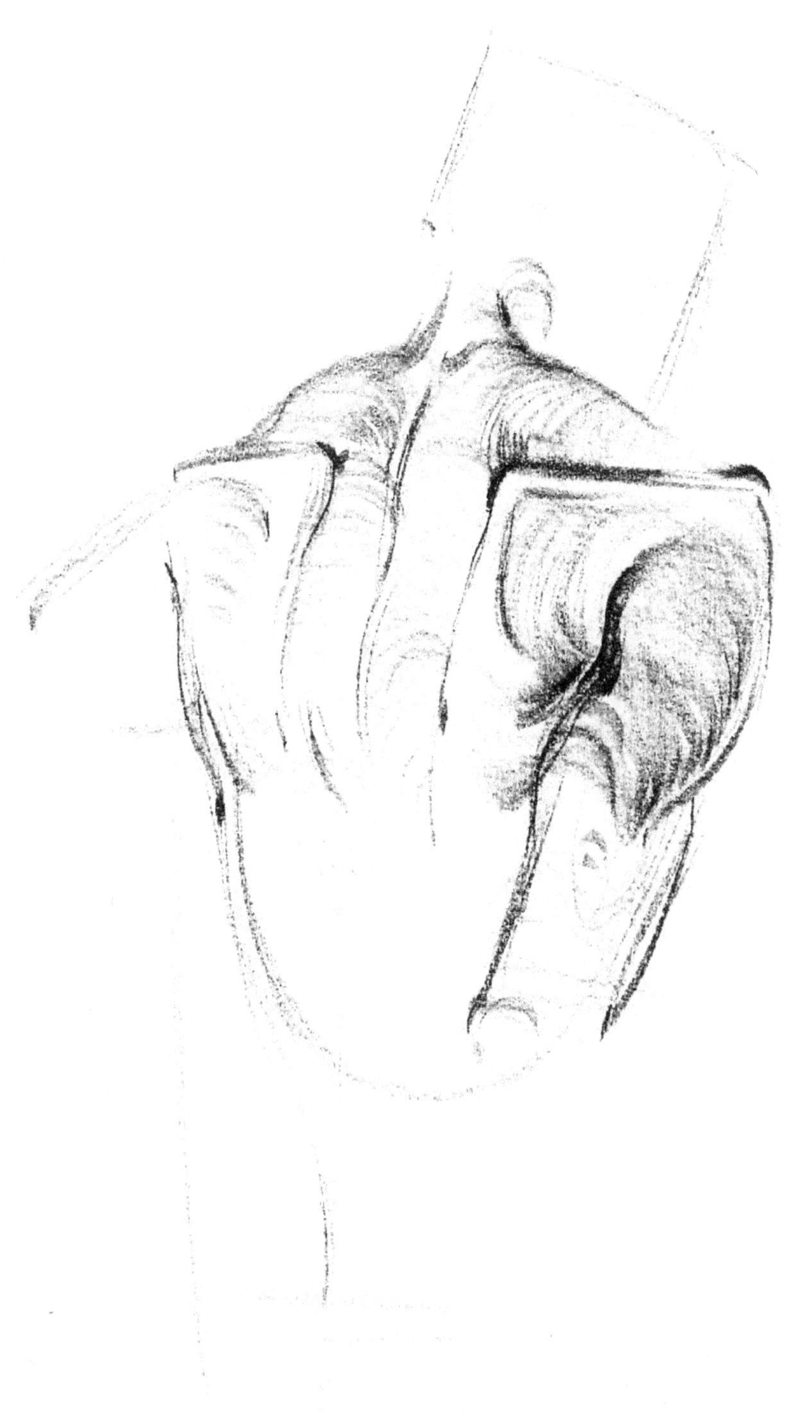

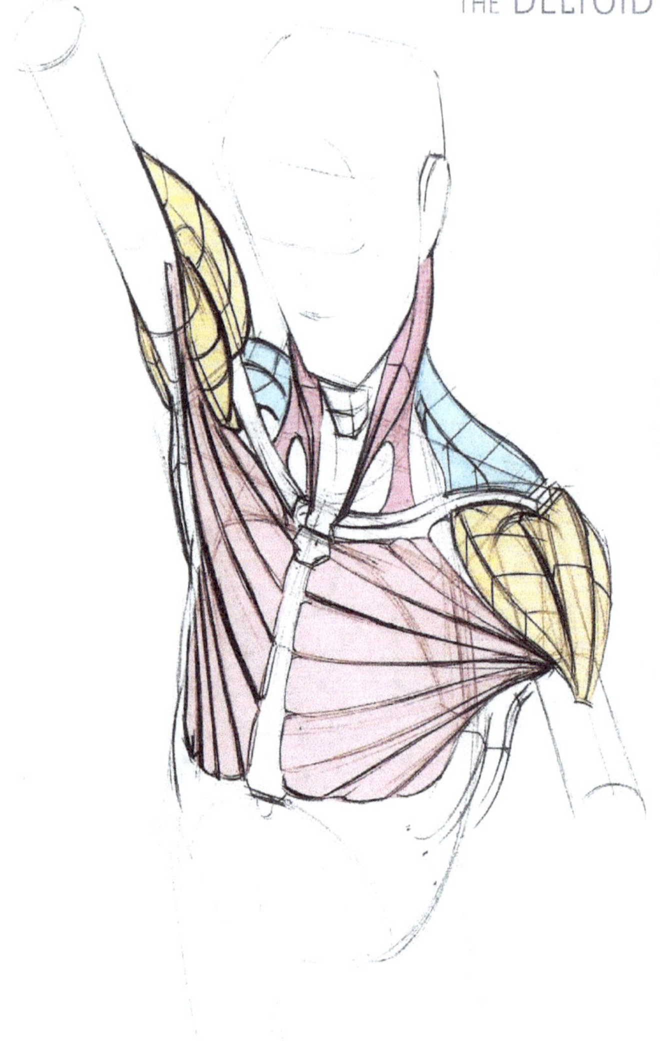

The deltoid consists of three separate heads. The first is the anterior portion, which raises the arm in front of the body. Second is the acromial portion, which pulls the arm away from the body. Third is the anterior portion, which pulls the arm backward.

The origin of the deltoid's shape is a continuous line passing along the last third of the clavicle, the border and top of the acromion of the scapula, and the lower edge of the spine of the scapula. The deltoid inserts outside of the humerus about half-way down the arm.

DELTOID – SHAPE

Seen from the side, the shape of the deltoid resembles an upside-down triangle. From the front or back, the deltoid still appears as a triangle — just a much thinner one.

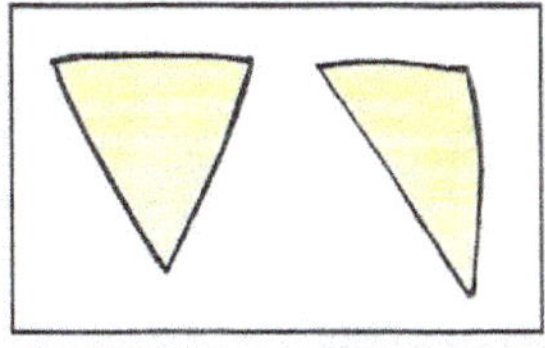

The red line in this diagram represents the area of connection for the deltoid.

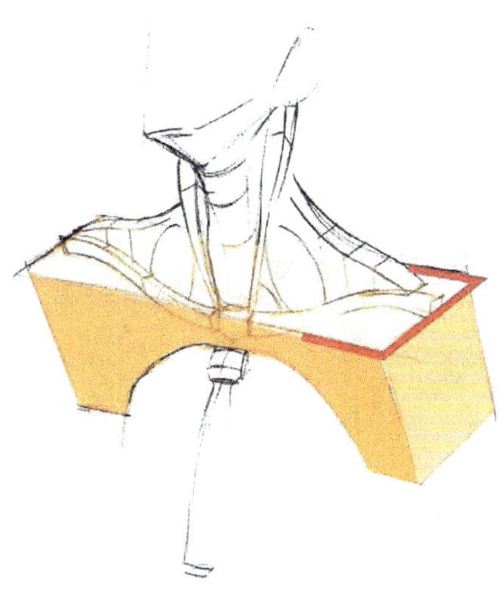

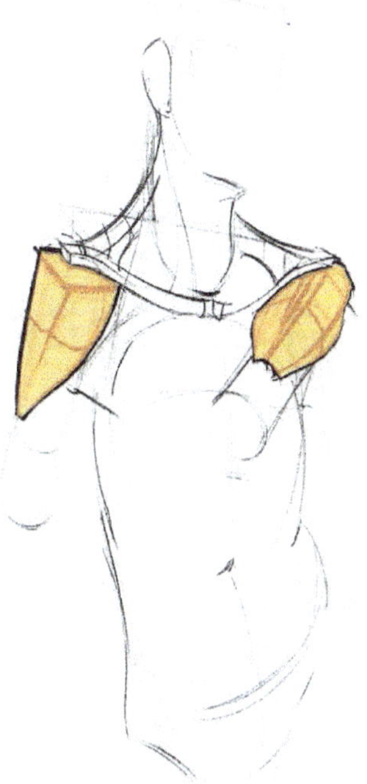

The above diagram represents a second perspective structure to describe the shoulder girdle. The shoulder girdle is the combination of the clavicle and scapula. This structure sits on top of the rib cage and acts as an incredibly useful tool for organizing/understanding the shoulders.

DELTOID - VOLUME

The most effective way to show the deltoid in perspective, with a strong sense of volume, is by wrapping its insertion point in the same perspective as the direction of the arm.

The deltoid should also reflect the perspective of the upper body as its origin pulls away from the shoulder girdle.

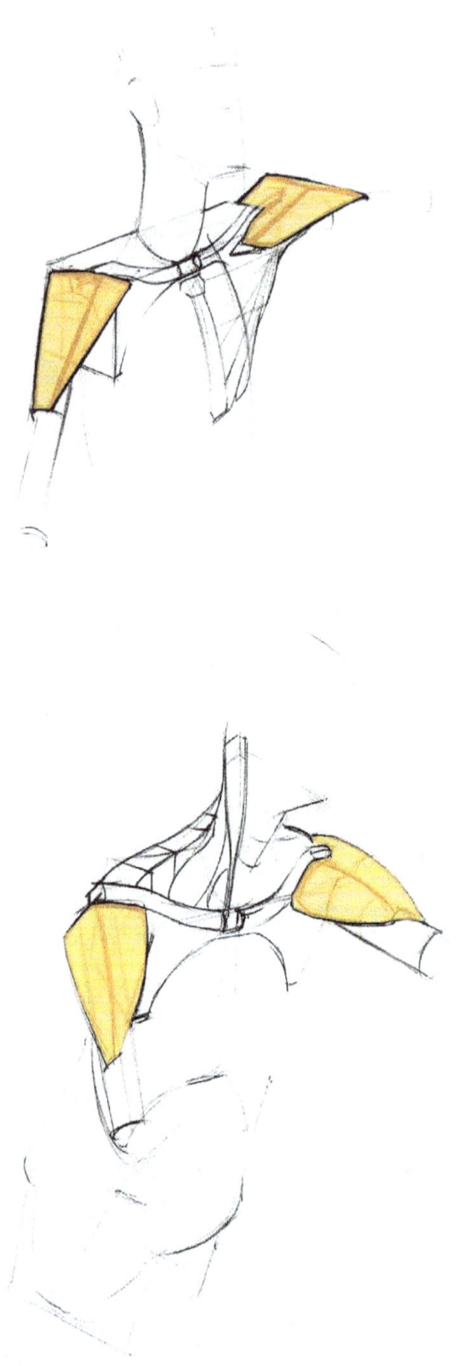

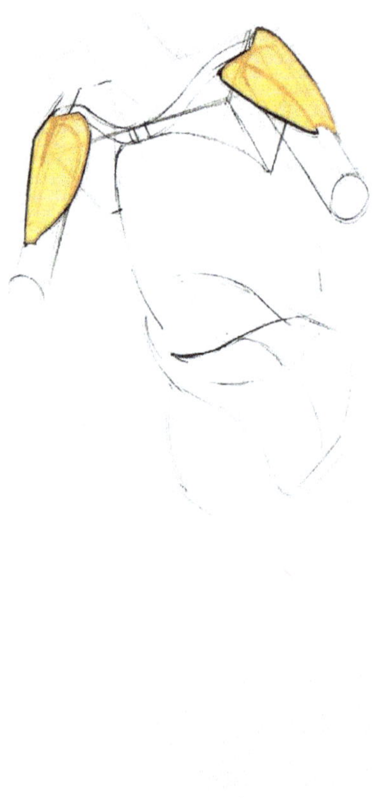

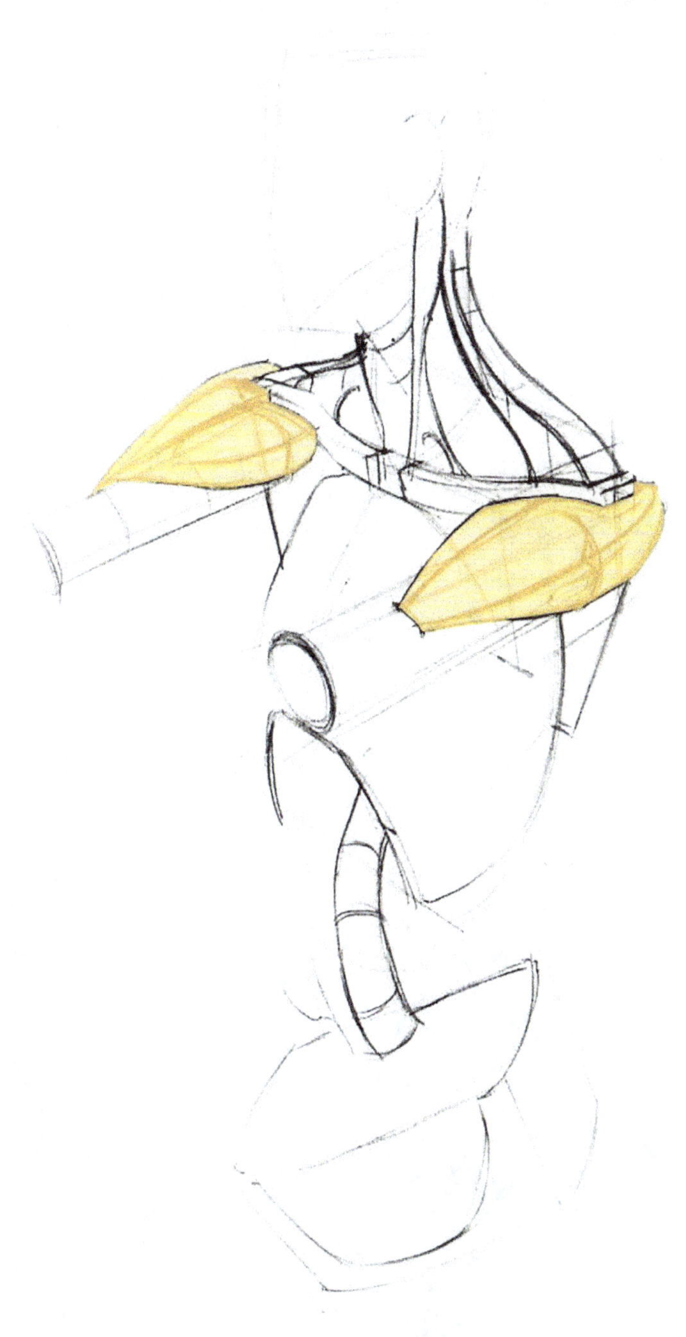

RECTUS ABDOMINIS – GESTURE

The rectus abdominis (or abdominal group) works to flex the trunk at the lumbar vertebrae. This muscle group begins at the base of the pubic bone and inserts into the surfaces of the fifth, sixth, and seventh ribs.

RECTUS ABDOMINIS – SHAPE

The abdominal group can be simplified into a shape resembling a bullet. The curved portion of the bullet fits into the pelvis, while the flattened end lies along the ribs above the thoracic arch. Within this shape, there are eight sections. Starting from a straight or horizontal line around the area of the belly button, these sections progressively rise to a peak (see diagram).

When the trunk moves front, back, or side to side, this shape can be shown pinching, stretching, or aiding in a twist.

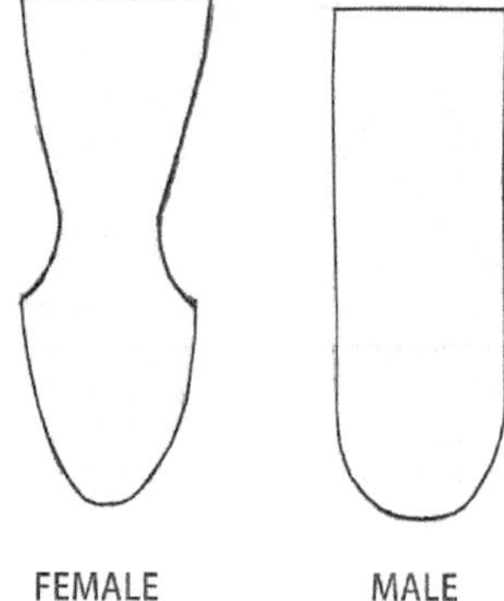

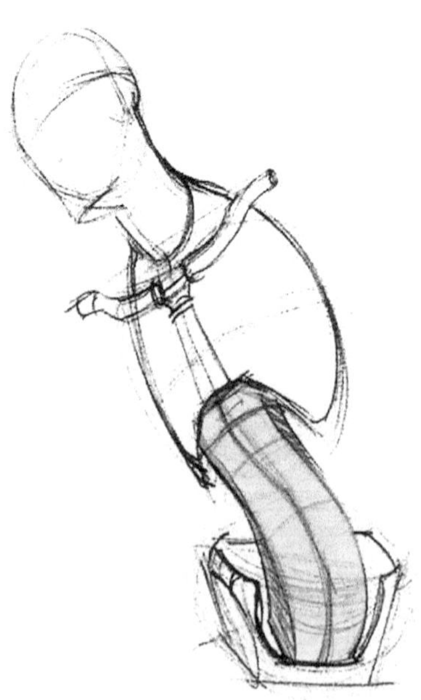

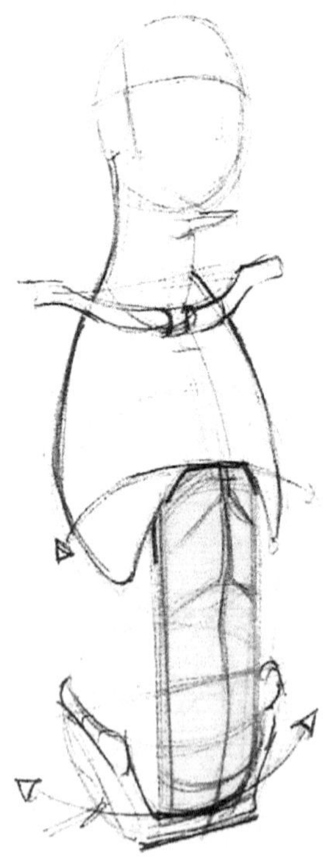

The volume of the abdominals should be shown with a very thin side plain. This gives the viewer an indication of the depth of this muscle group. Ultimately, this volume will resemble a flattened box or rectangle.

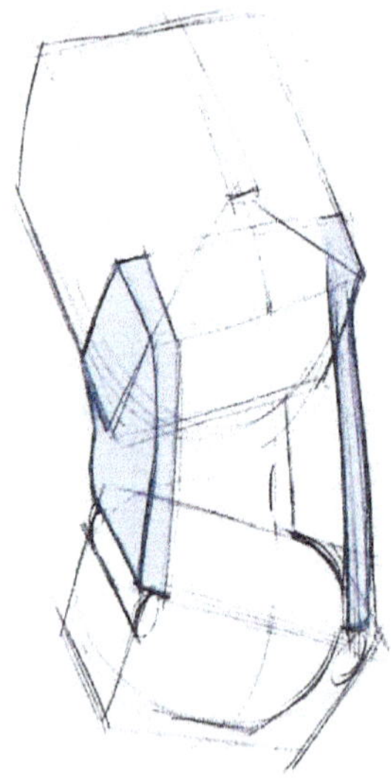

The external oblique attaches to the lower eight ribs of the rib cage, running downwards and to the back, inserting into the pelvis.

The function of this muscle is to bend the trunk laterally, to twist, and when both sides are used simultaneously the rib cage will be pulled down towards the pelvis.

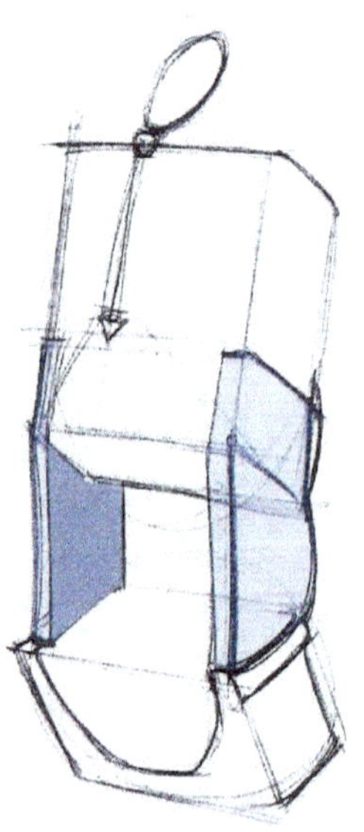

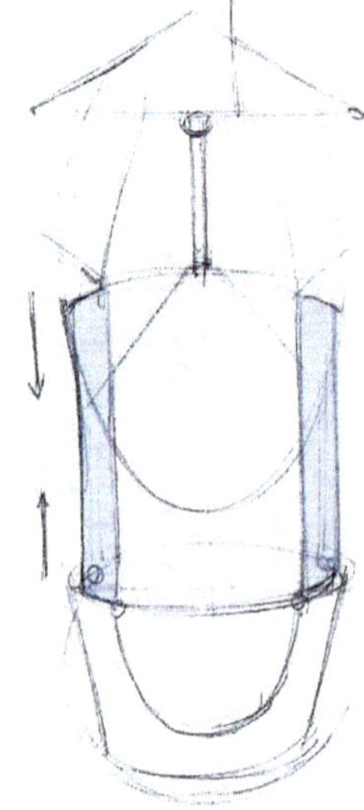

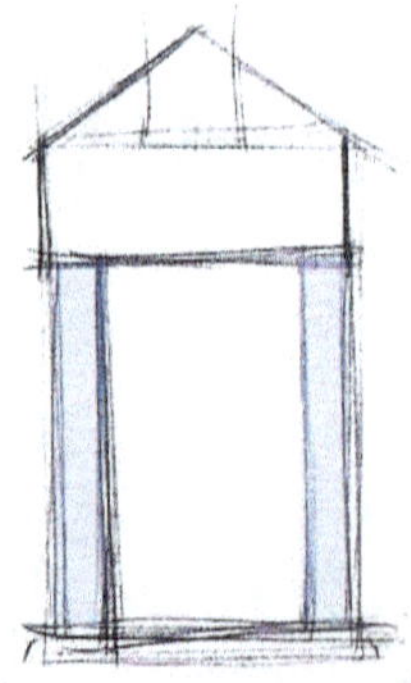

OBLIQUES – SHAPE

The shape of the obliques can be thought of as two elongated rectangles, similar to two columns supporting the rib cage over the pelvis.

When drawing this shape, attach the obliques to a wrapping line lower down on the form of the rib cage, while relating them from the side to the back.

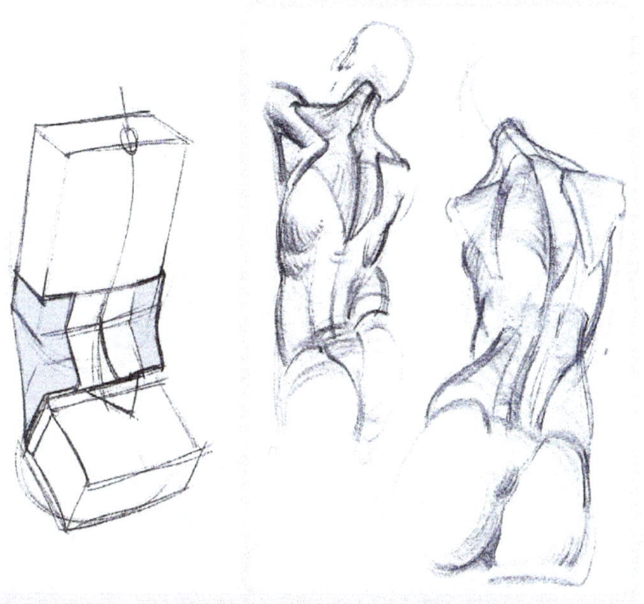

As discussed earlier in the section on Connections, the goal after wrapping these shapes to their corresponding perspectives is to design them with a "C" or "S" depending on the activity they are engaged in.

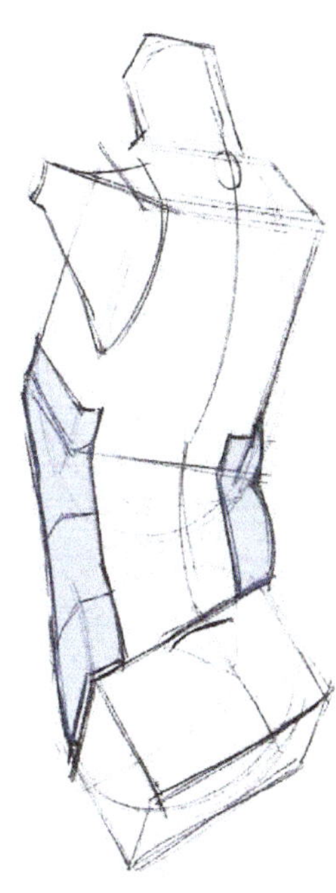

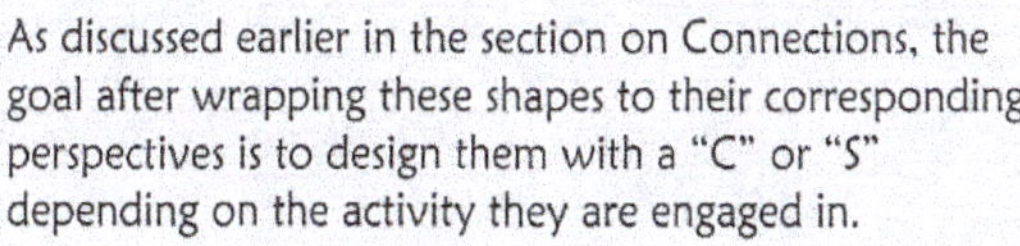

OBLIQUES – VOLUME

When assigning volume to the obliques, make the shape wrap around the existing forms, but also think of this shape as having a depth similar to a flattened box.

SERRATUS ANTERIOR - GESTURE

The serratus slides the scapula forward and aids in elevating the shoulder and the arm. It originates on the outer surfaces of the first eight or nine ribs and inserts into the underside of the scapula.

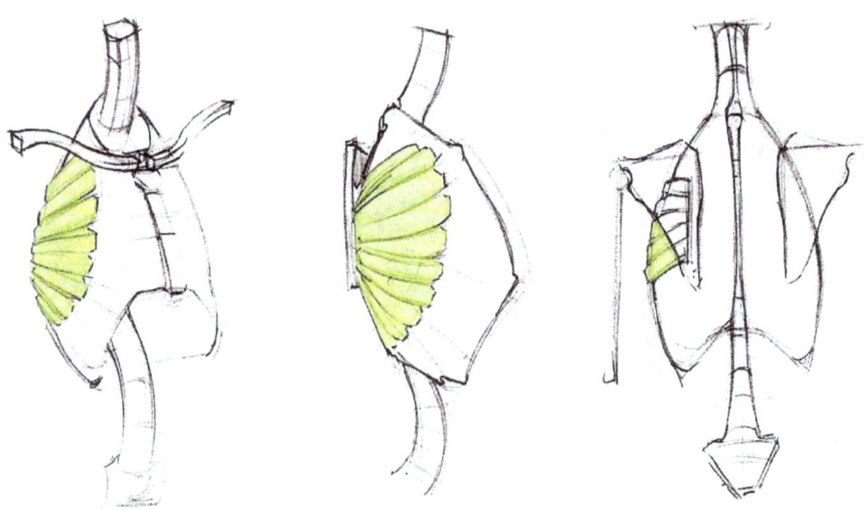

SERRATUS ANTERIOR - VOLUME

The volume of the serratus should show the muscle as a sphere or ovoid form. Additionally, this muscle should appear to be a smaller form sitting on the larger form of the rib cage.

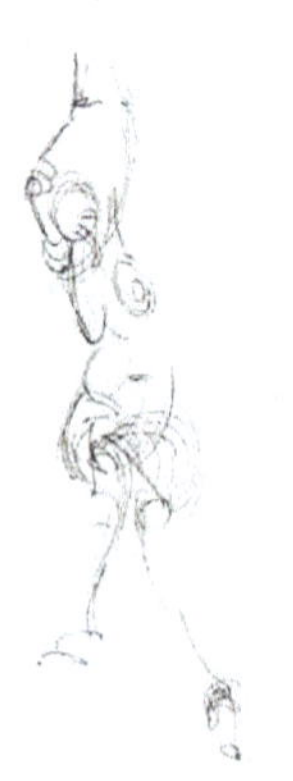

SERRATUS ANTERIOR - SHAPE

The shape of the serratus can resemble an egg form with three legs or tabs coming out from the bottom. The top portion of the egg represents the muscle pulling under and to the bottom of the scapula, while the feet or tabs shows the muscle pulling into and between the ribs. Additionally, these feet can be shown connecting into the upper portion of the obliques in order to develop a strong connection between the two anatomical shapes.

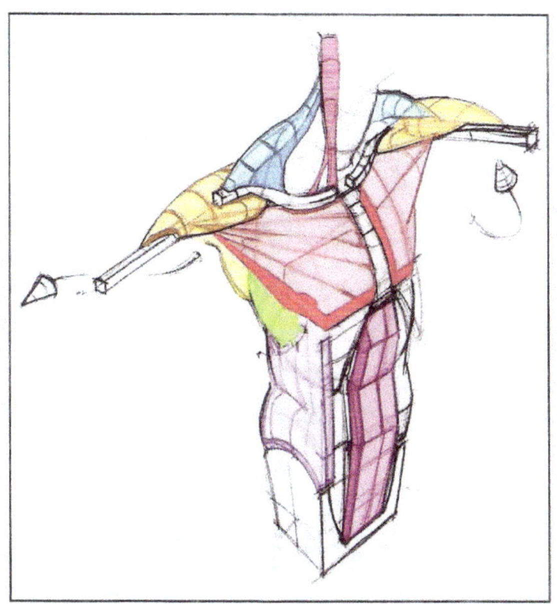

Study the diagram for the shapes and planes of the anatomical features covered so far. Strong knowledge of the anatomical planes is the best tool for creating believable light and shadow in tonal drawings.

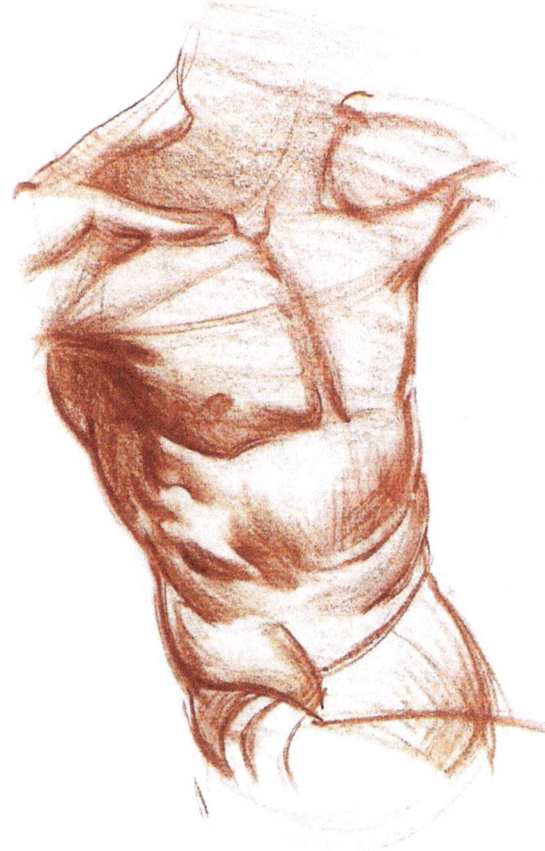

ERECTOR SPINAE - GESTURE

The erector spinae (or sacrospinalis) is a muscle group that extends the vertebral column and head, as well as aiding in flexion and rotation. This muscle group begins on the inner surface of the iliac crest and ends into numerous points on the back of all twelve ribs, and eventually into the skull.

ERECTOR SPINAE - SHAPE

The erector spinae can be simplified into a shape that resembles two corn dogs side-by-side, or two ovoid shapes above two cylinders. Depending on the activity taken by the figure, these simple shapes can easily be shown to pinch (by contracting their shape) or stretch (by elongating the forms).

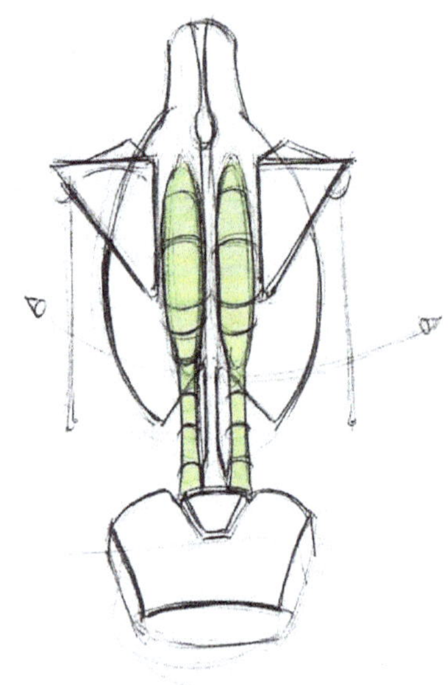

ERECTOR SPINAE - VOLUME

The volume of the erector spinae should be shown following the perspective and volume of the spine, rib cage, and pelvis.

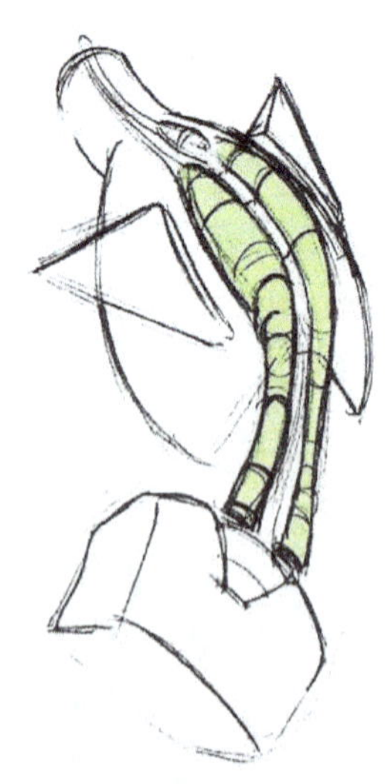

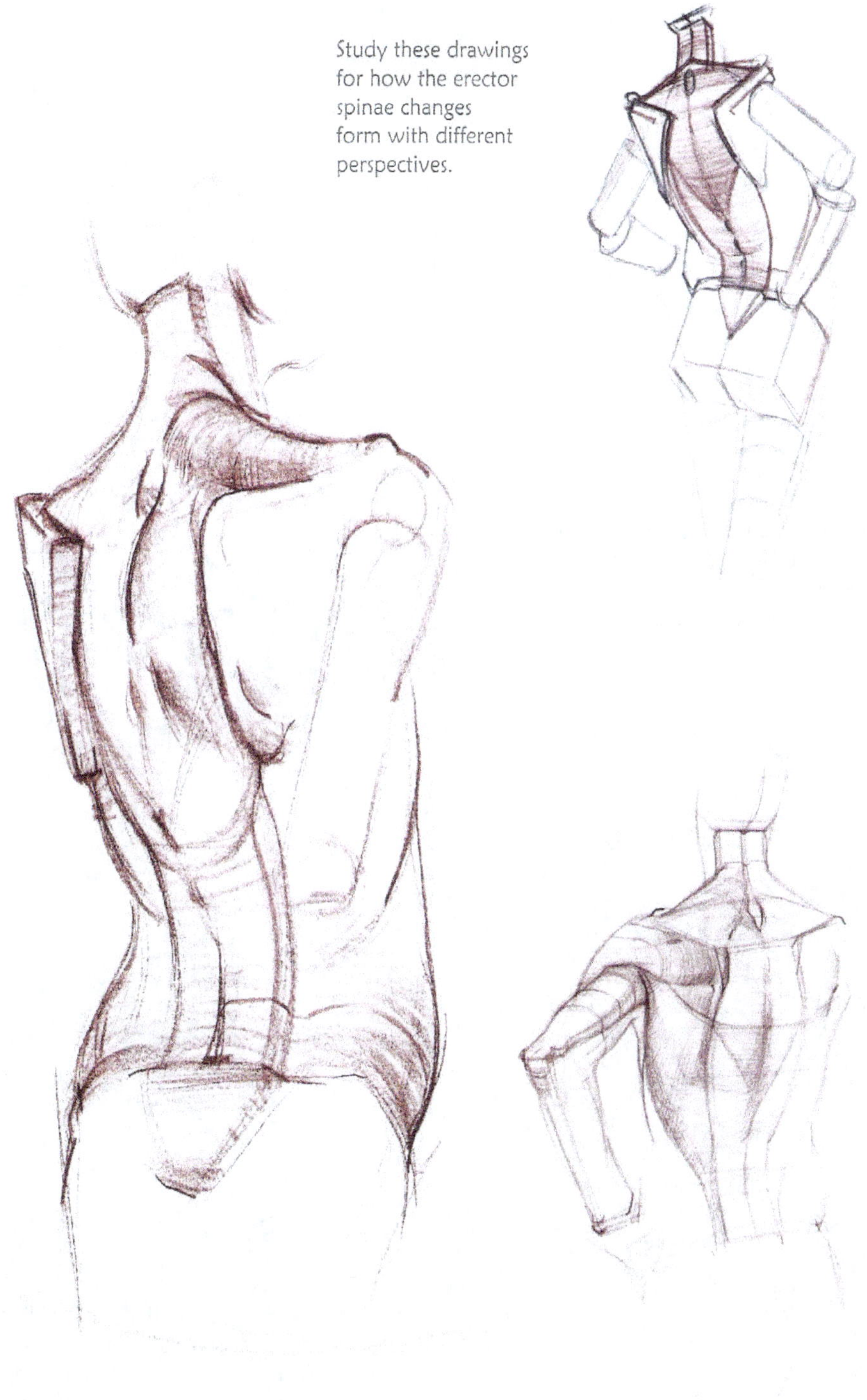

Study these drawings
for how the erector
spinae changes
form with different
perspectives.

LATISSIMUS DORSI - GESTURE

Adducts, extends, and rotates the humerus. Additionally, the latissimus pulls the arms down (like when swimming) or lifts the body up (as in climbing).

The latissimus originates along the lower sixth thoracic vertebrae and the crest of the ilium. It inserts on the front of the humerus higher than the attachment of the pectoralis.

LATISSIMUS DORSI - SHAPE

The shape of the latissimus resembles a bowl with arms and legs. The arms on the sides of the bowl are small triangles. The design of these arms suggests that the shape of the latissimus, like the pectoralis, unfolds when it is stretched.

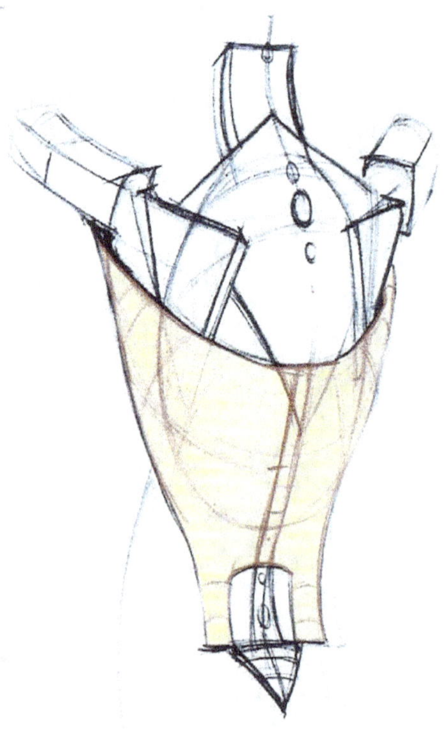

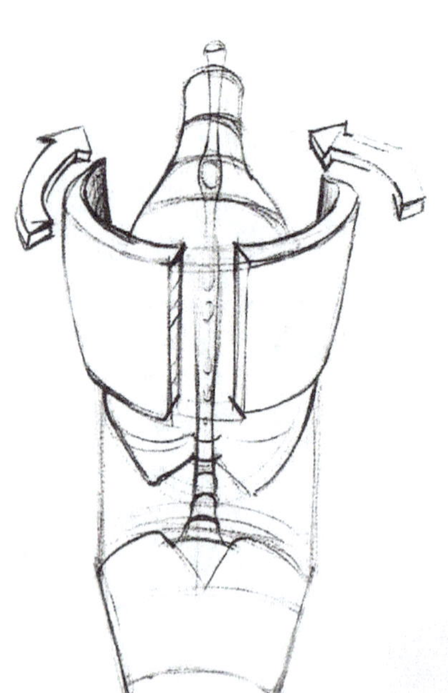

LATISSIMUS DORSI - VOLUME

The latissimus can be given perspective and a feeling of volume by wrapping the top of the shape with the perspective of the bottom of the rib cage (from behind). As the bottom of the latissimus ends on the pelvis, it should be drawn to sit on the perspective of the pelvis.

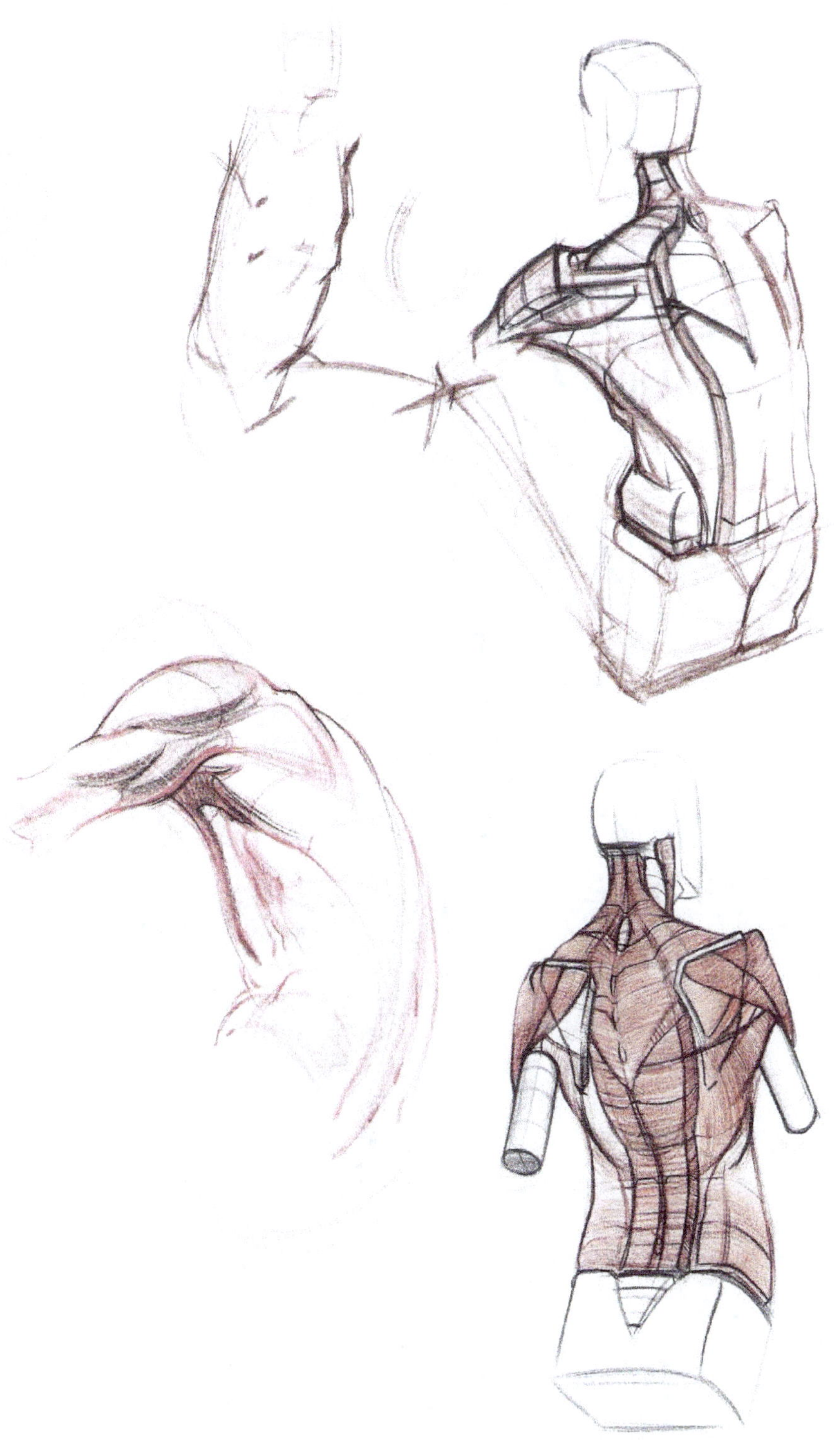

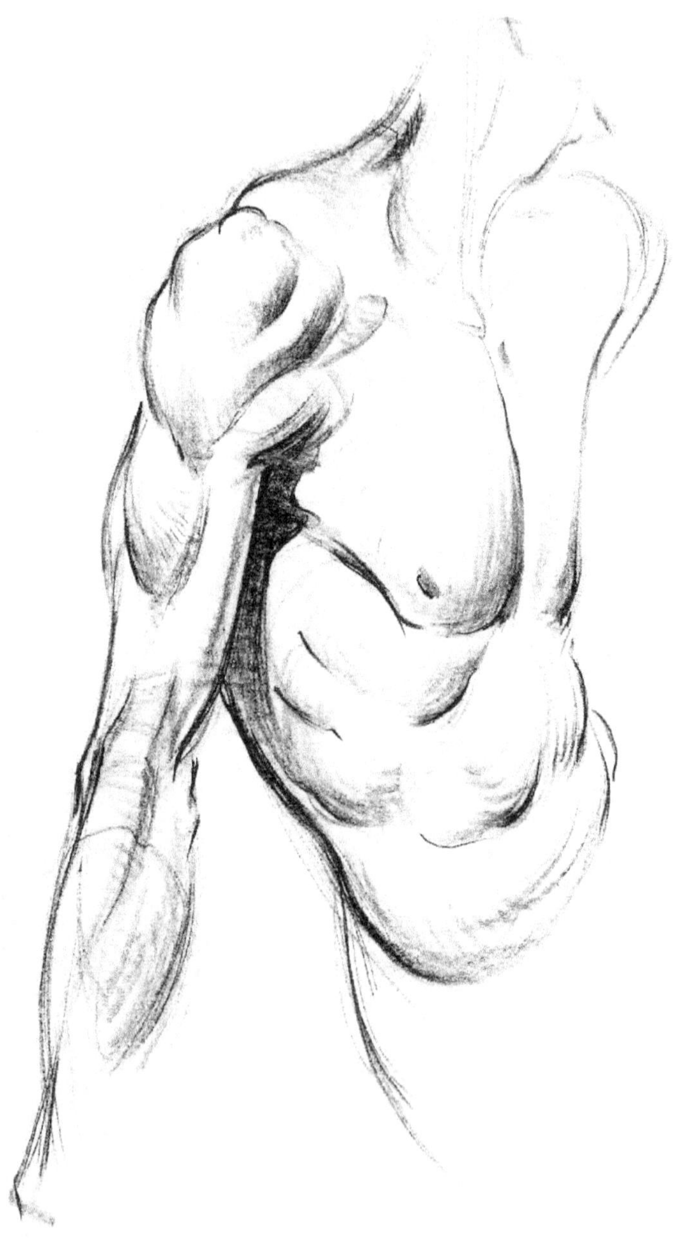

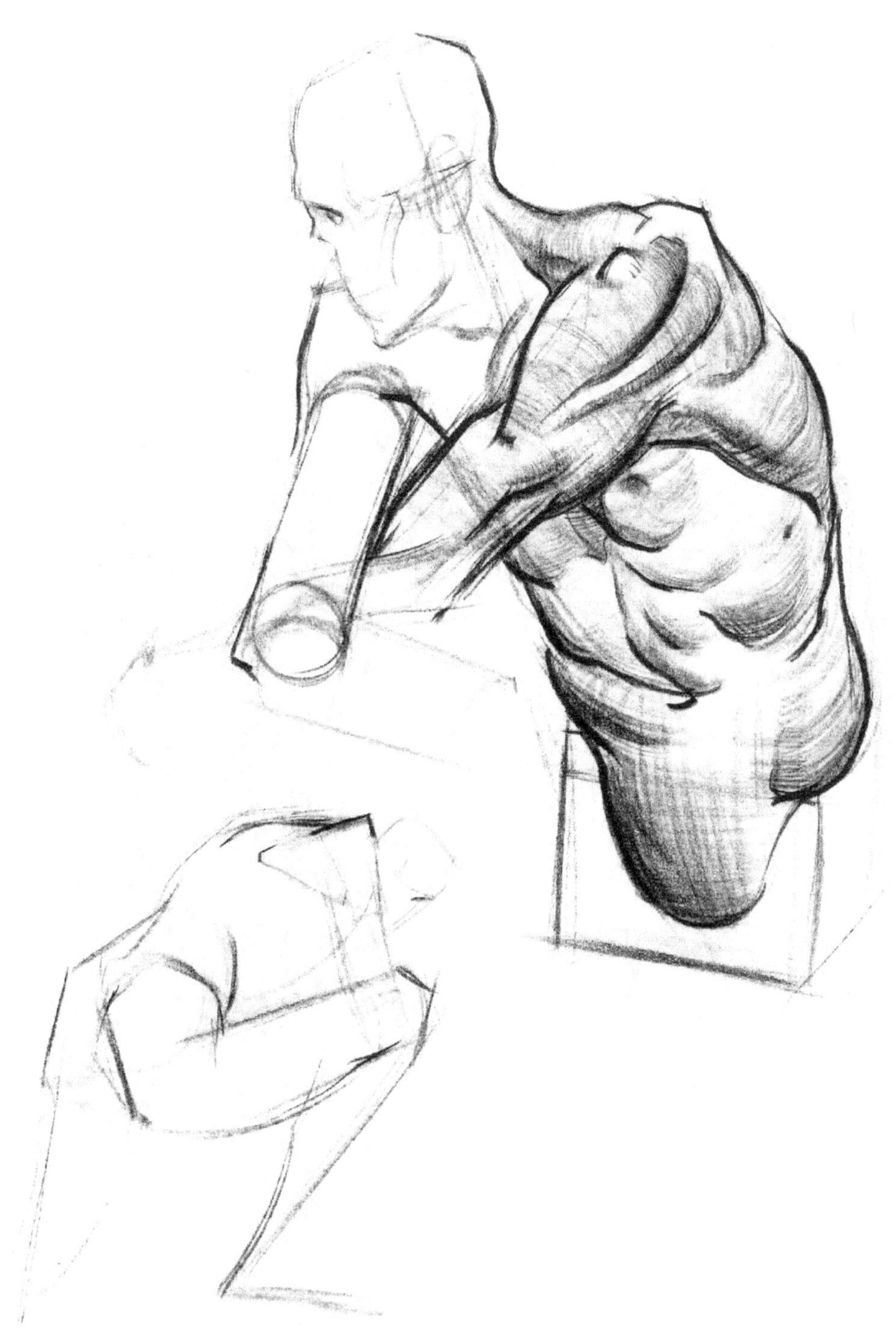

ANATOMY and ARCHITECTURE

Many artists throughout history have let their studies of the figure inform their architectural designs (and vice versa). One of the many benefits of this is the sharing of strong perspective and spatial principles.

Additionally, thinking of the anatomical shapes we covered in a more geometric, architecturally informed way can result in endless design solutions or more memorable shapes and patterns.

The following drawings exaggerate the use of perspective in order to present architectural influences in the study of the figure.

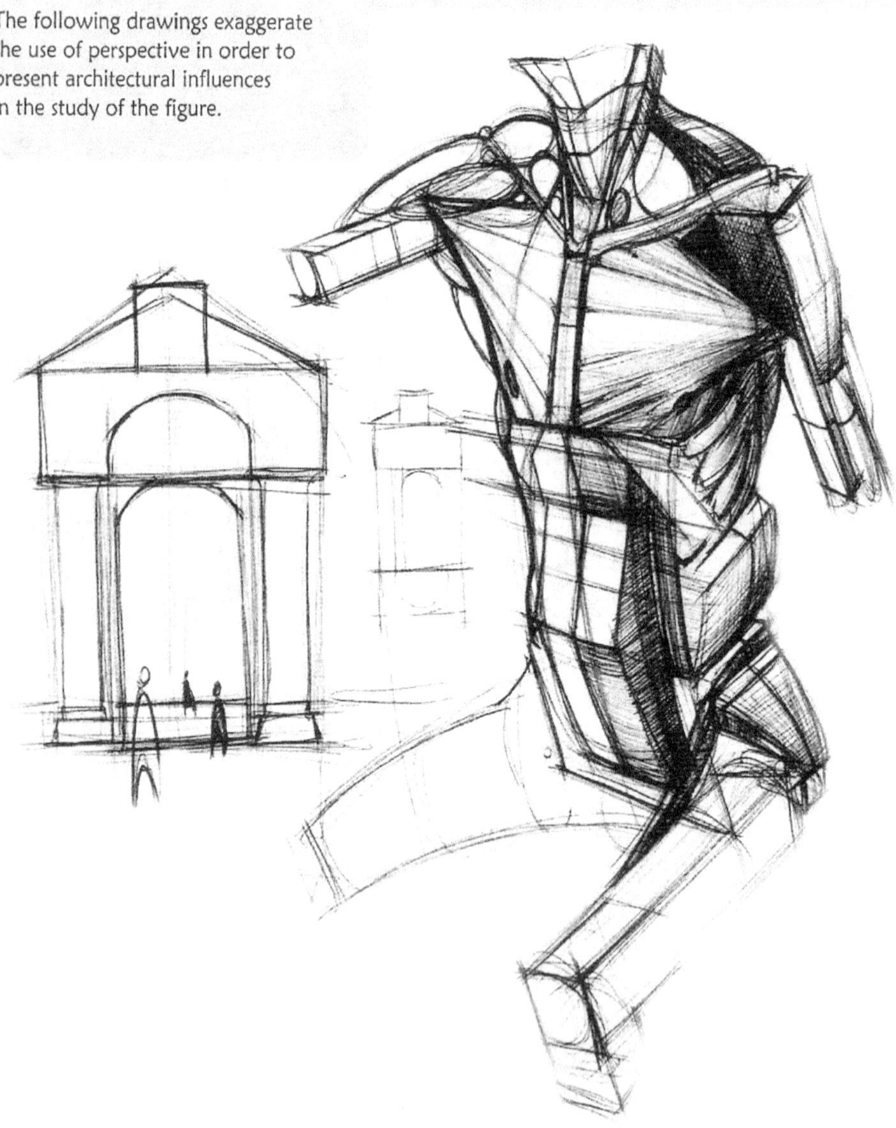

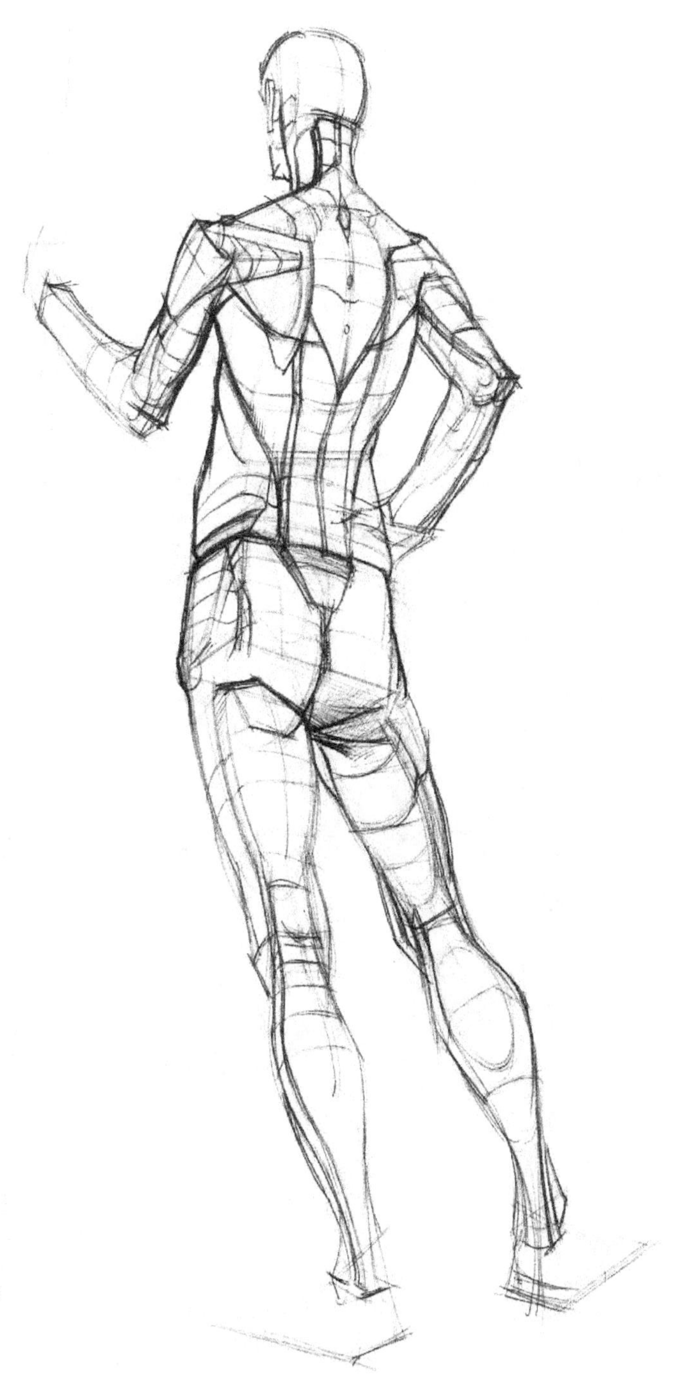

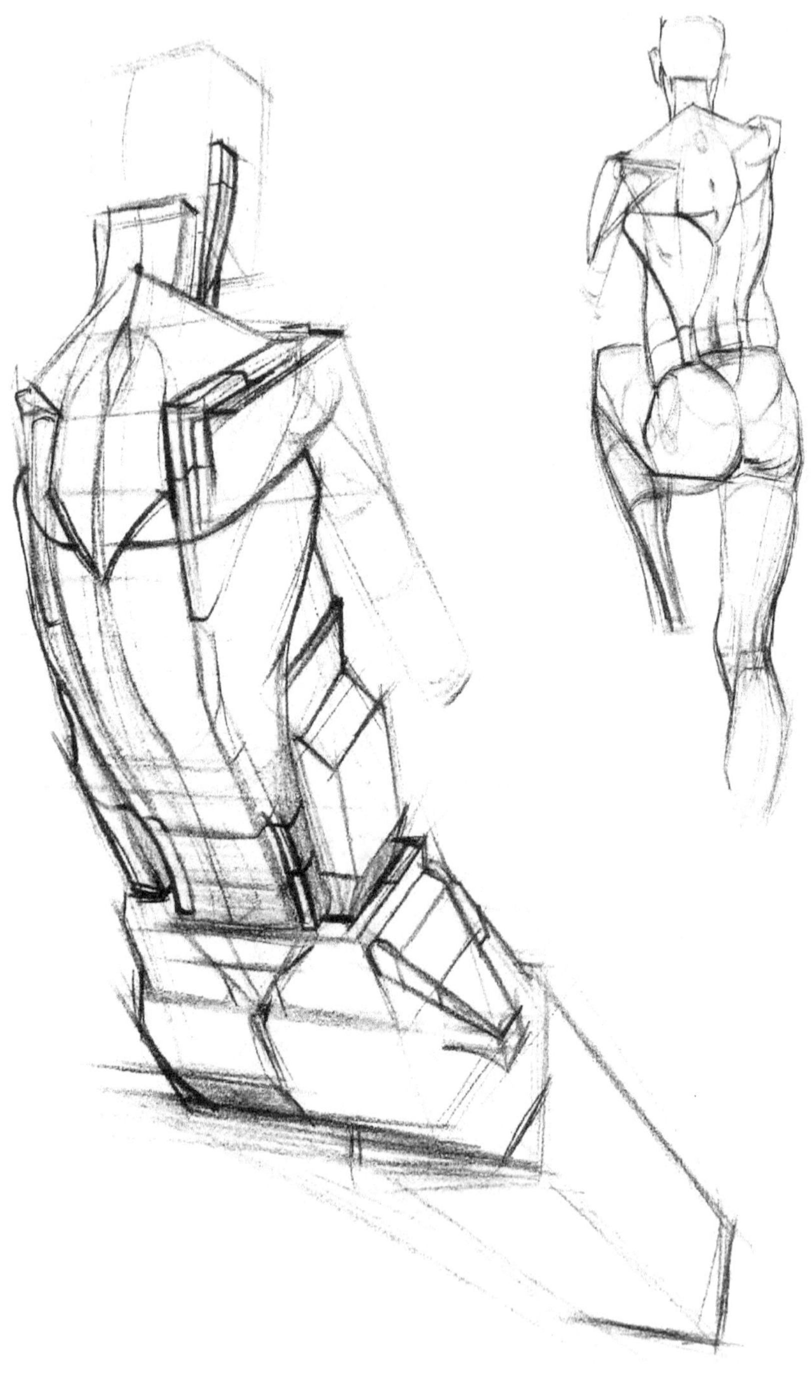

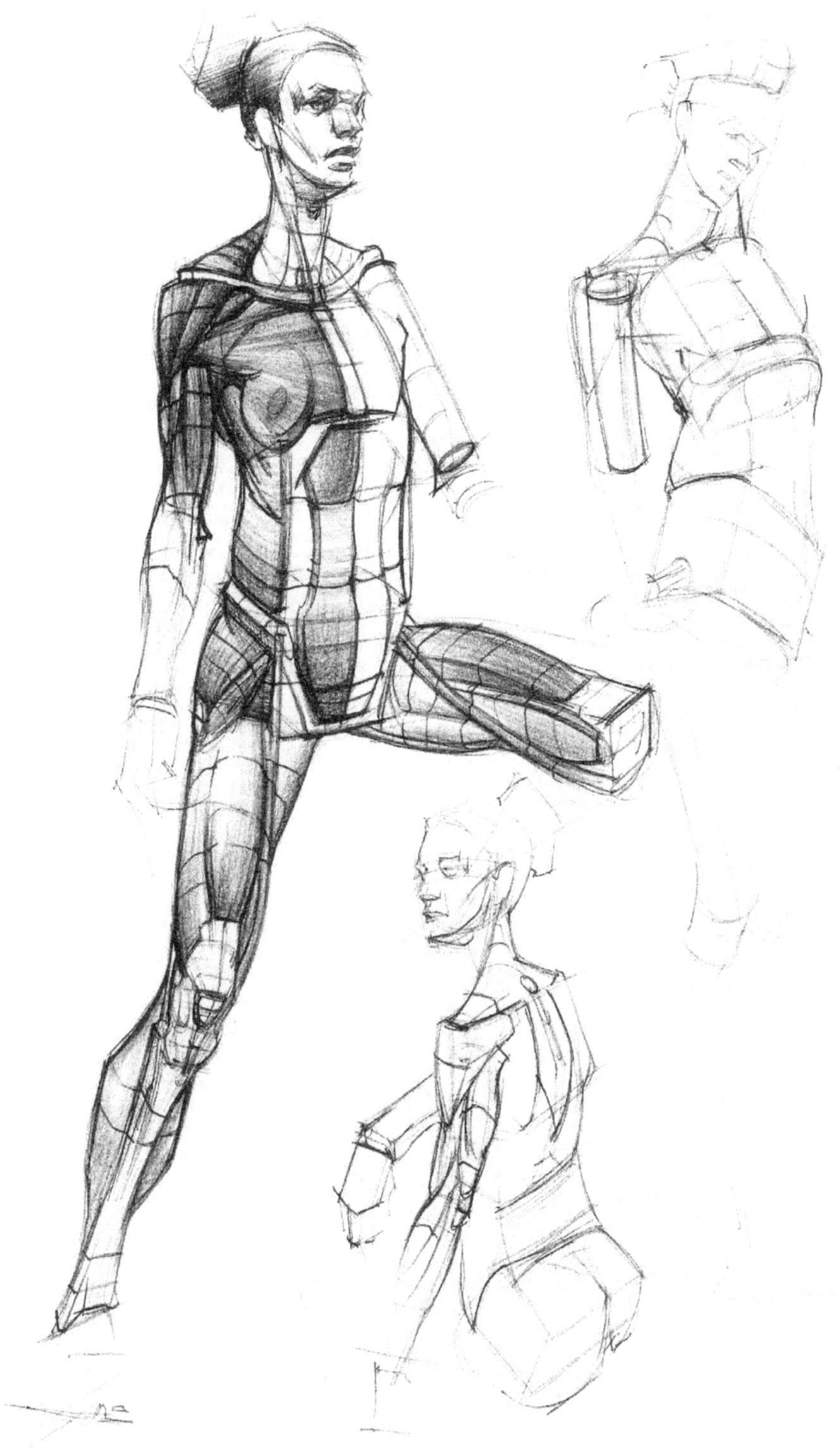

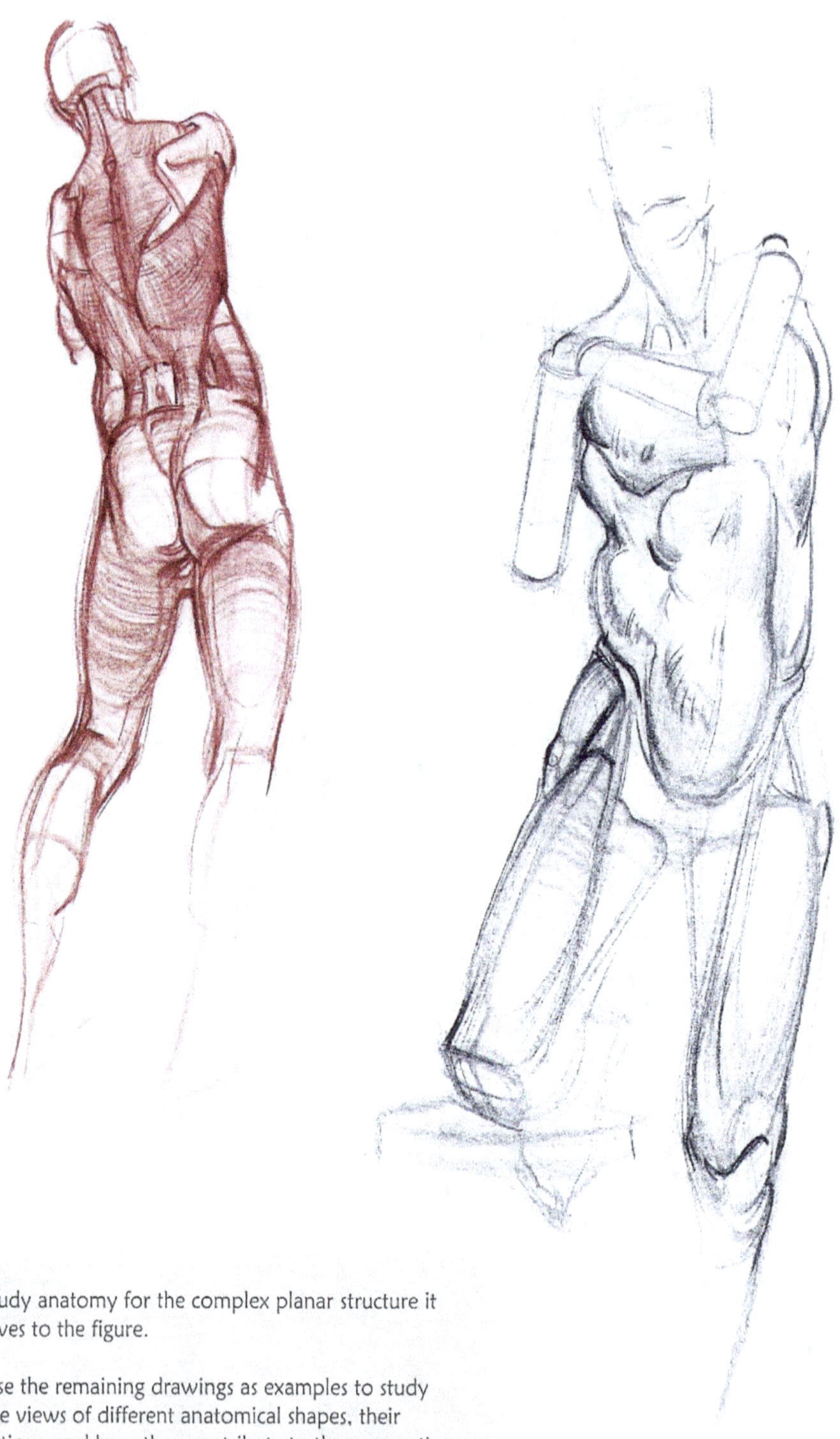

Study anatomy for the complex planar structure it gives to the figure.

Use the remaining drawings as examples to study the views of different anatomical shapes, their actions, and how they contribute to the perspective.

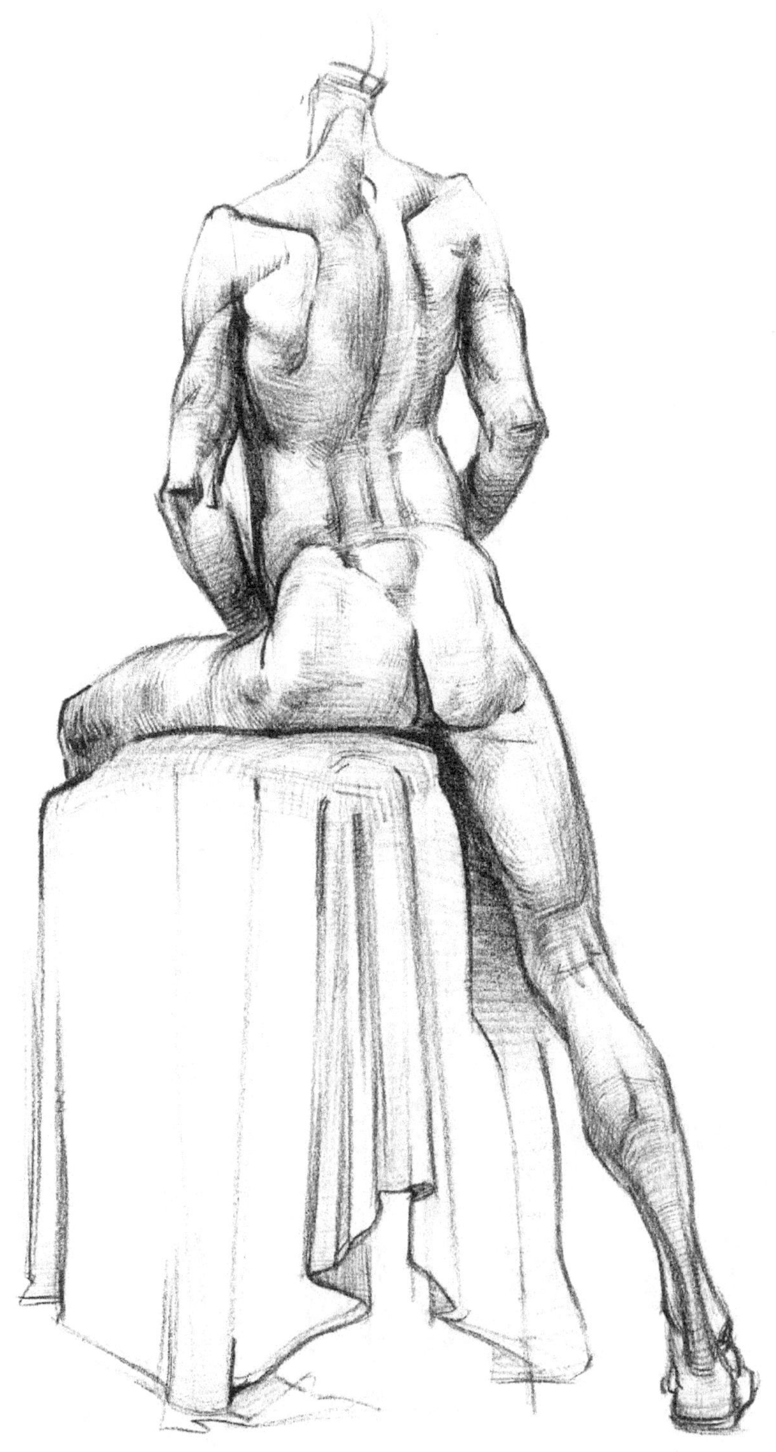

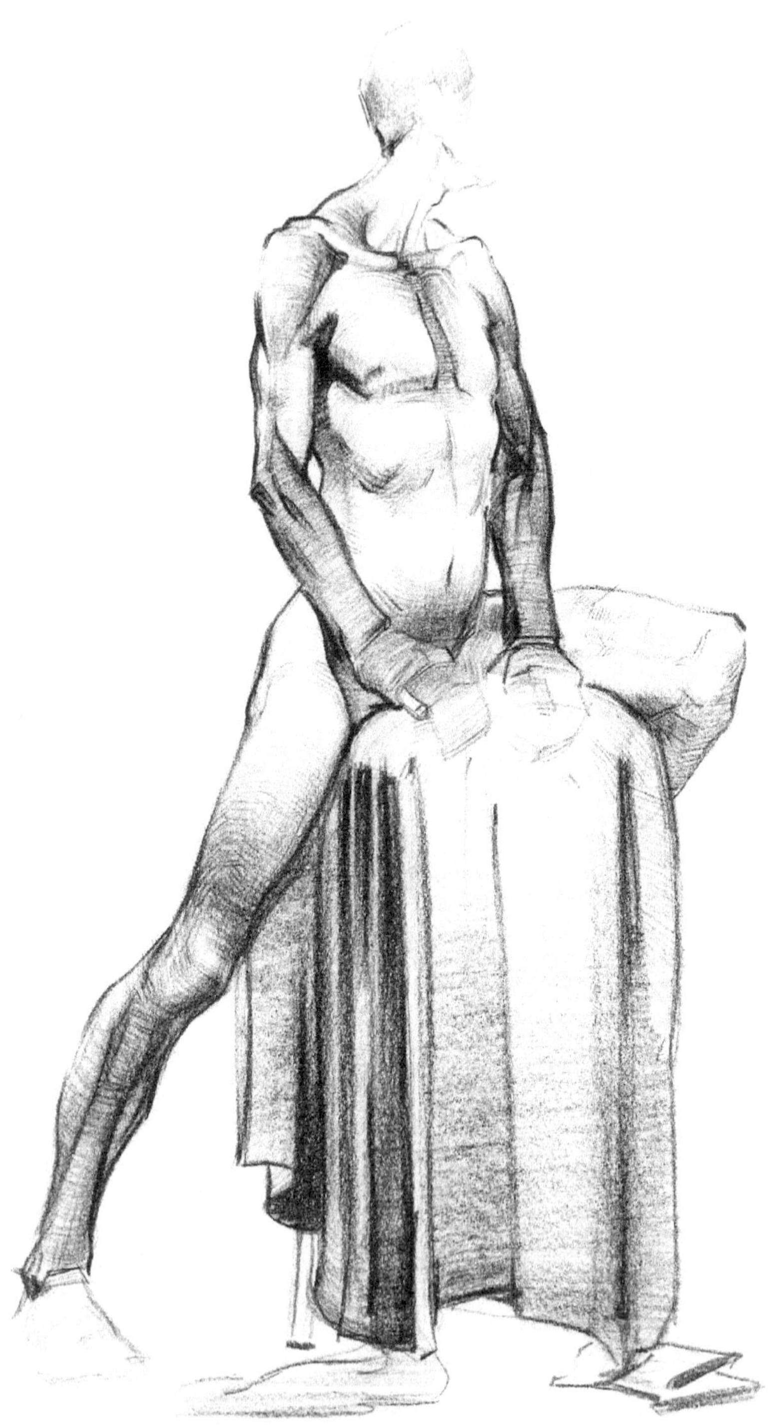

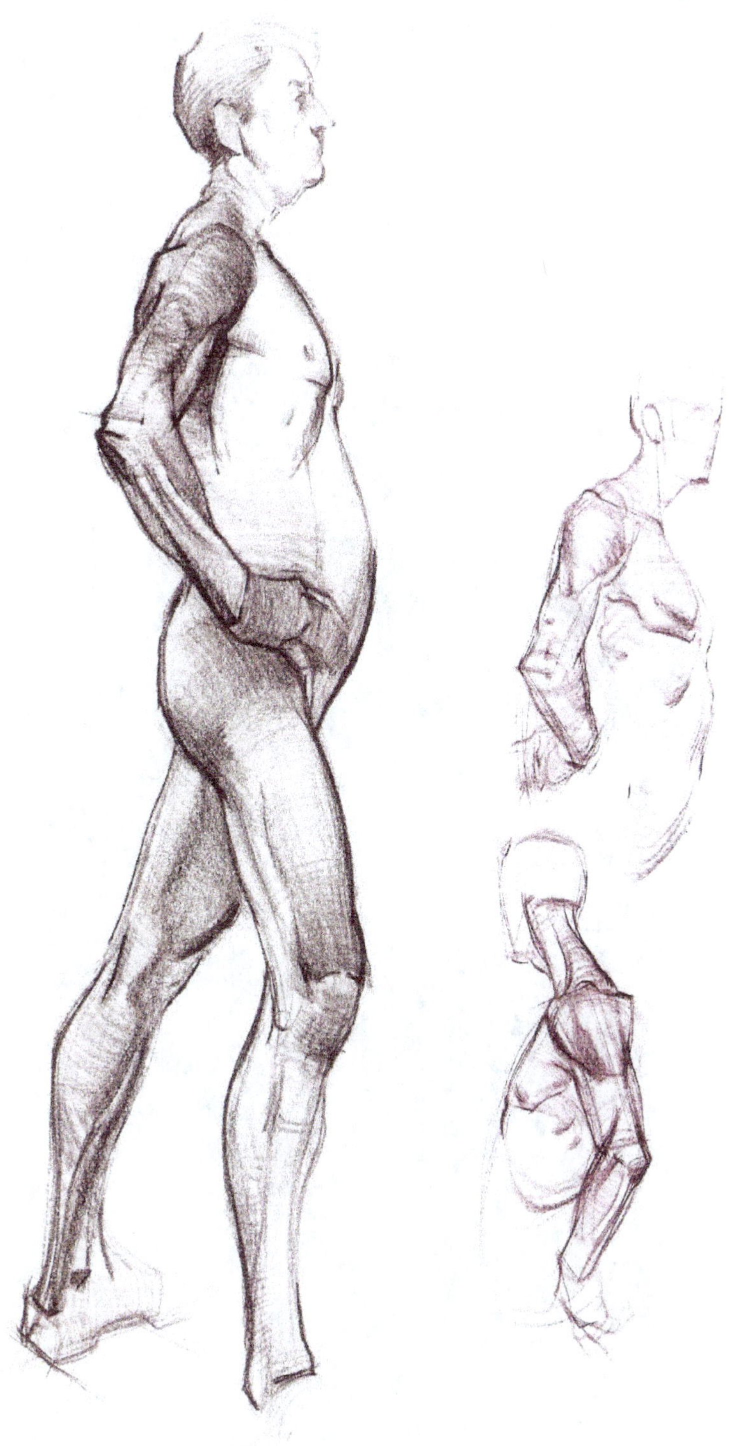

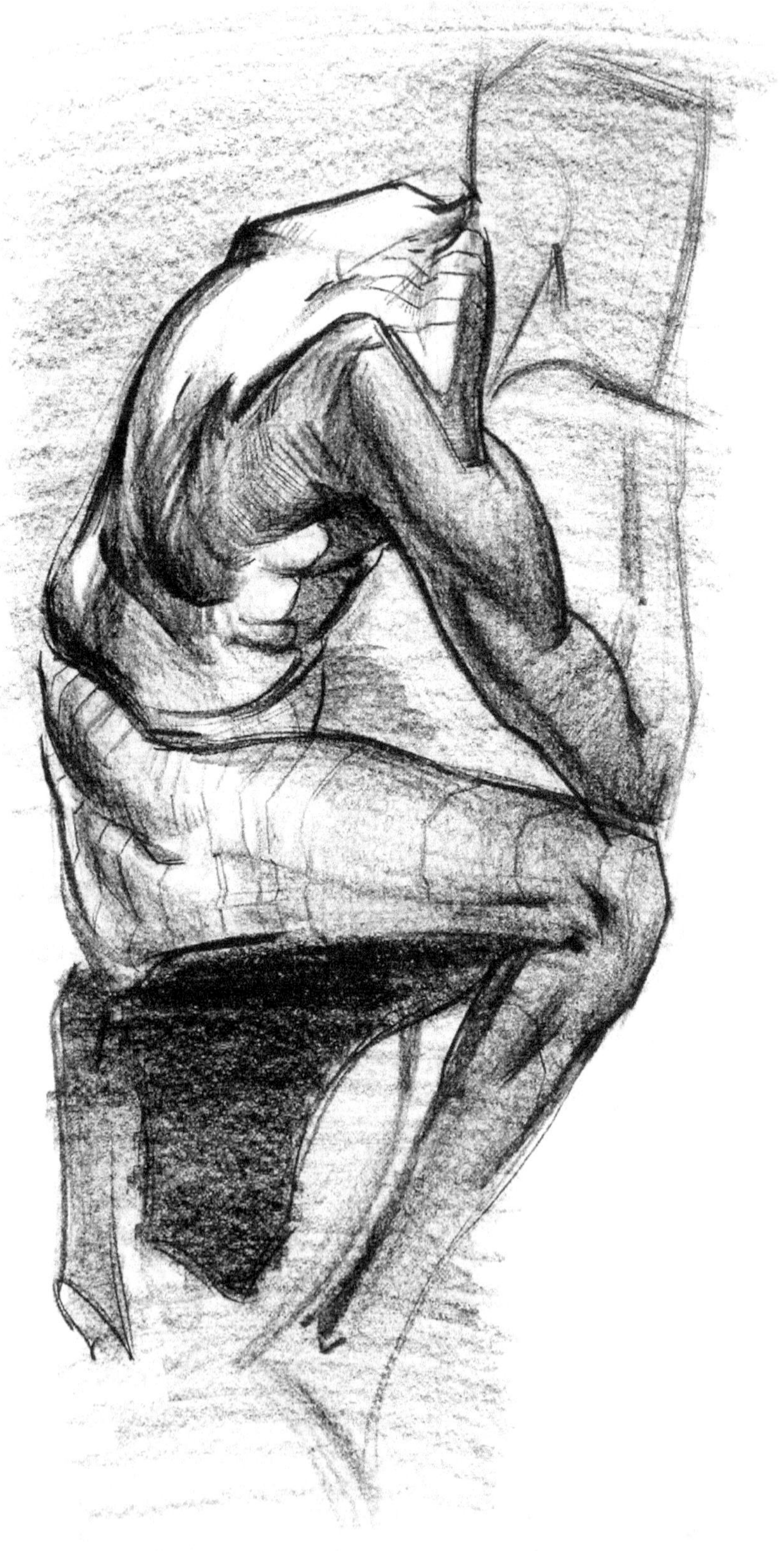

THE ARM

When drawing the arm, the first thing to consider is its relationship with the large form of the rib cage. The structure of the shoulder girdle, introduced in the anatomy of the upper torso, should serve as the transitional form between the rib cage and the arm.

The following information contained in this chapter should be added to this structure.

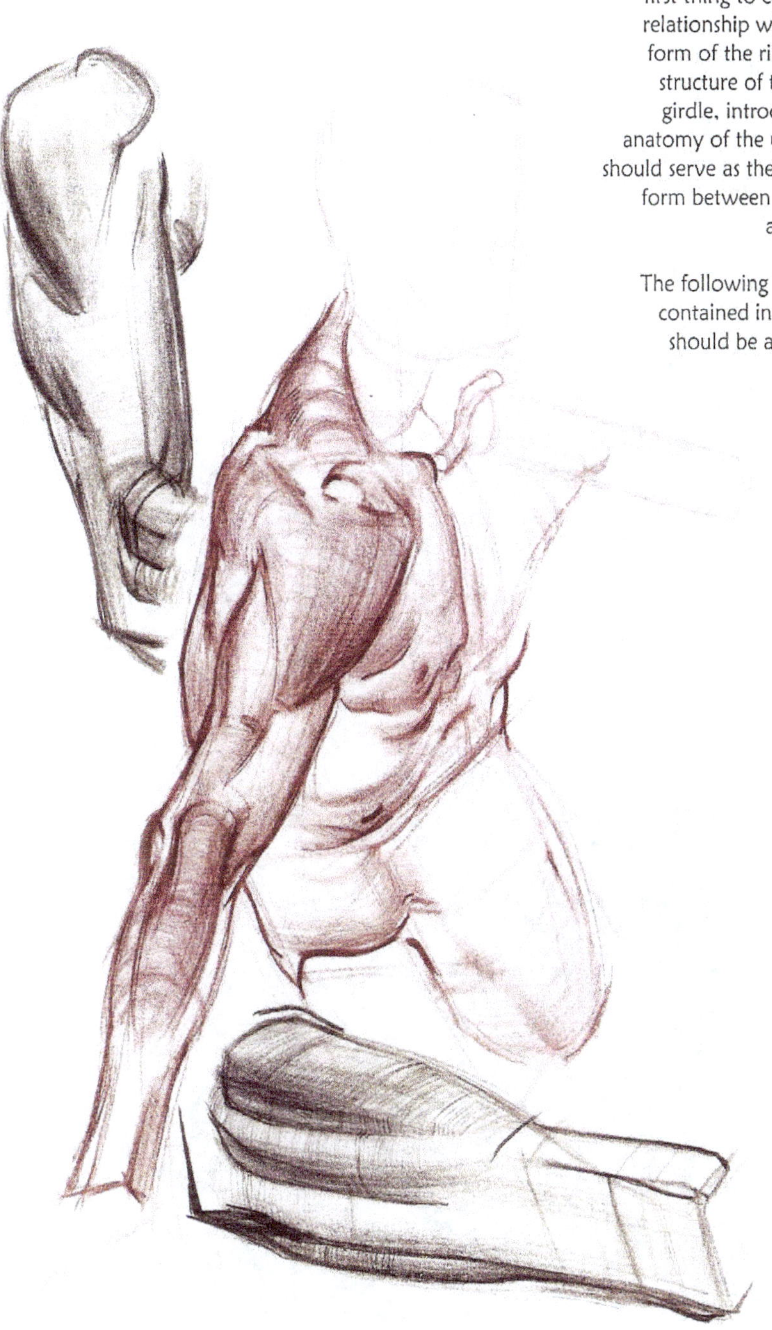

The shoulder girdle is a perspectival unit made up of the clavicle (front) and scapula (back). The importance of having a separate form for these bones is to allow them to move independently of the rib cage, as well as adding planar elements. When working with the arms, first pay attention to the development and placement of the shoulder's perspective. Think of this structure as being similar to the shoulder pads worn by football players.

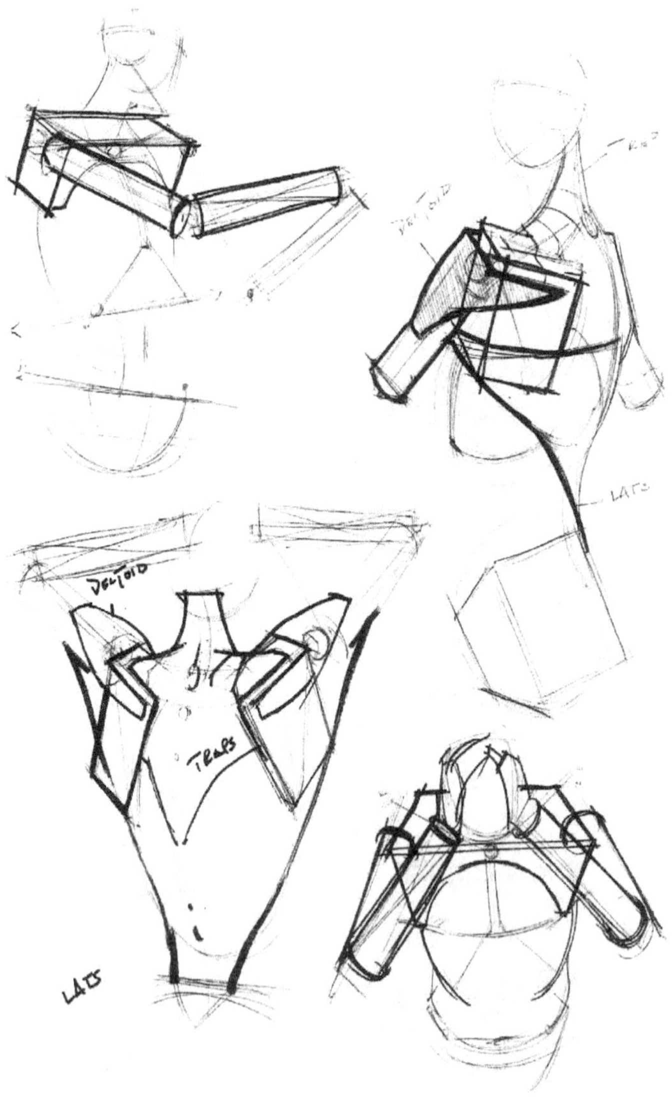

The points to look for in the construction
of the shoulder girdle consist of:

- One line through the top of the pit of the
  neck. This should be your starting line
  because it will also determine the tilt of the
  shoulders.

- One line, of equal length to the one above,
  drawn through the 7th cervical vertebrae.

- At the ends of the clavicle, two lines that are
  perpendicular to the first two and construct
  a top plane.

- At the bottom of the pit of the neck, a "C"
  curve that folds over the surface of the rib
  cage and joins the two structures.

- Parallel lines that angle the structure back
  and join the top plane to the side and front.

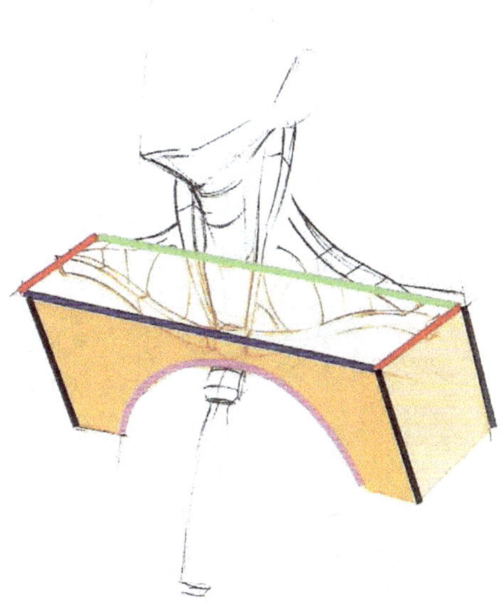

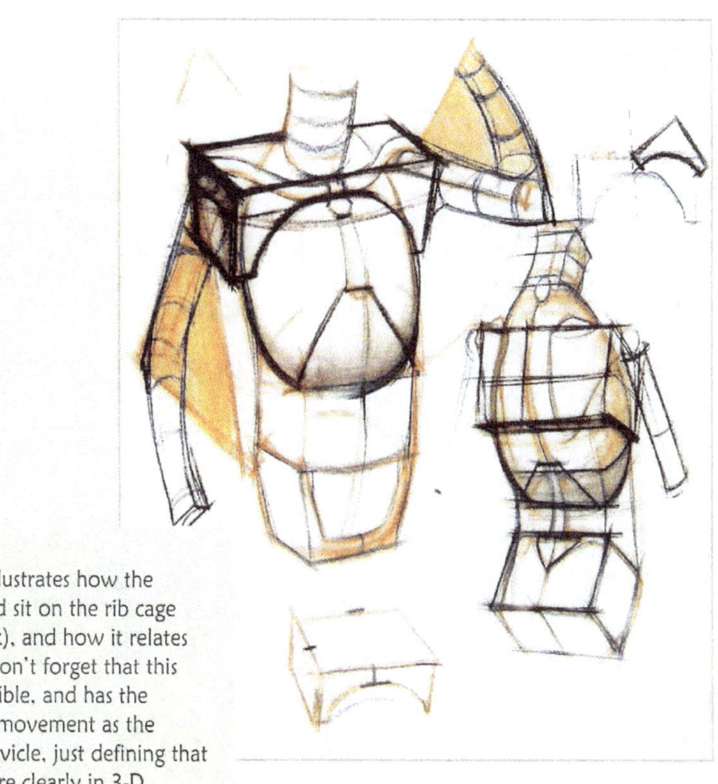

This drawing illustrates how the
shoulder should sit on the rib cage
(front and back), and how it relates
to the arms. Don't forget that this
structure is flexible, and has the
same range of movement as the
scapula and clavicle, just defining that
movement more clearly in 3-D.

PROCESS FOR THE DEVELOPMENT OF THE ARMS & LEGS —

1. LINE OF ACTION > (OF ARMS OR LEG)
2. PERSPECTIVE
3. GESTURE OF ANATOMICAL SHAPES.

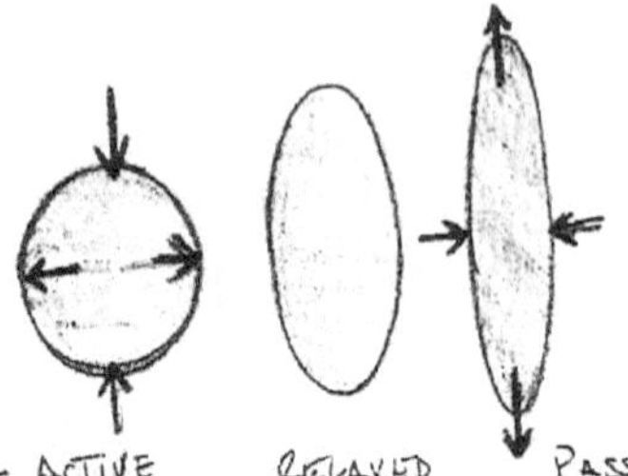

- MOST OF THE ANATOMY IN THE ARMS & LEGS
  CAN BE REPRESENTED AS ELLIPSES.
- THE GESTURE OF THESE ASYMMETRICAL
  ELLIPSES CAN BE SHOWN BY SQUASHING OR
  STRETCHING THE SHAPE.

SHAPE STILL
RETAINS SAME VOLUME
AS THE NORMAL RELAXED
ELLIPSE BUT HAS IT DISPLACED
INTO A SHORTER WIDER
APPEARANCE.

THE PASSIVE OR
STRETCH SHAPE IS PULLED
AT THE ENDS & THINNED
IN THE CENTER.

4. CONNECTION OF ANATOMY TO UNDERLYING PERSPECTIVES.

① "T" OVERLAPS   ② CONNECTING ALL LINES
                    INSIDE A FORM TO A WRAPPING
                    LINE

Above is a diagram that gives a process for drawing the arms and legs. This is similar to the anatomy and motion diagram in the previous chapter. This process should look very familiar by now, as it is the same one used for just about everything so far. If you are having difficulty in your drawings, return to this process and double-check your approach. If you notice your drawings are weak in a specific step, focus all your practice time on improving that one area.

The challenge in drawing the arms and legs is that all the anatomical shapes are essentially spheres or ellipses, and that they need to be developed on top of a difficult, complex surface (the cylinder).

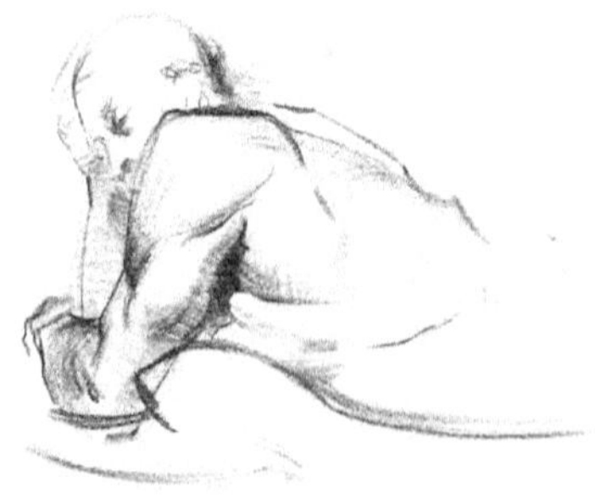

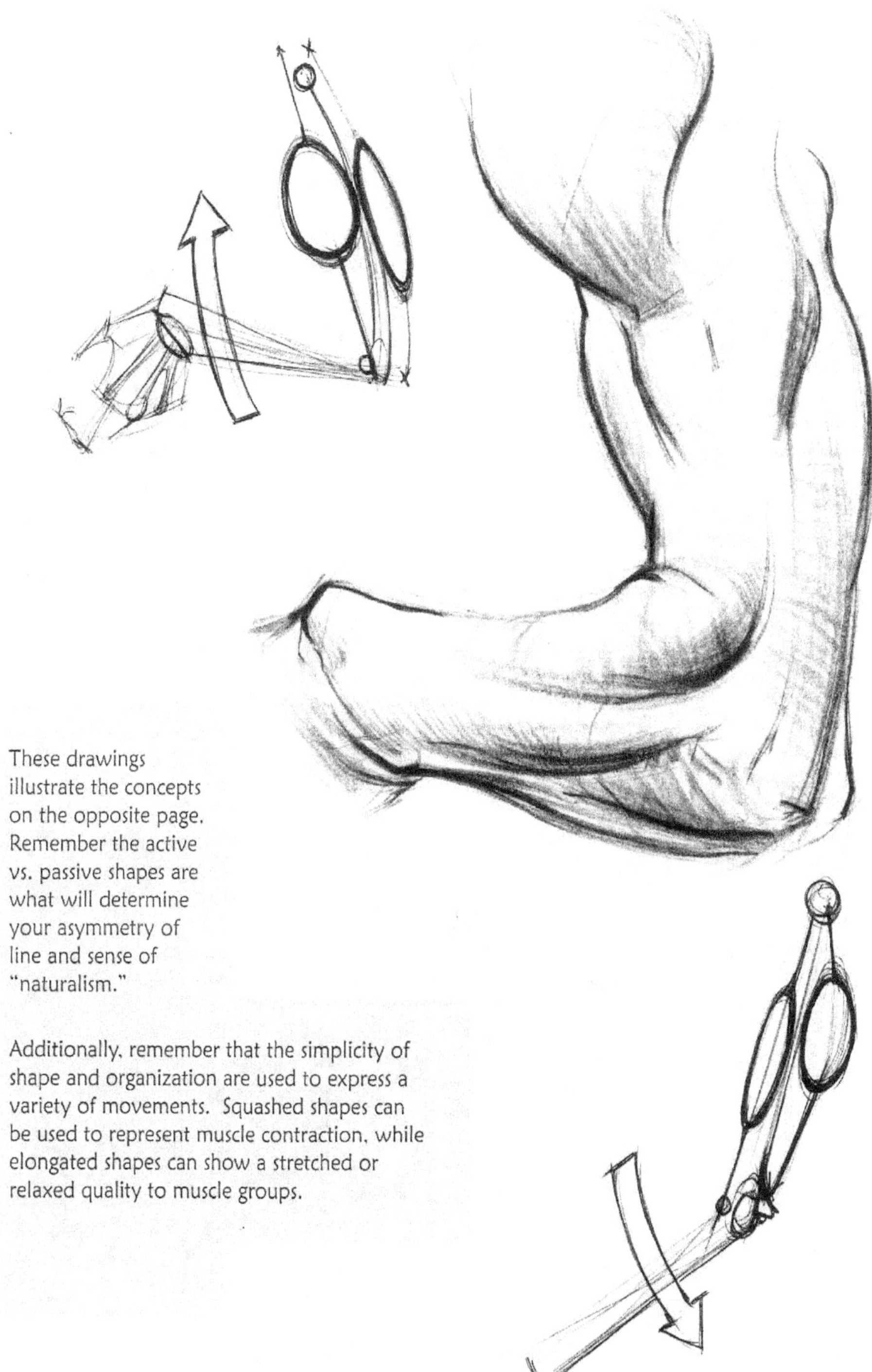

These drawings
illustrate the concepts
on the opposite page.
Remember the active
vs. passive shapes are
what will determine
your asymmetry of
line and sense of
"naturalism."

Additionally, remember that the simplicity of
shape and organization are used to express a
variety of movements. Squashed shapes can
be used to represent muscle contraction, while
elongated shapes can show a stretched or
relaxed quality to muscle groups.

We will begin studying the anatomy of the arm by looking from the scapula into the head of the humerus. You may want to review the scapula shape, its movement, etc. as a refresher before jumping right into the anatomy.

Looking at the scapula, there are some very important muscle shapes to be aware of in order to accurately define the arm in a number of positions and activities. These muscles are often referred to as the rotator cuff, and aid in the rotation of the humerus. Become familiar with the shapes in the diagram below, as they will provide the most accurate way of providing a believable transition from the scapula to humerus.

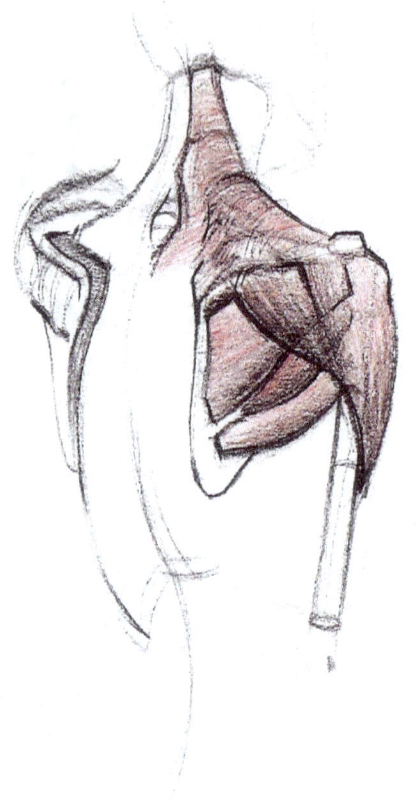

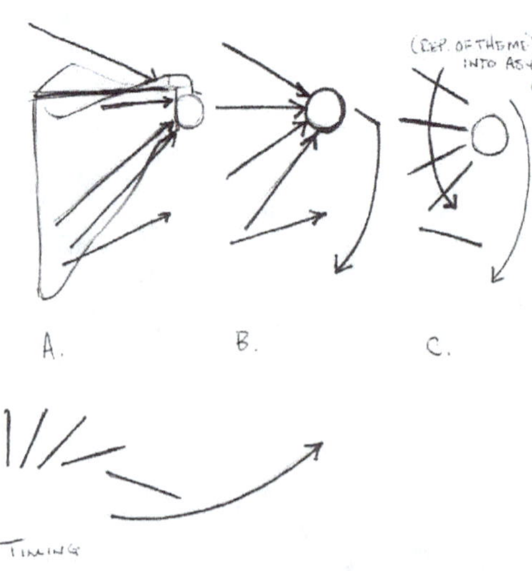

This diagram shows a simplified design for the placement and positioning of the muscles, borrowing from ideas of rhythm and asymmetry, discussed in the gesture chapter. For complex areas of anatomy, think of simple design solutions to clarify and provide rhythmic passage from one form to the other. Use the diagram as a suggested layout, but also be encouraged to design your own, based off of skeletal anatomy.

The elliptical shapes on the larger triangular form of the scapula represent:

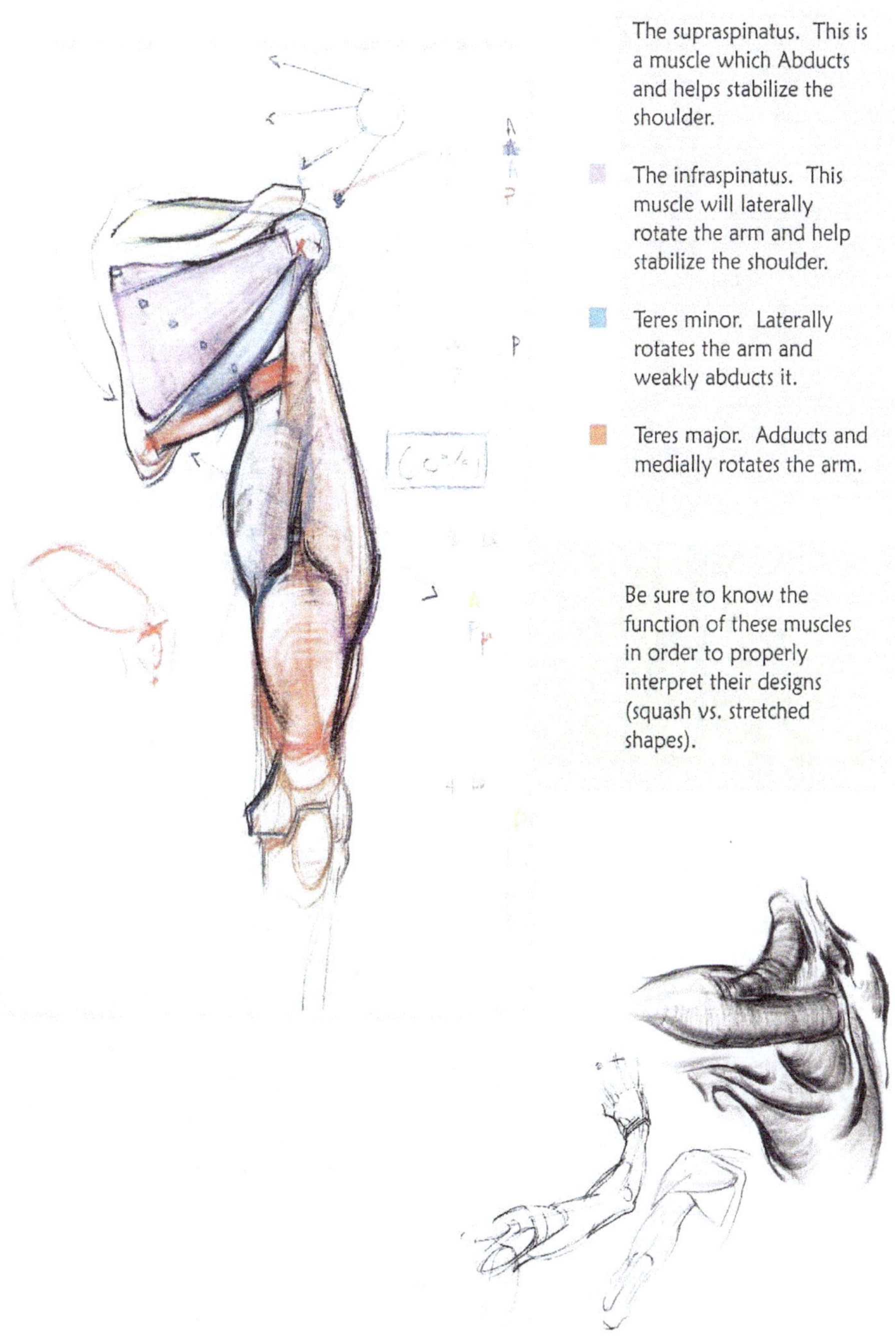

The supraspinatus. This is a muscle which Abducts and helps stabilize the shoulder.

The infraspinatus. This muscle will laterally rotate the arm and help stabilize the shoulder.

Teres minor. Laterally rotates the arm and weakly abducts it.

Teres major. Adducts and medially rotates the arm.

Be sure to know the function of these muscles in order to properly interpret their designs (squash vs. stretched shapes).

The previous section described the basic process for working out the arm. Here we will look at the anatomy as ellipses in a generalized design. These ellipses should be treated in the same manner as the rest of the anatomy discussed, in order to project as feeling of realistic movement. The muscles briefly outlined below are the components included in basic designs of the arm.

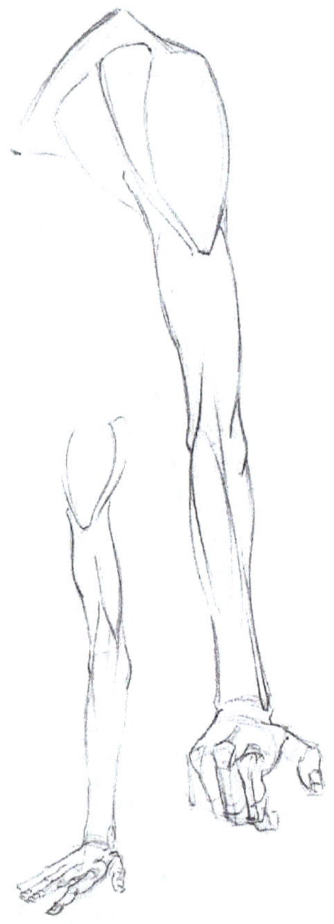

Brachialis. Starts about half way down the humerus and connects to the ulna. This muscle is used to flex the forearm.

Biceps brachii. This muscle begins at the top of the arm/shoulder and ends at the radius. The biceps is used to flex and supinate the forearm.

Triceps. There are three heads to the triceps — lateral, medial and the long head. These muscles work to extend the arm at the elbow.

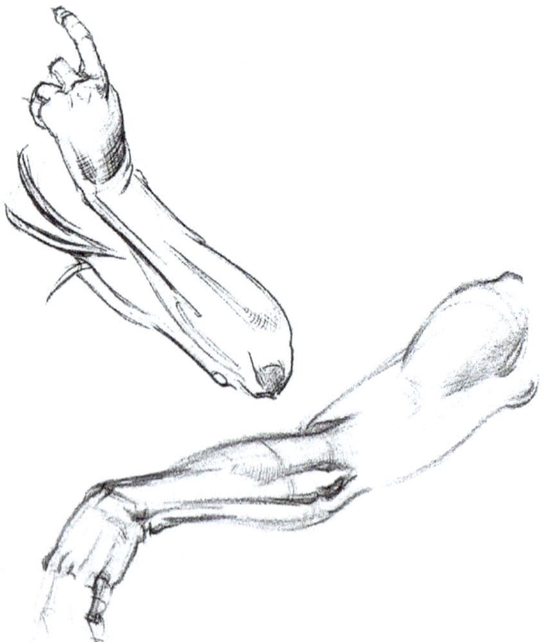

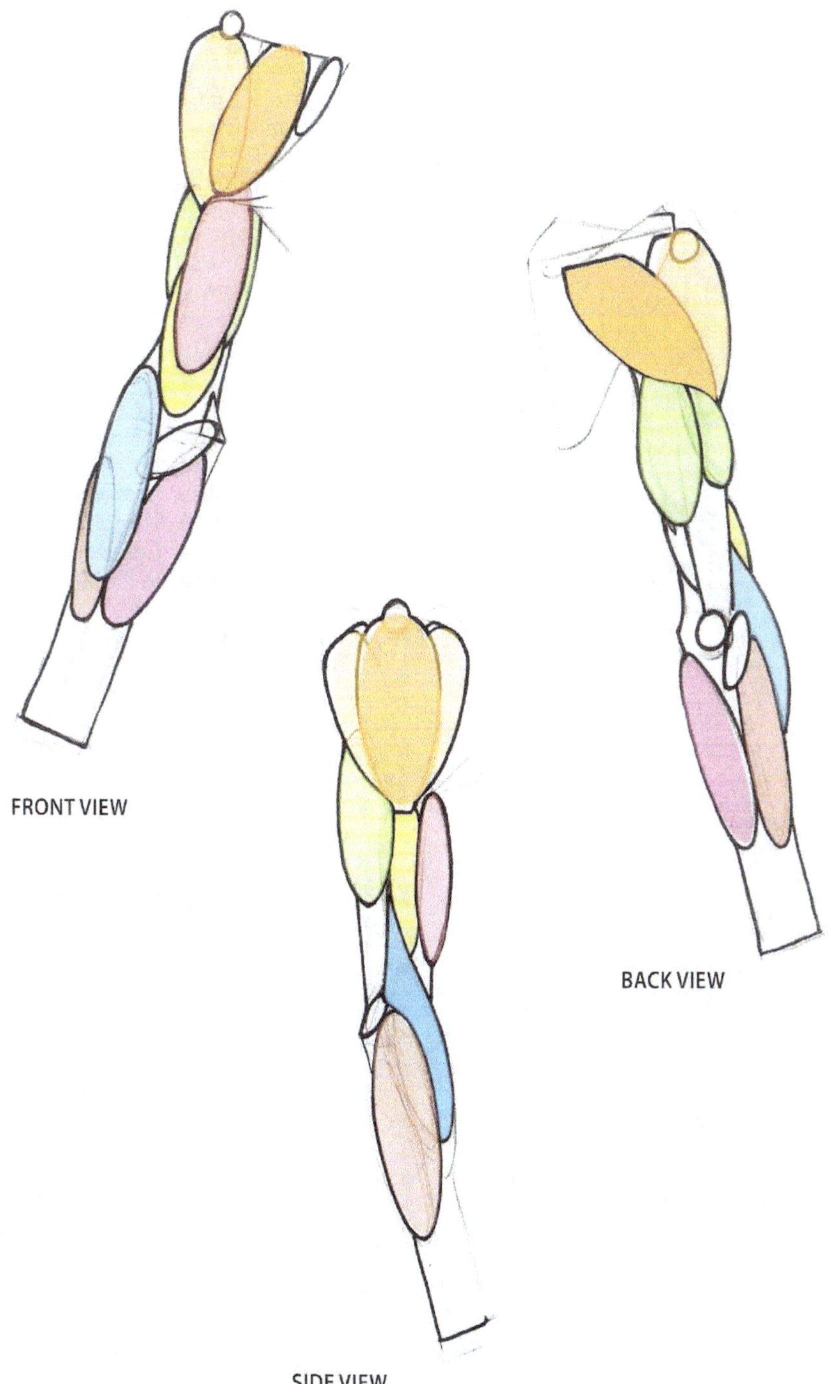

FRONT VIEW
SIDE VIEW
BACK VIEW

The forearm is an incredibly complex weaving of anatomy that will ultimately deserve more time and study then given here. I have taken great liberties to simplify the anatomy in order to integrate these basic ideas into a process. Again, this approach highlights a working process and should under no condition be used as an anatomy reference. I highly recommend that this approach be supplemented with additional anatomical texts.

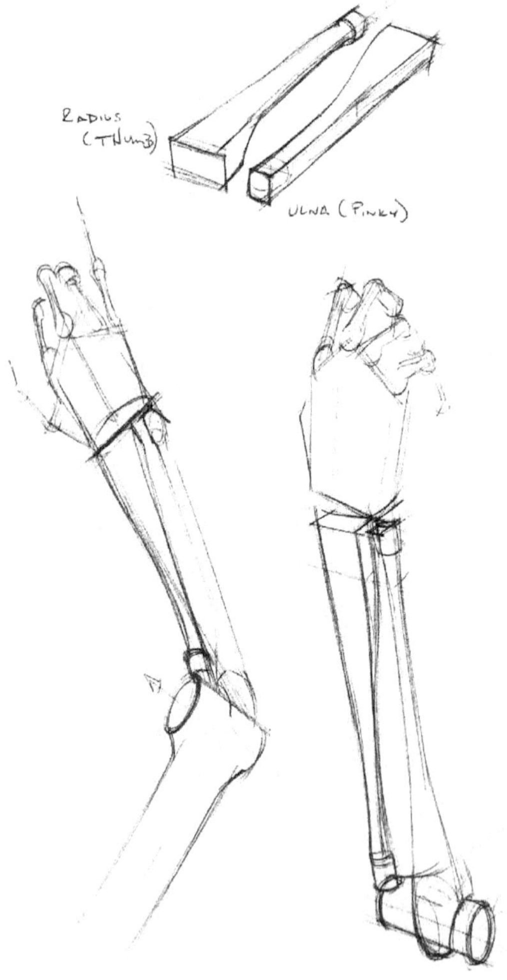

When drawing the forearm, the first, most important step is to determine the location of the radius and ulna during pronation and supination. The ulna remains static, connected to the humerus by a hinge joint, while the radius rotates around the ulna.

Notice that the radius and ulna are similar in shape: both are elongated triangles.

The ulna is wider at the top, as it fits into the humerus, and smaller towards the wrist. The ulna always appears on the pinky-finger side of the wrist.

The radius is opposite in shape to the ulna: a
triangle that is fat at the bottom and skinny
at the top. The radius always appears on the
thumb-side of the wrist.

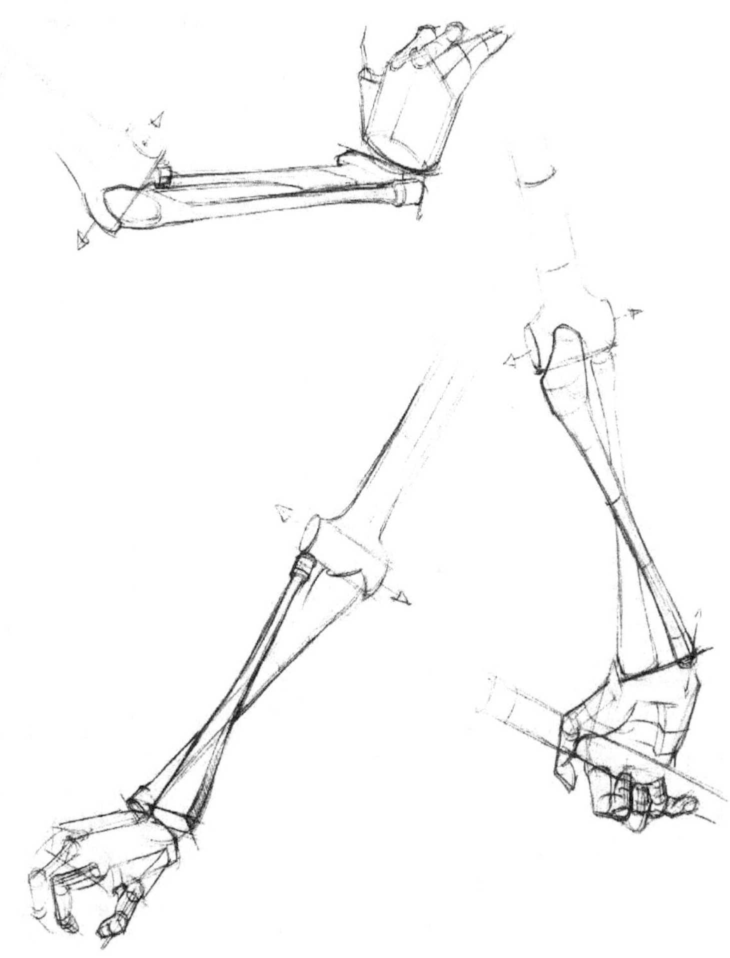

The design of the forearm can be reduced into three basic shapes/ellipses: the flexors, extensors, and ridge muscles. While we will add to these three, these are the major shapes to focus on. These muscles can be seen in the diagrams on page 137.

The flexors are made of three separate muscles on the medial half of the forearm, beginning at the medial epicondyle of the humerus, and continuing to the palmar (inside) region of the hand. The flexors are a more powerful muscle grouping than the extensors or ridge muscle, and are actively seen when the hand is gripping, making a fist, or pronating the arm.

The ridge muscles are composed of two individual muscles. They aid in flexing the forearm at the elbow joint, and their shape can be seen on the outside of the forearm.

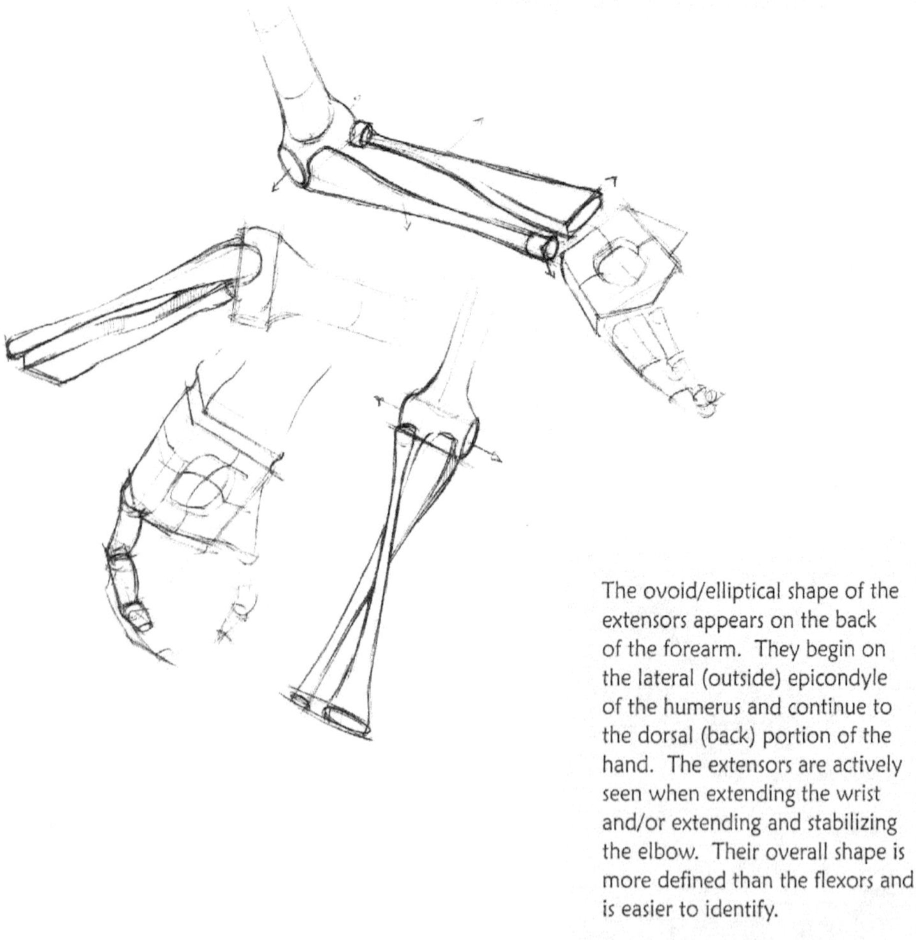

The ovoid/elliptical shape of the extensors appears on the back of the forearm. They begin on the lateral (outside) epicondyle of the humerus and continue to the dorsal (back) portion of the hand. The extensors are actively seen when extending the wrist and/or extending and stabilizing the elbow. Their overall shape is more defined than the flexors and is easier to identify.

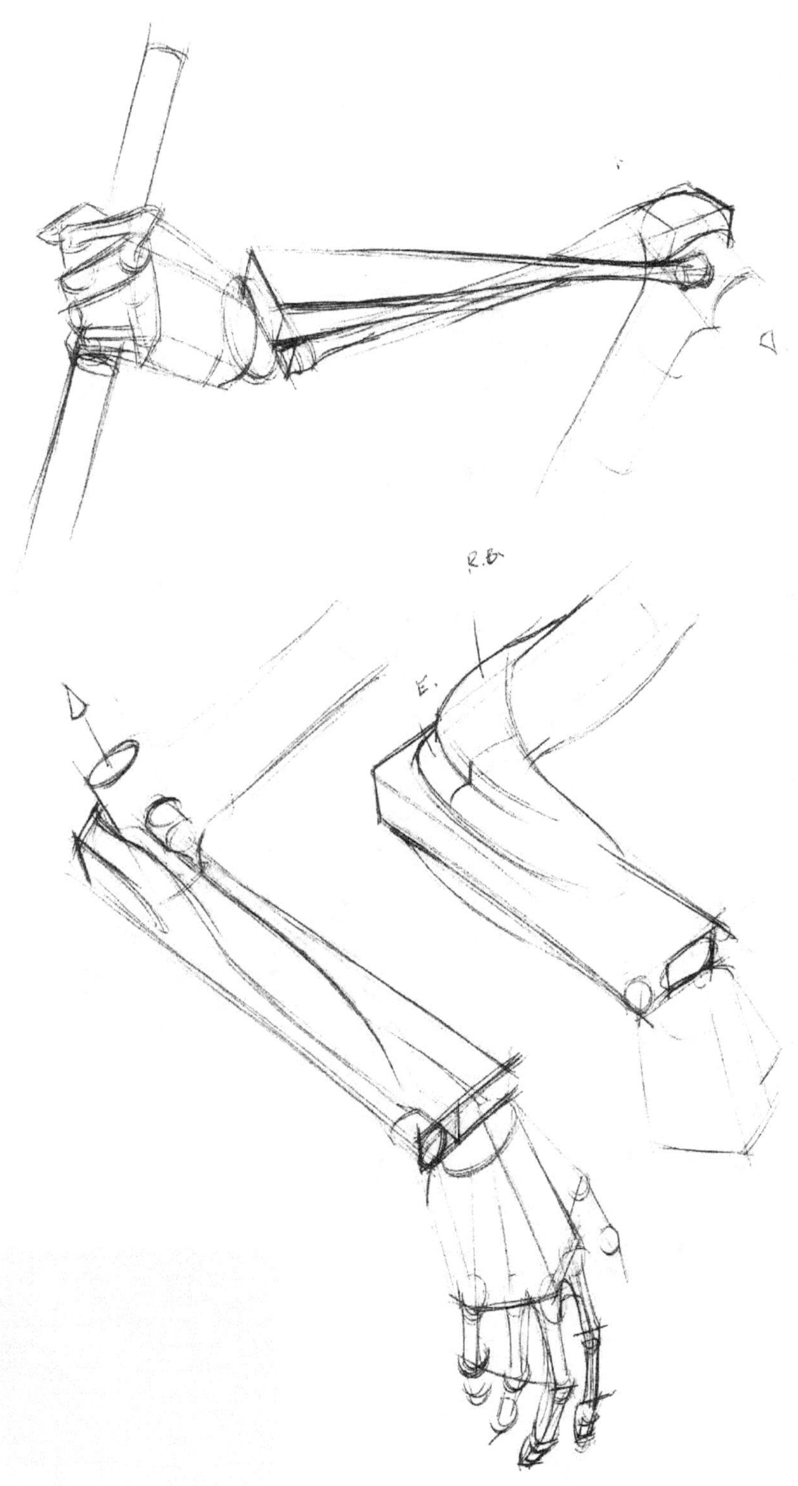

R.B
E.
D

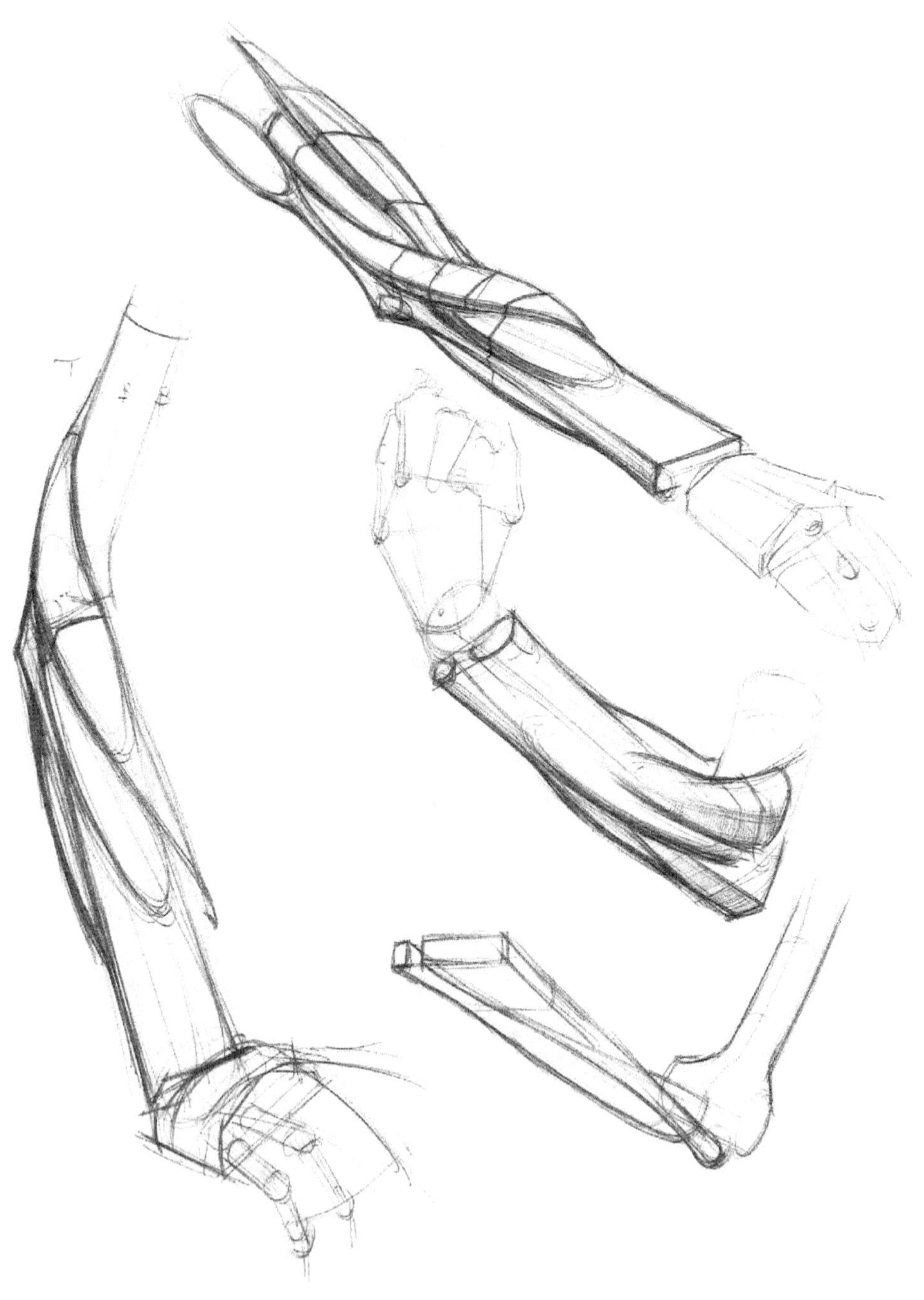

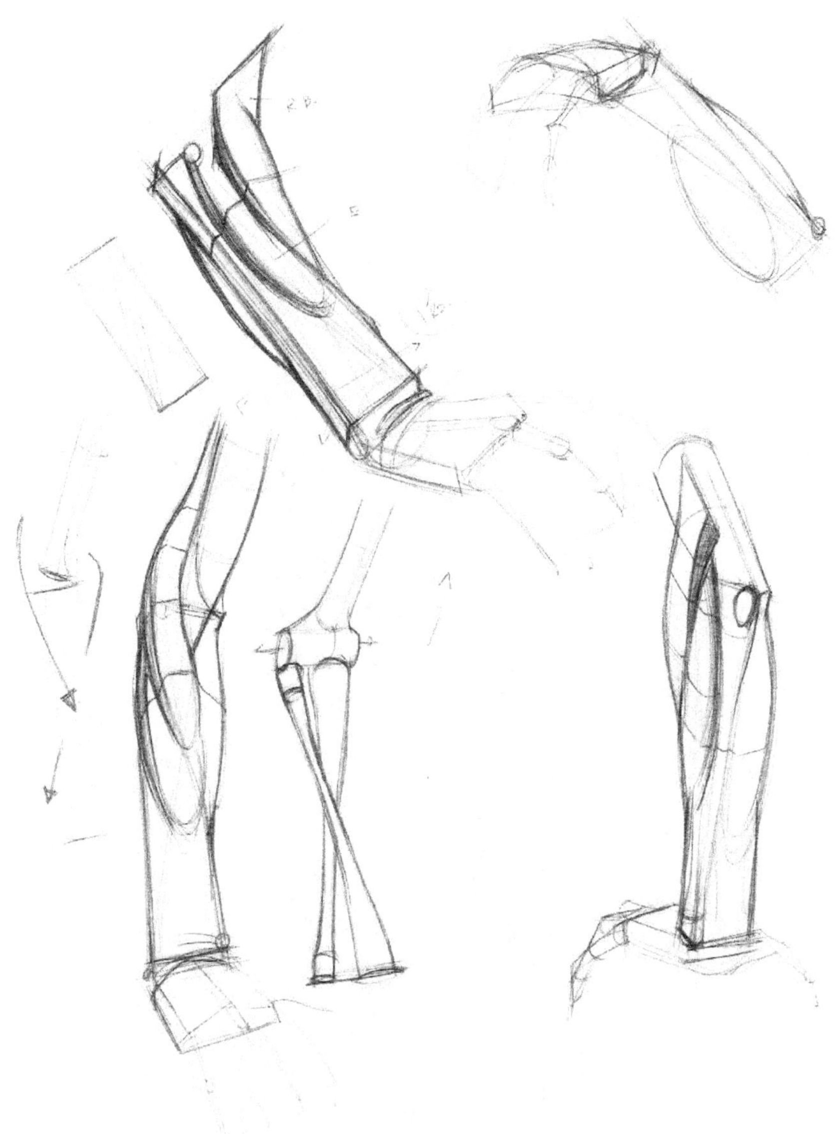

Successful understanding of the forearm's anatomy will help you depict pronation and supination of the arm.

In pronation, the radius crosses over the ulna. In terms of anatomy, pronation involves the pronator teres, pronator quadratus, and flexor carpi radialis.

In supination, the bones of the radius and ulna lie parallel to one another. Supination is more powerful than pronation. Supination involves the biceps brachii and supinator.

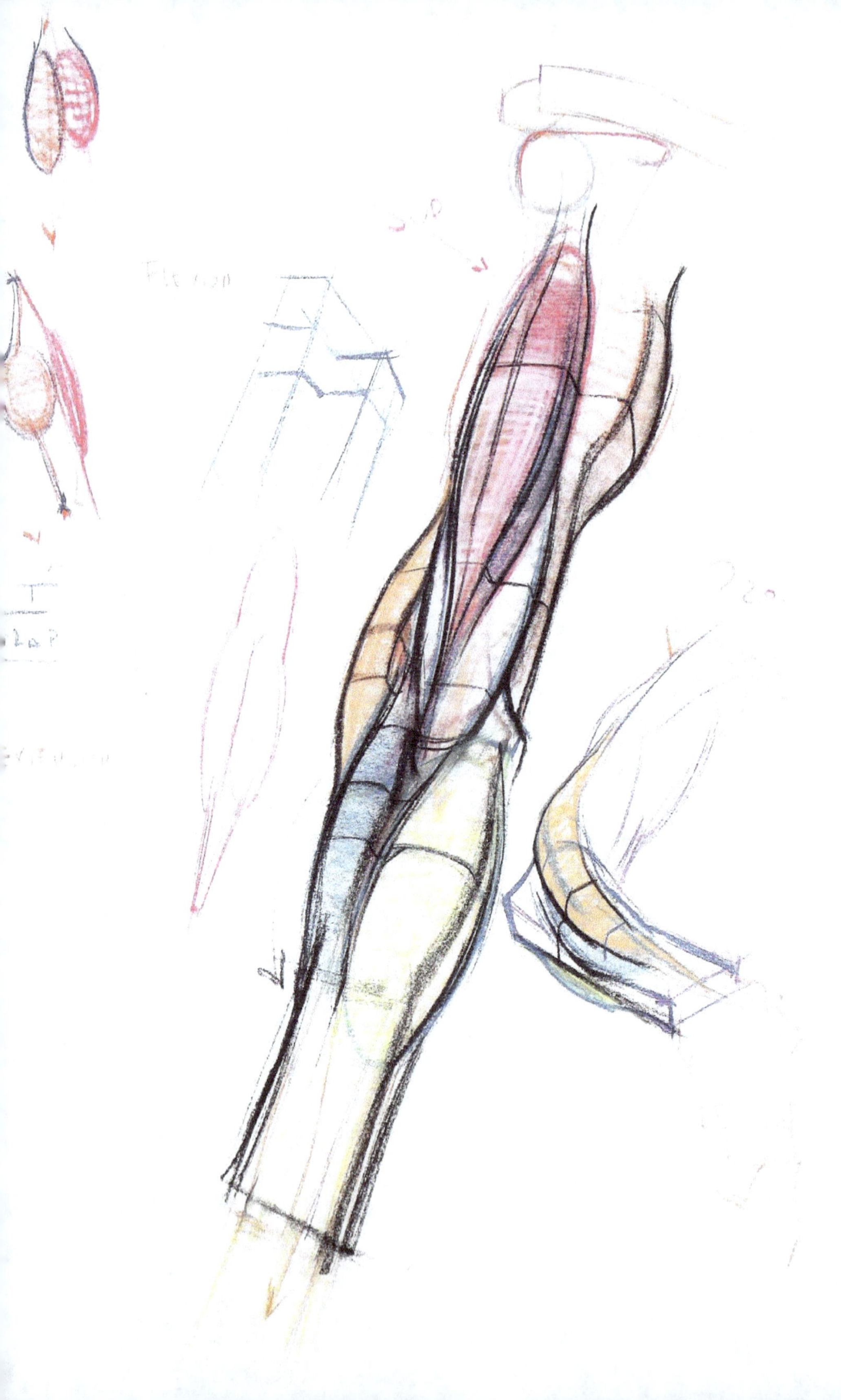
Flexion

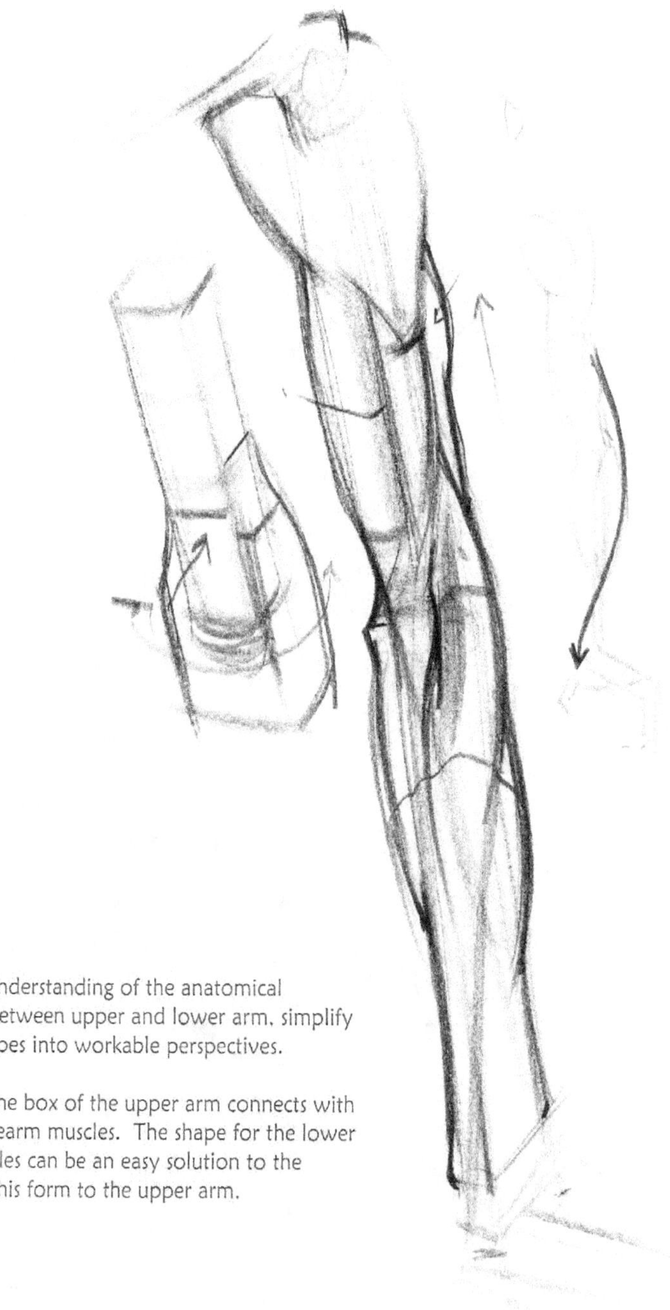

To aid your understanding of the anatomical relationship between upper and lower arm, simplify the larger shapes into workable perspectives.

Notice how the box of the upper arm connects with the lower forearm muscles. The shape for the lower forearm muscles can be an easy solution to the swiveling of this form to the upper arm.

When working on the arm itself, it is again very important to follow a process. This process will help you organize the most important qualities first, and will aid in invention.

1. To start the arm, it is first most important to position/work out the gesture. On top of the more lyrical, overall gesture (demonstrated in Chapter 1), place a straight line to give a strong feeling on the bones as well as 2-D position.

2. Second, build your perspectives on top of the straight. These cylinders (demonstrated in the Landmarks chapter) are the most important element to provide a believable sense of space and form. The anatomy will need to be wrapped around these cylinders using the two main ideas of transitional volumes.

3. Place the asymmetrical design of the anatomy of the arm on top of the cylinders.

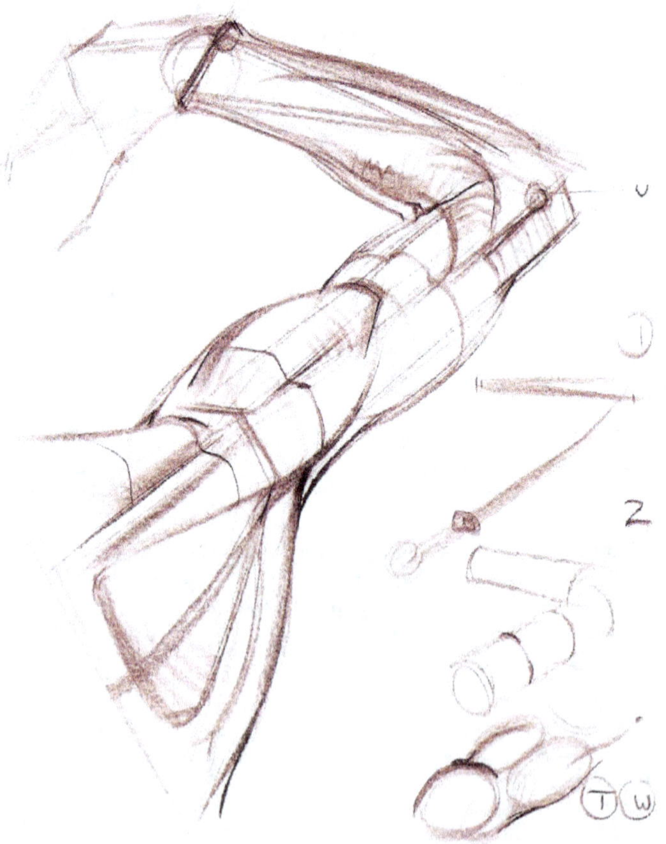

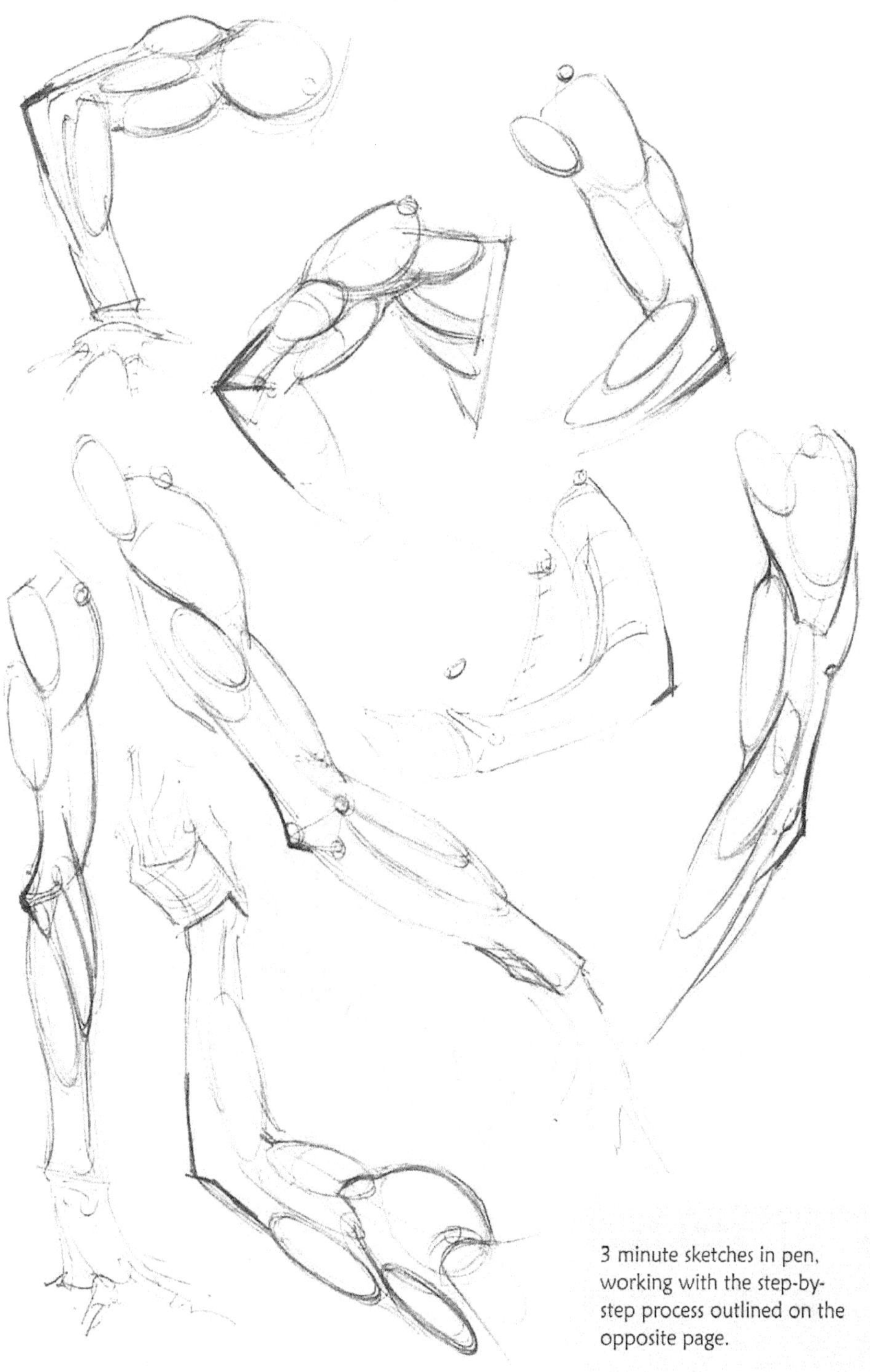

3 minute sketches in pen, working with the step-by-step process outlined on the opposite page.

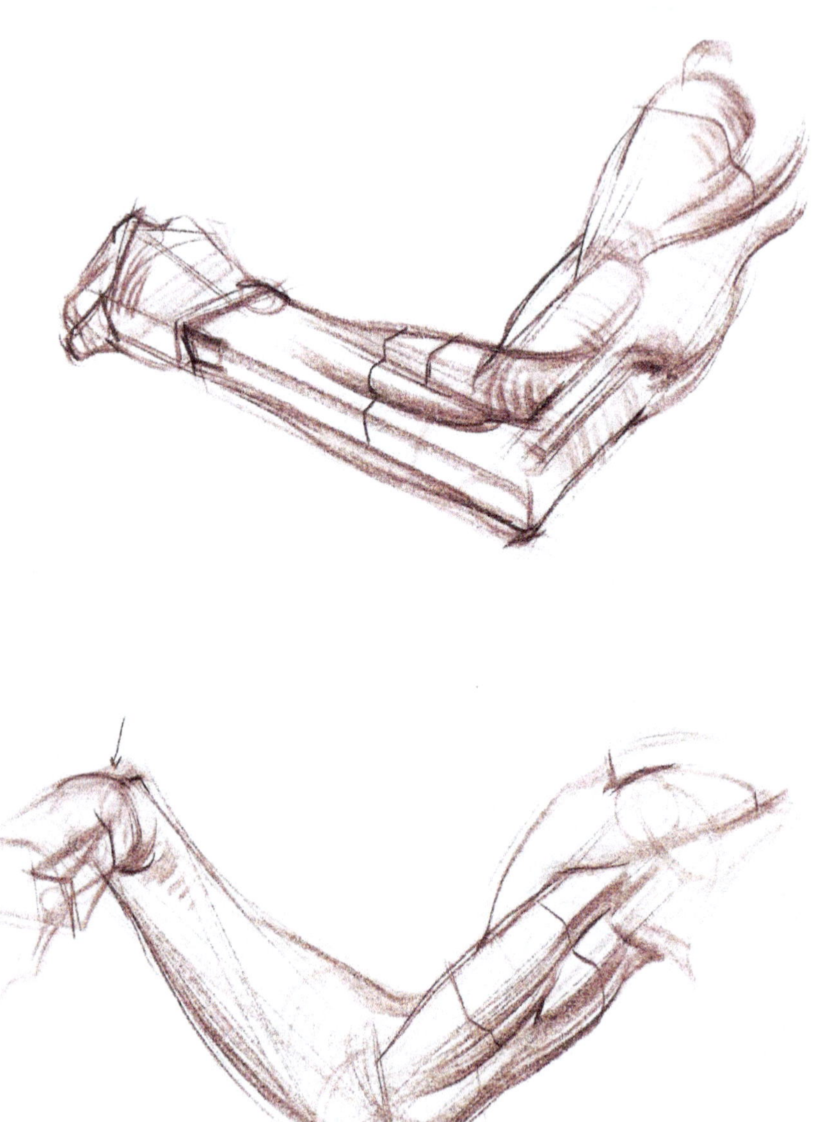

BRACHIALIS

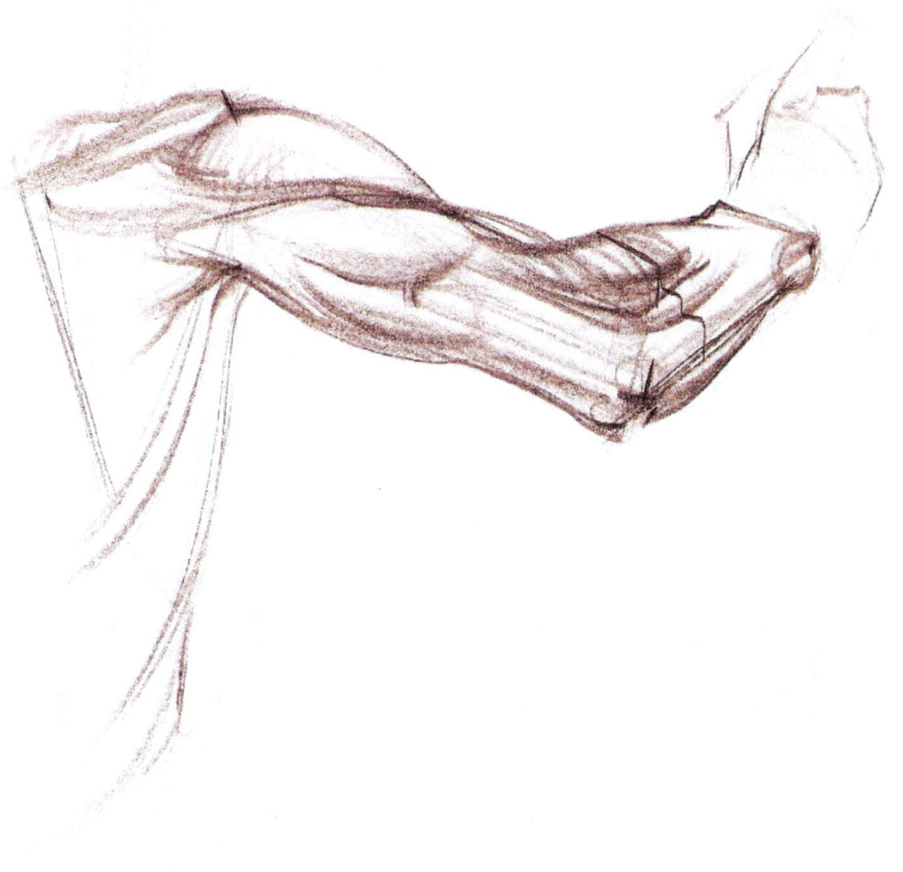

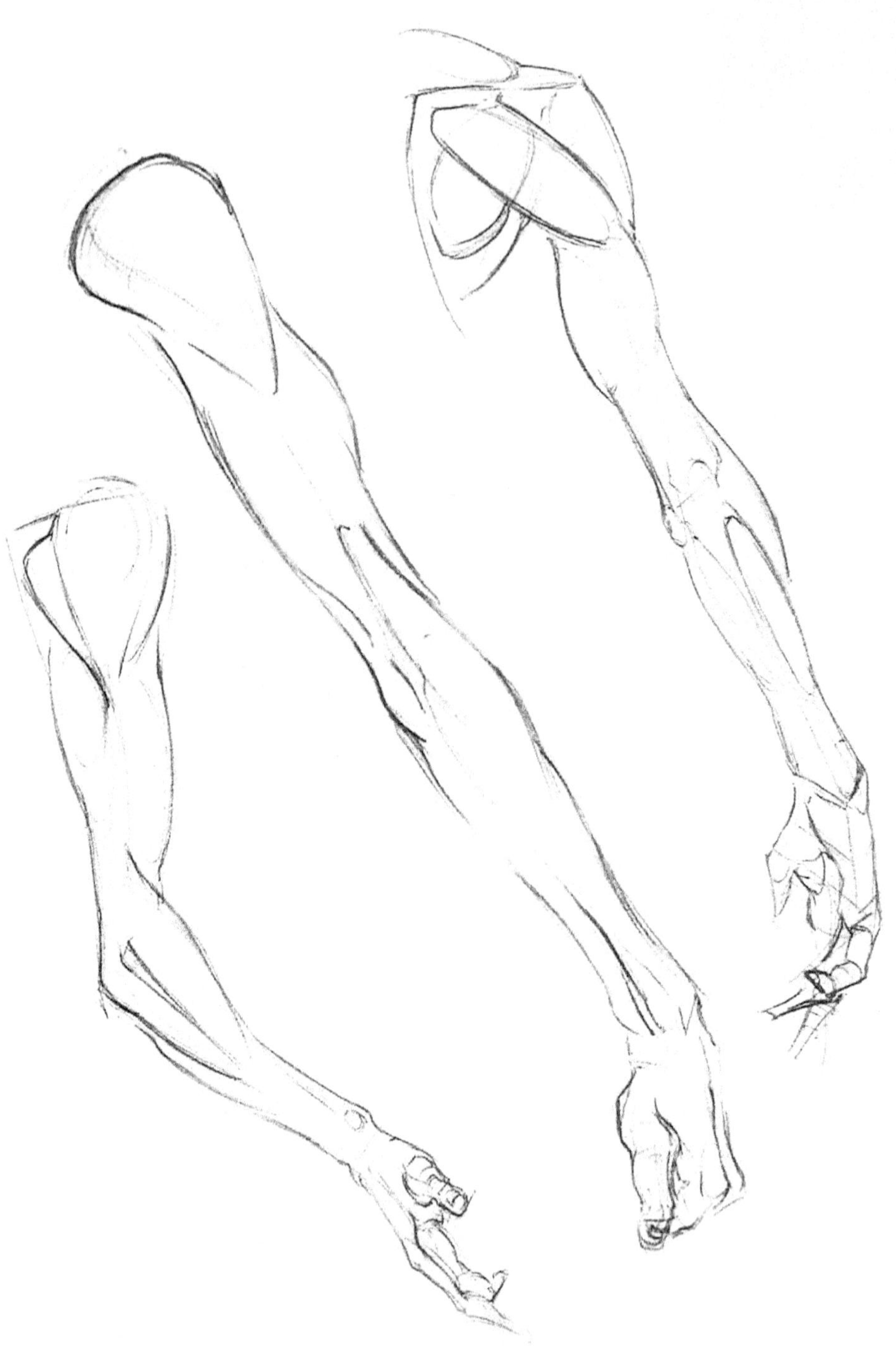

The drawings on the following pages are included to give a number of different views and positions of the arms for study. In many cases, the arms have been isolated, so remember to always include the working of the rib cage and shoulders.

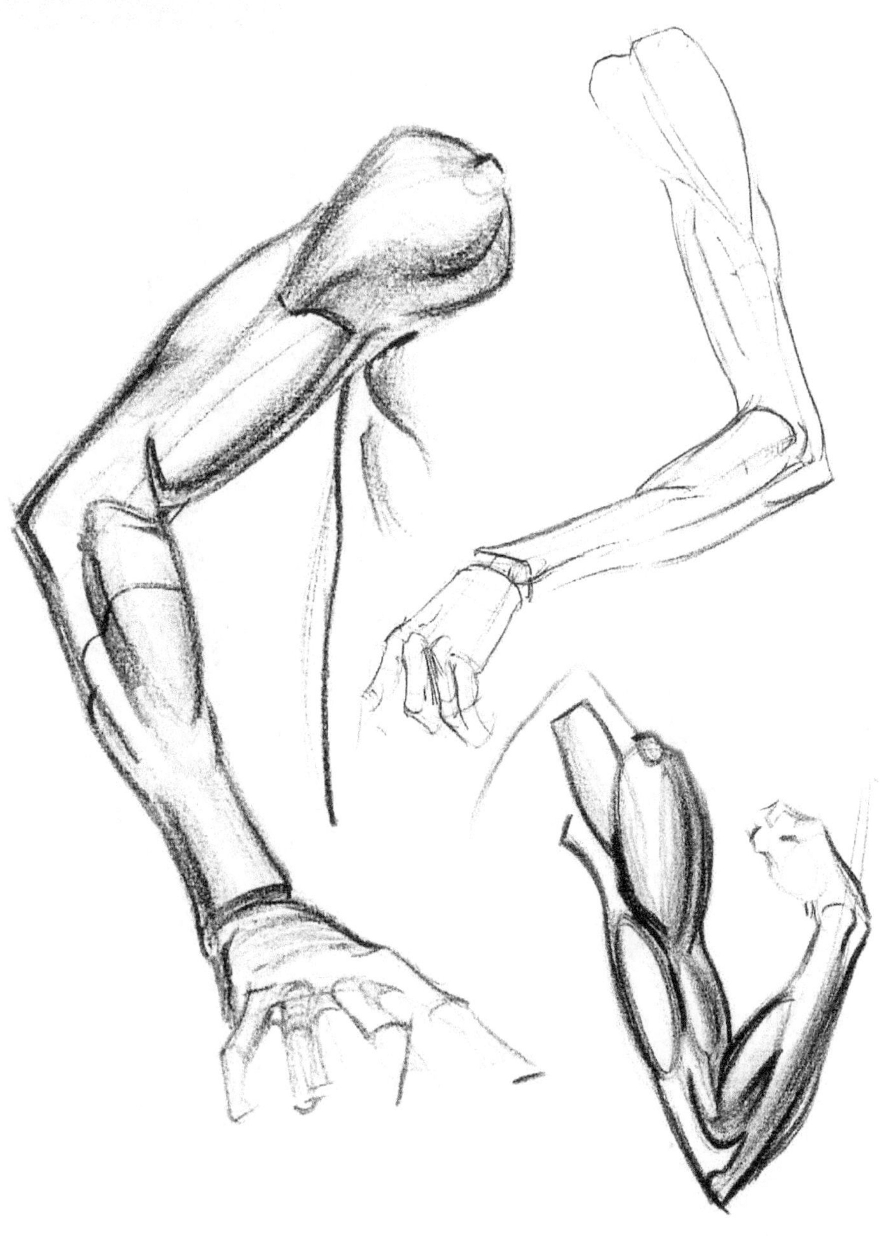

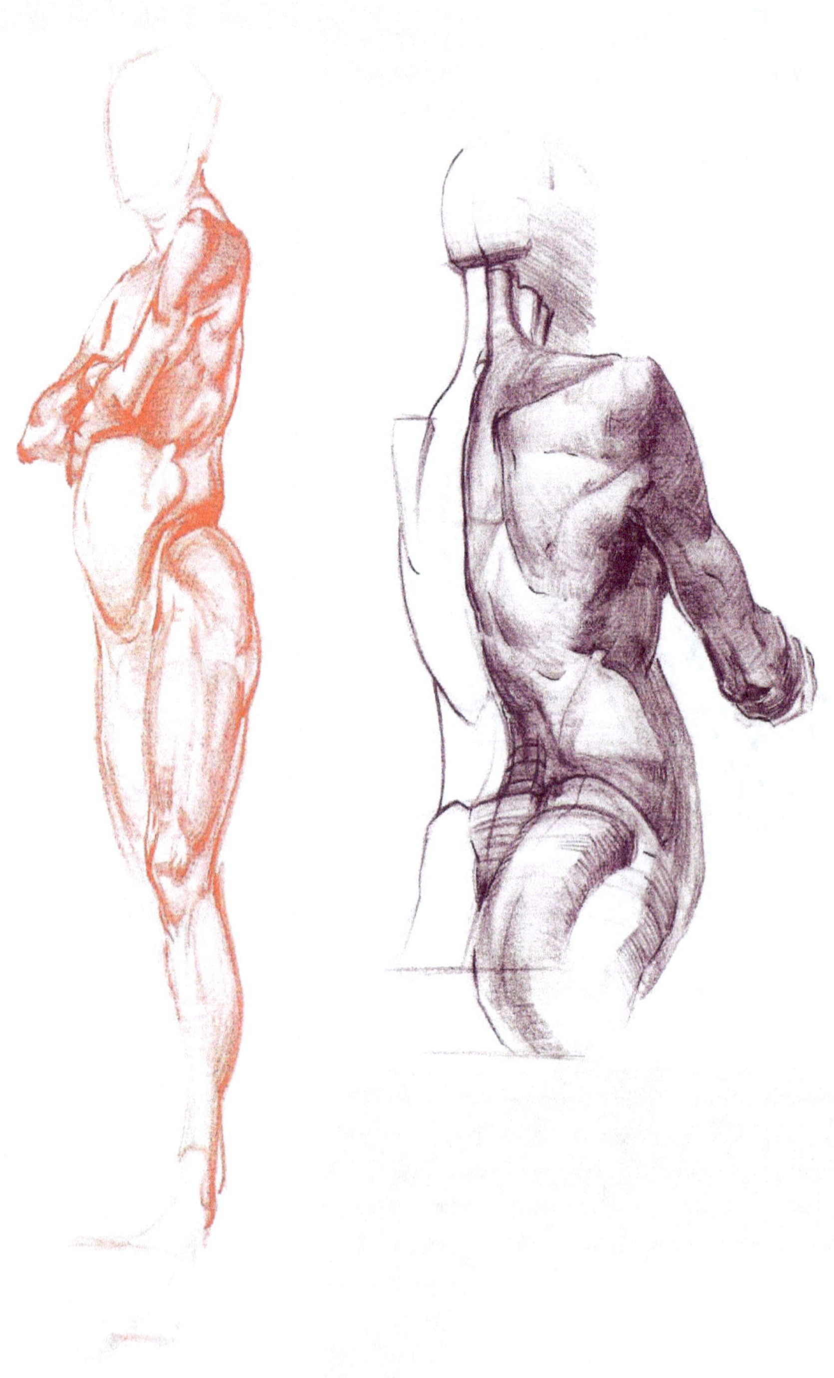

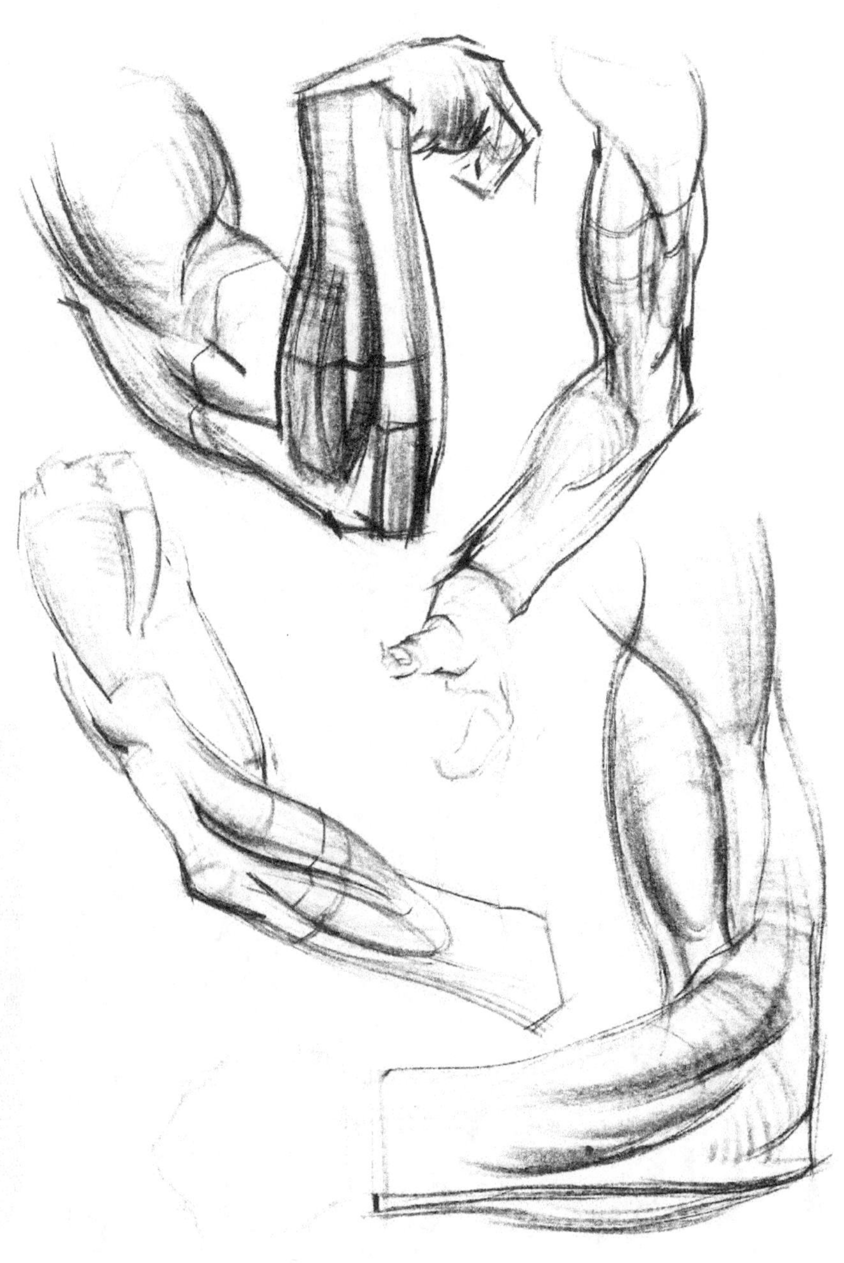

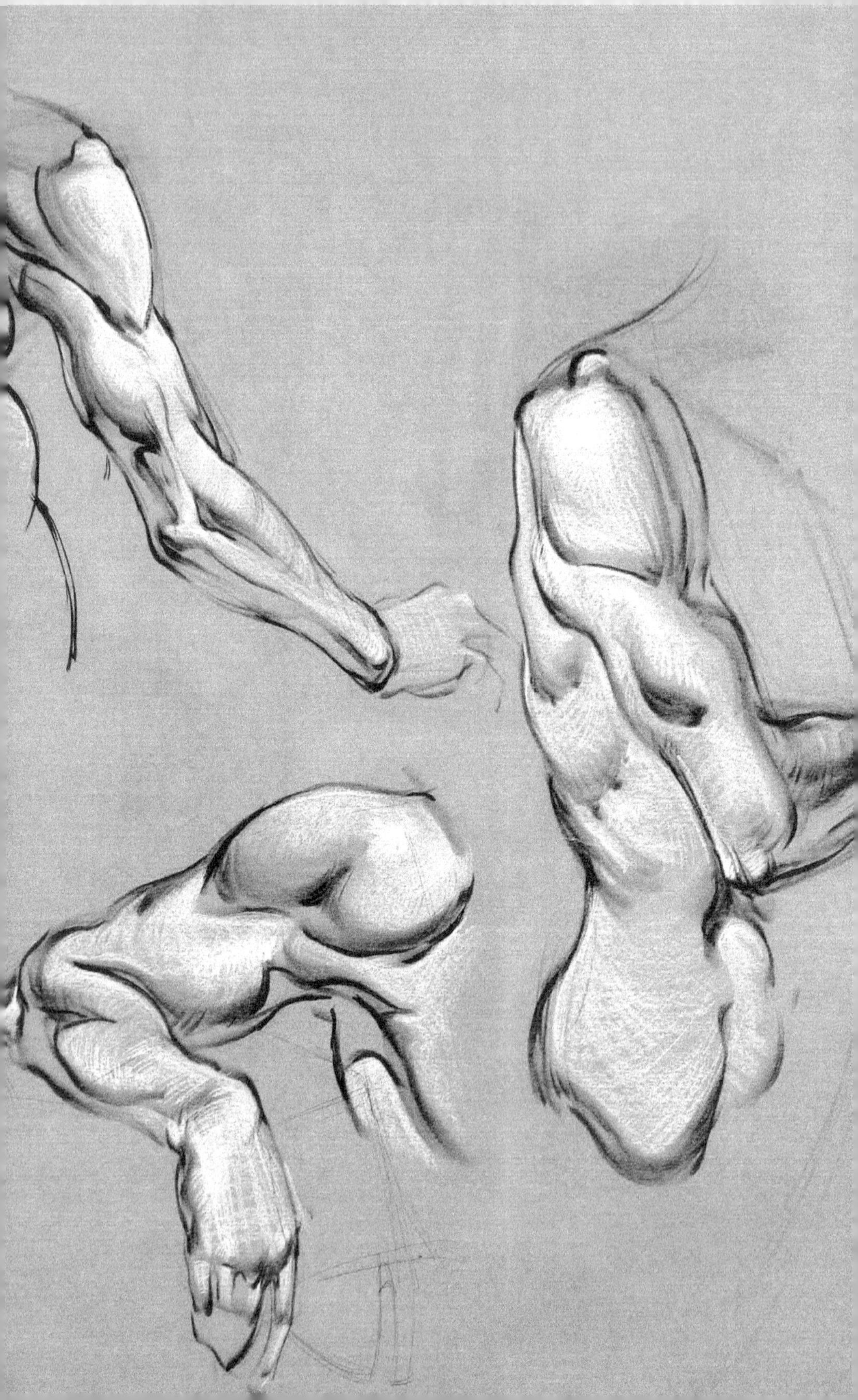

The benefit to only using spheres and ellipses is that lighting,
or inventing a light source, becomes much easier.

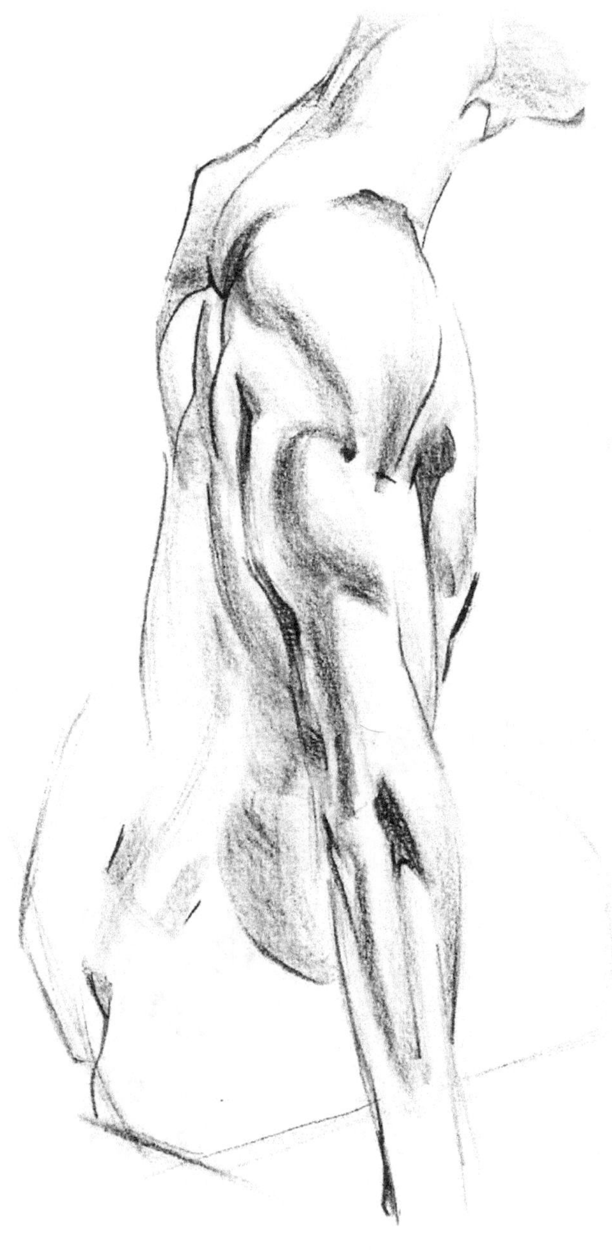

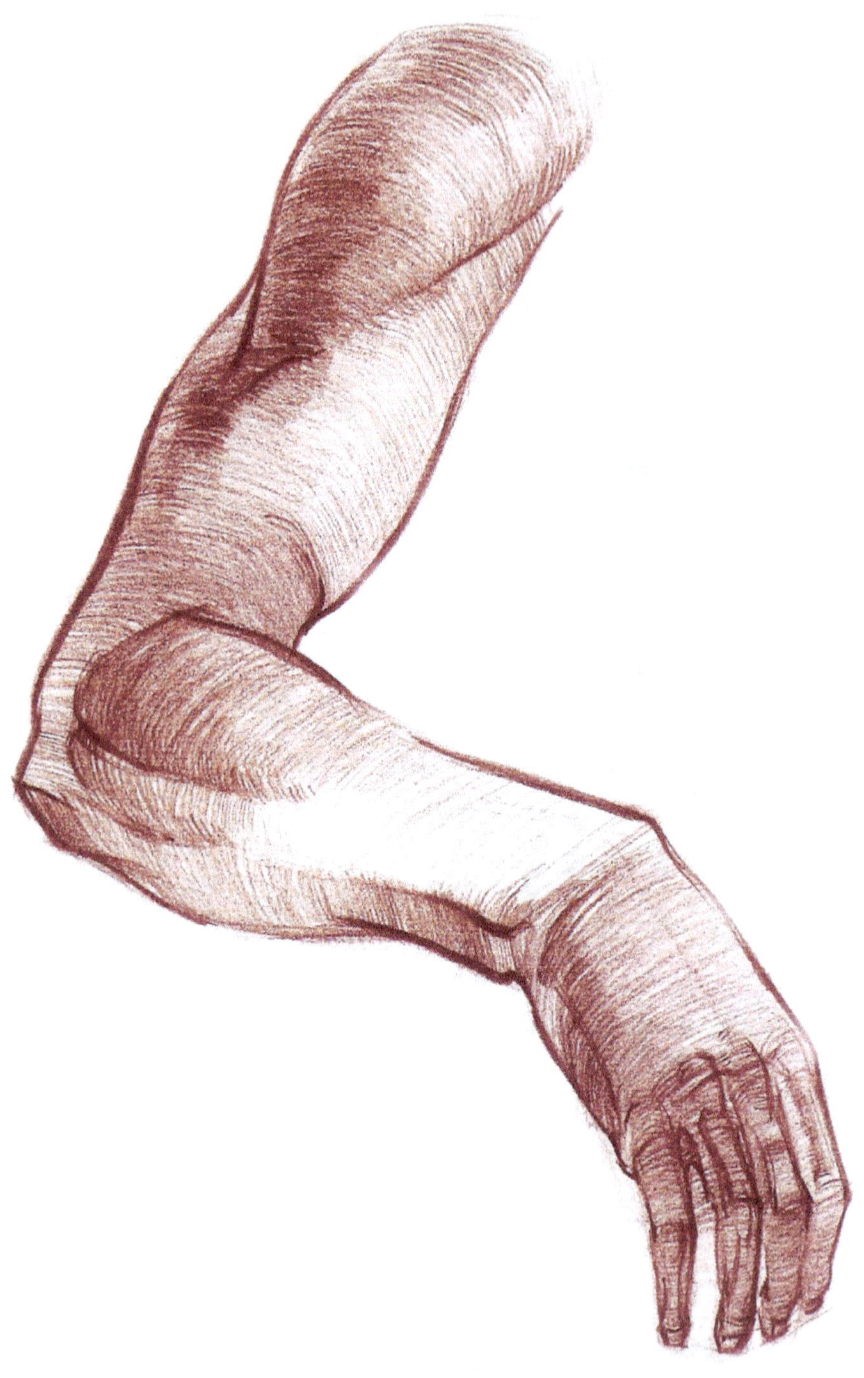

A different exercise, used here, is
the development of form and value
through the use of only ellipses. Study
the drawings, paying attention to the
surfaces they depict.

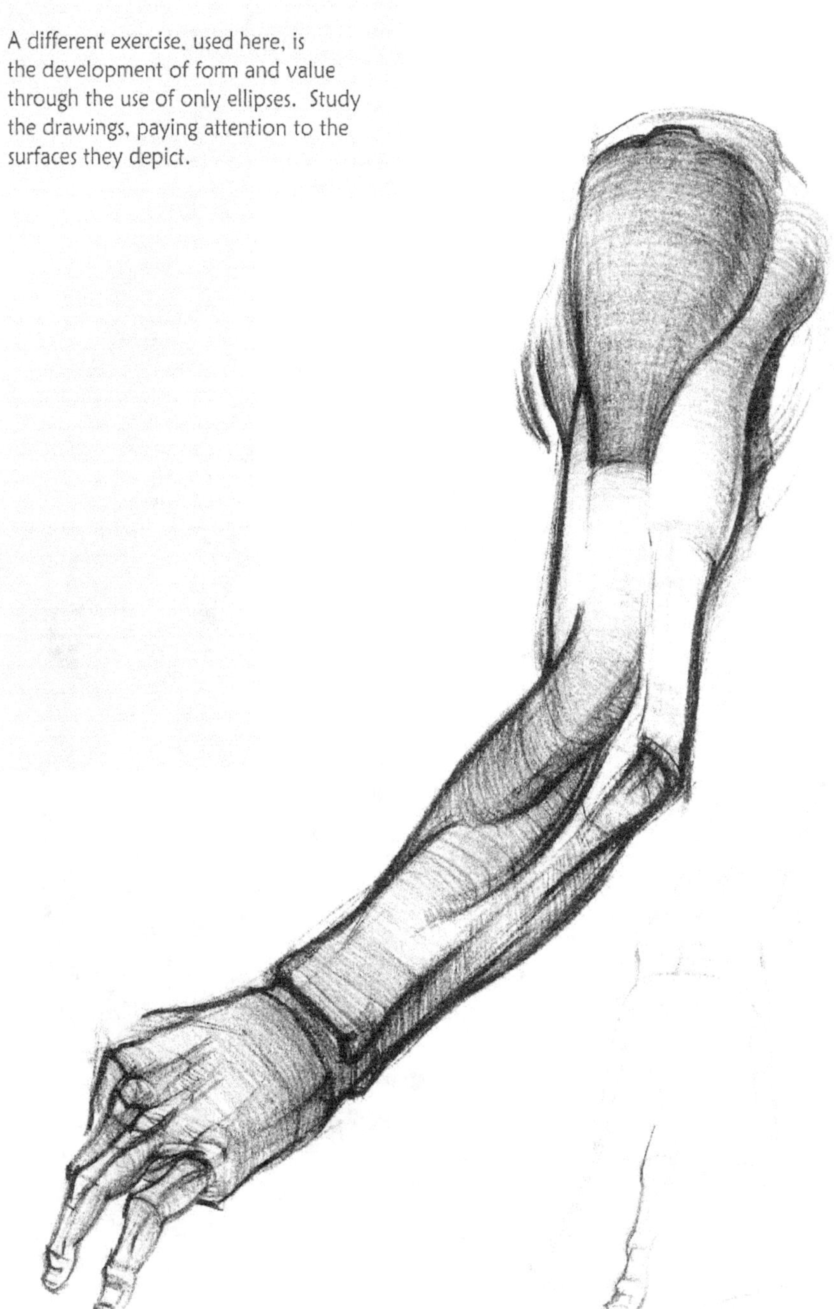

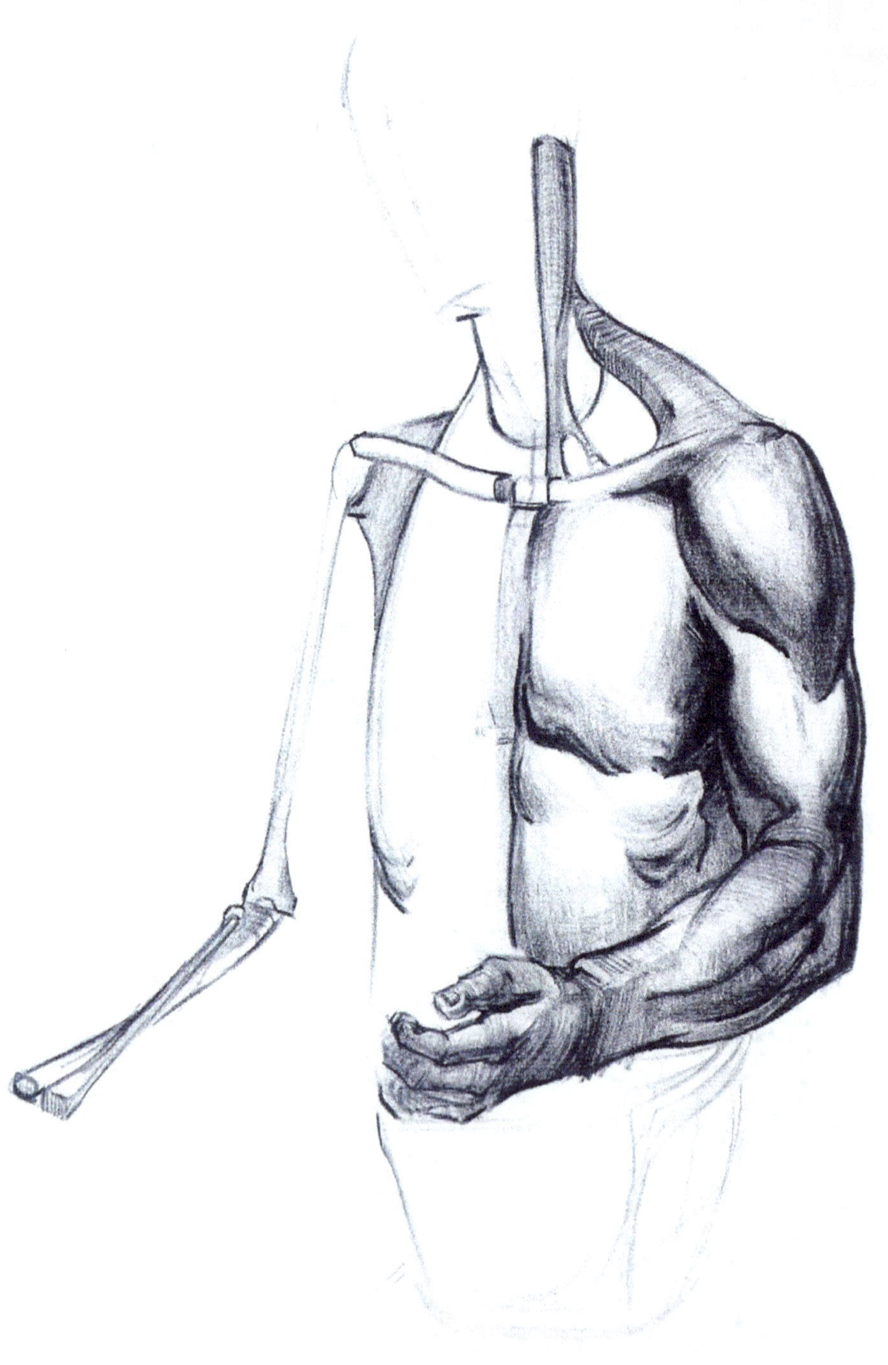

These two drawings show
the evolution of the process
and working method.

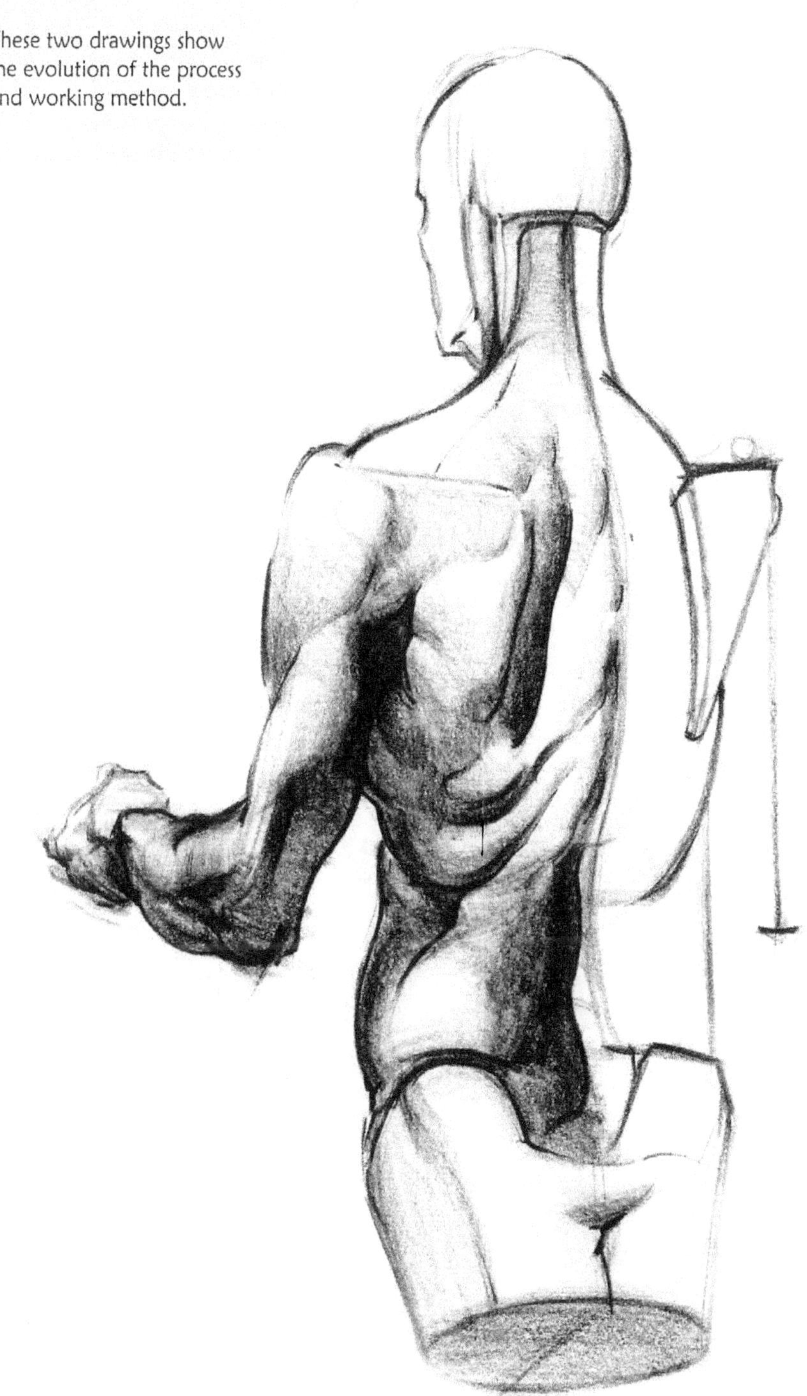

THE HAND

When drawing the hand, you will follow
a similar process to the one used when
drawing at the head, and this same process
will again be used again when looking at
the foot in a future chapter.

To review, you will begin by looking at
the skeleton, proportion, and anatomy,
in order to develop designs that can
be effective in the process of drawing.
Next, take those informed designs and
render them with variations of the box,
cylinder, and sphere in order to create a
believable effect of space. Finally, you
should create contours that can be worked
over the forms to present a more organic
description.

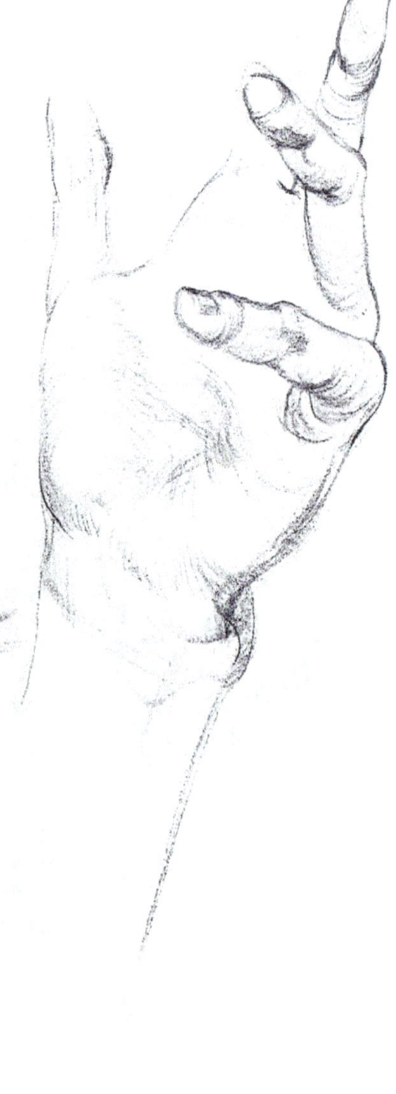

HAND STRUCTURE and PROPORTION

By looking at the skeleton of the hand, you can take away important proportional information. The hand is primarily made up of three groups of bones: the carpus group, the metacarpus (the bones of the palm), and the phalanges (the bones of the fingers).

The drawing below is a diagram of the bones in the hand. The circle shown near the top of the hand represents the carpus group. The carpus group is a collection of eight bones organized in a bridge-like form.

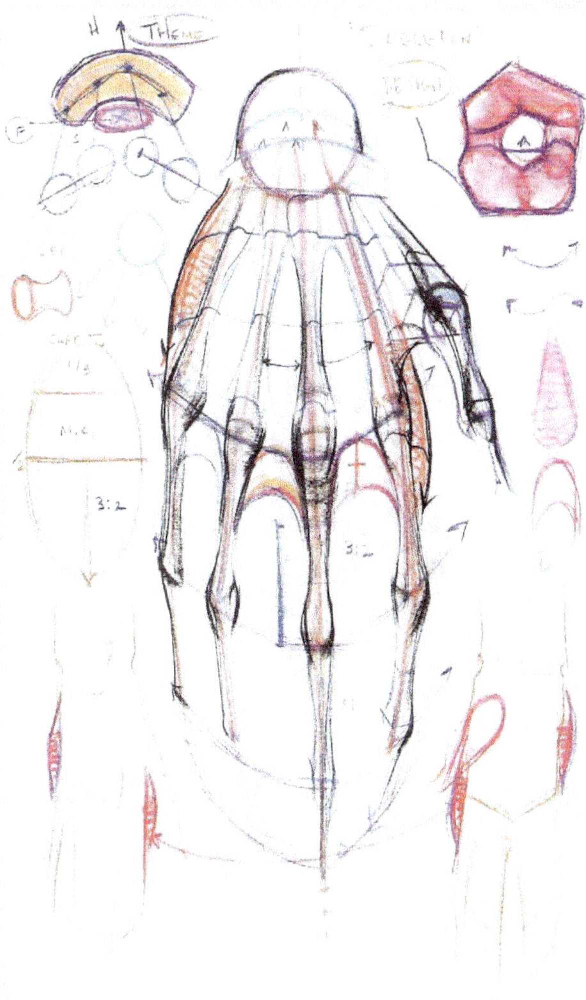

The diagram below shows this bridge seen as if the fingers were pointing directly at you. The carpus group is carried through every stage of the hand drawing. This area, and shape shown below, are the design theme for the hand. Because the shape of the hand is dependent on this form for its effective use and function, you should integrate its effects into your depiction of the hand at every stage, from proportion to the design of perspectives.

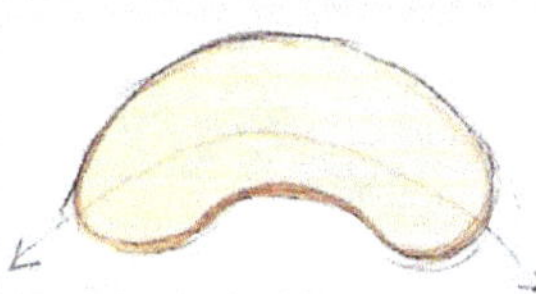

Note that the wrapping line drawn over the carpus group (orange circle) suggests an apex or high point to the form. Because of the bridge-like quality of the carpus group, a vacancy is created beneath. This area is known as the carpal tunnel, and is occupied by tendons that branch out into the fingers.

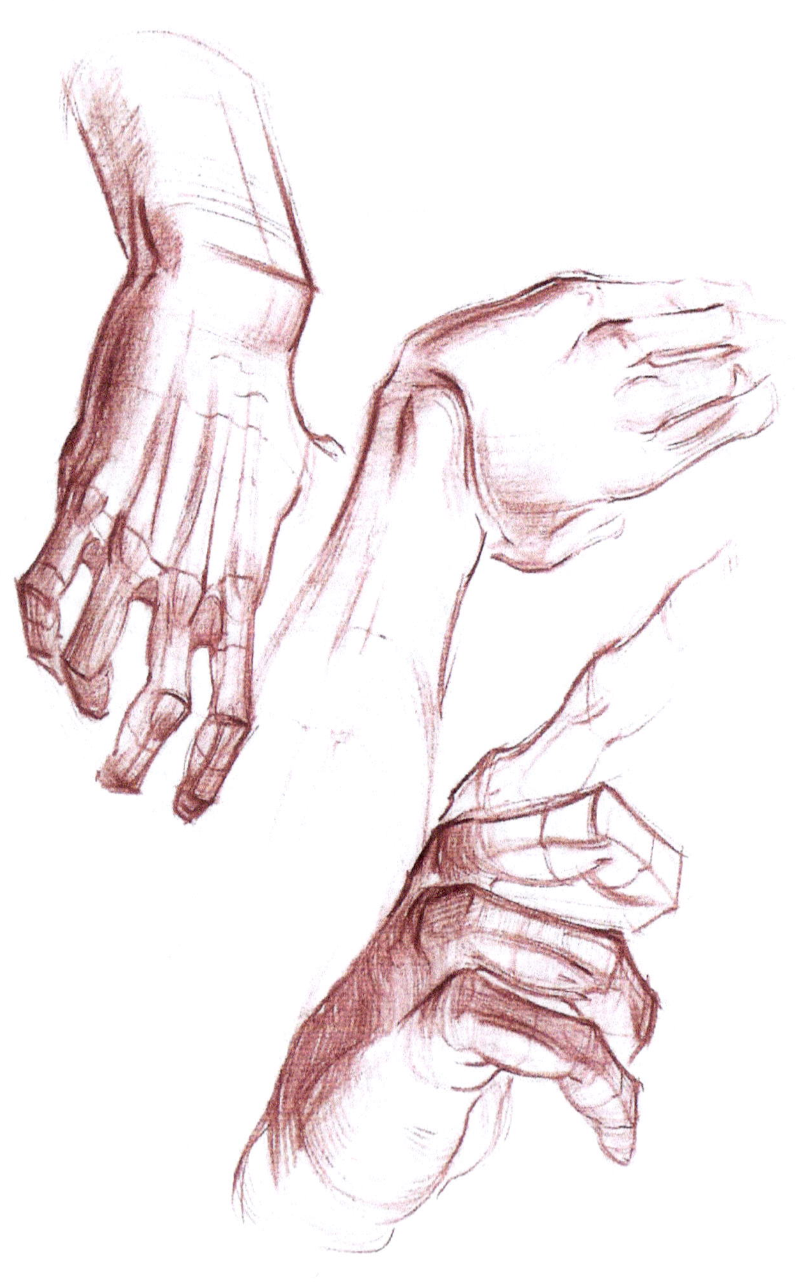

The diagram to the right shows the proportional relationship between these groups. The entire oval shown is a simplified version of the diagram on page 161. That oval is split at the half-way point to represent the end of the metacarpals (knuckles). The upper half of the hand is made up of one-third carpus group and two-thirds metacarpals. The lower half of the hand is made up entirely of the fingers/phalanges.

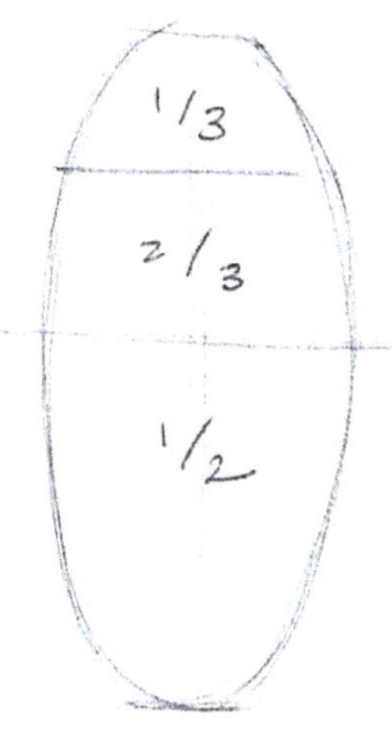

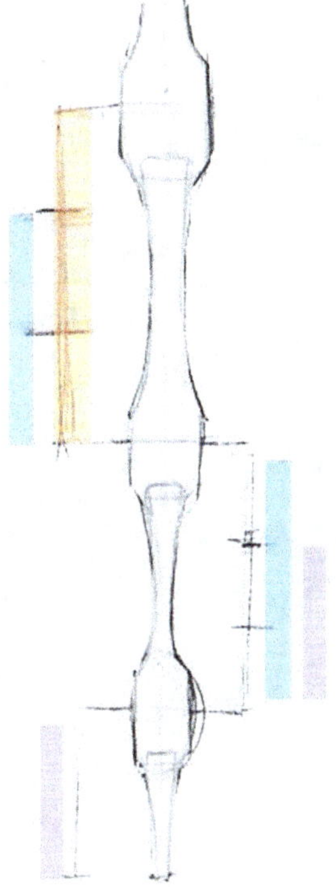

The bones in the fingers are proportionally-based on a 3:2 ratio. The diagram to the left shows the relationship between the three bones in the finger: the proximal (closest to the palm), middle, and distal (furthest bone or tip of the finger) phalanx.

Note that the proximal phalanx as a whole (shown in orange) has been broken into three sections. Two of these sections are shown in blue. These two parts are the length of the next bone, the middle phalanx. The same approach is then applied to the middle phalanx: the overall length (shown in blue) is broken into three equal parts. Two of these three parts (shown in purple) are the length of the last bone in the finger, the distal phalanx.

Remember that these proportions are meant to give you a general sense of the relationship between the sizes of the bones in the finger. You will not always have the opportunity to execute an exact measurement. This is not necessarily because techniques of measurement aren't helpful, but rather that many poses won't allow for this specific view, enough time to measure, or, in the case of an invented pose, a subject to measure.

HAND ANATOMY

In addition to being familiar with the hand's skeletal structure, understanding the anatomy of the hand will contribute to an understanding of gesture and volume. The anatomy shown below represents a simplified understanding of the major muscle groups shown in the hand. Remember, this is meant to aid in the development of a working process. An in-depth study of anatomy requires additional resources.

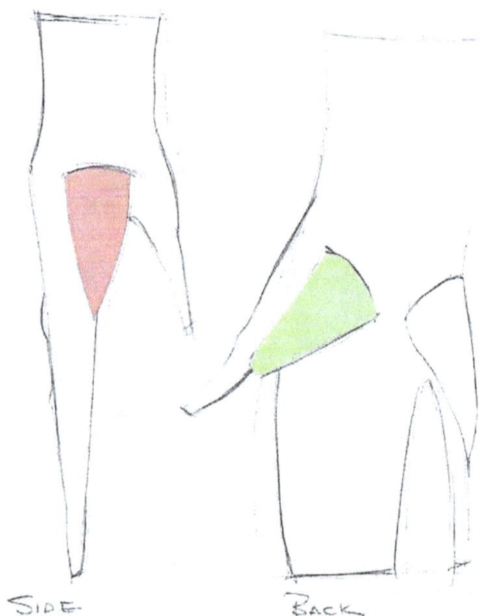

The upper portion of the hand is primarily dominated by bone, whereas the palm is fleshy. The diagram shows a side view (to the left) and a back or palm view (to the right).

There are three major muscle groups in the hand, which can all be simplified into a simple teardrop shape.

These muscles include the thenar eminence (shown in green), the hypothenar eminence (shown in pink) and the first dorsal interosseous (shown in red). While the shape of these muscles may vary depending on the view and position of the hand, they will always be a variation on the teardrop.

Additionally, depending on the movement of the fingers, these shapes will be where you will find a pinch or a stretch taking place. For example, if the thumb moves against the index finger, the shape of the first dorsal interosseous will be compressed and pinched.

Studying the anatomical shapes in the hand is at this point to help foster a better sense of your relationship to the forms. Remember these simple shapes in order to later expand them into more complex surfaces.

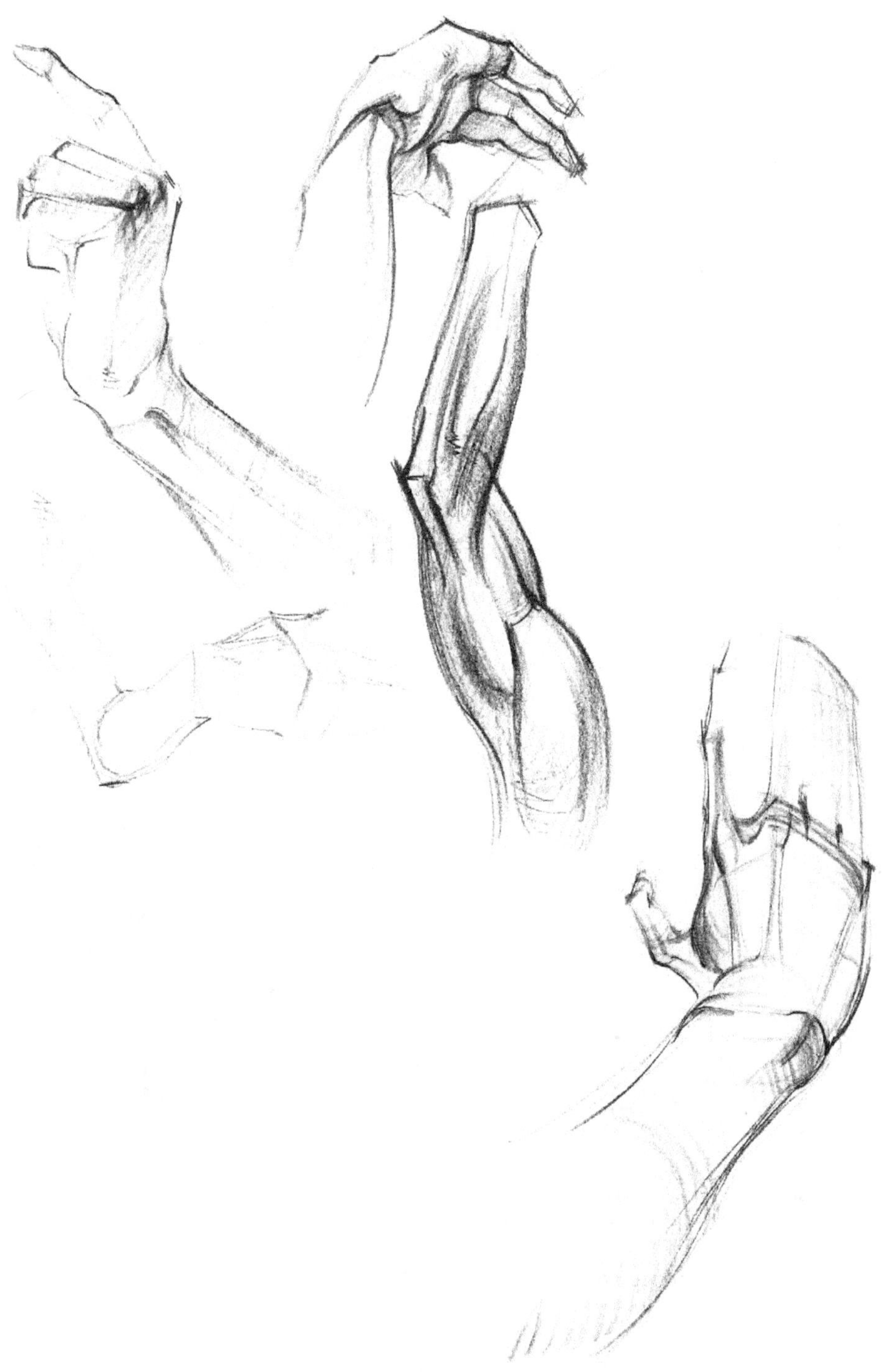

PERSPECTIVE

The second step in developing
the hand is to conceive of
the separate elements in a
perspective-based way.

In constructing the palm, all
of the information taken from
the skeletal structure and
anatomy should be applied.
Notice that the basic form of
the palm is a simple box with
a few adjustments.

The most important
adjustment continues the
major design theme of the
carpal tunnel. Notice that
the top of the palm is raised,
or comes to an apex, roughly
around the area of the middle
finger. The top of the box is
more like the roof on a house,
having its peak at the knuckle
of the middle finger.

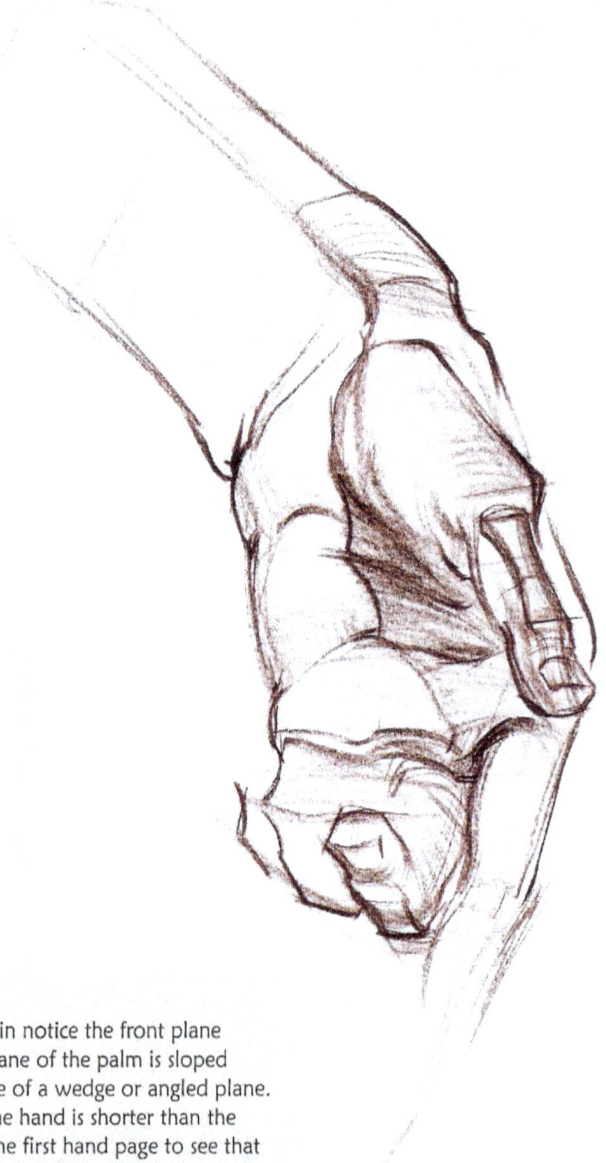

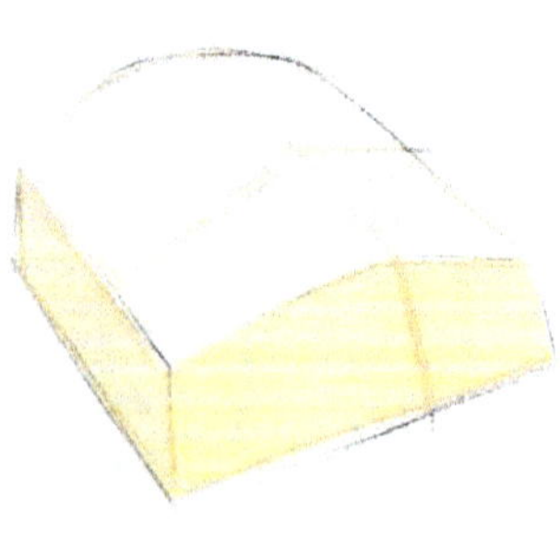

Instead of being a basic box, again notice the front plane
(shown in orange). The front plane of the palm is sloped
forward, creating the appearance of a wedge or angled plane.
This is to show that the top of the hand is shorter than the
palm. Look at the diagram on the first hand page to see that
the webbing on the palm side (shown in green) continues the
palm further than the top portion of the hand.

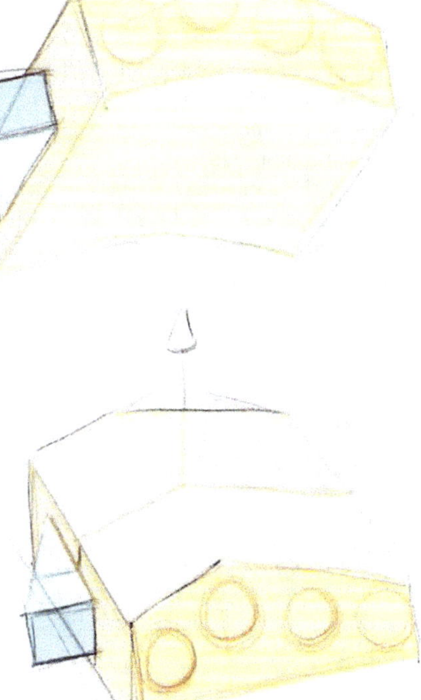

The diagram to the right shows how the underside of the palm is affected. This gives the structural description of the palm a feeling of naturalism.

The last structural element based on bone structure is the way the thumb is added. Notice in the diagram at the beginning of the chapter that all of the fingers (excluding the thumb) move, more or less, in the same direction.

In order to show that the thumb moves in a completely different direction, this separate form is added on to the side of the palm. The structure (shown in blue) resembles a door wedge. Additionally, keep in mind that the thumb has only a proximal and distal phalanx.

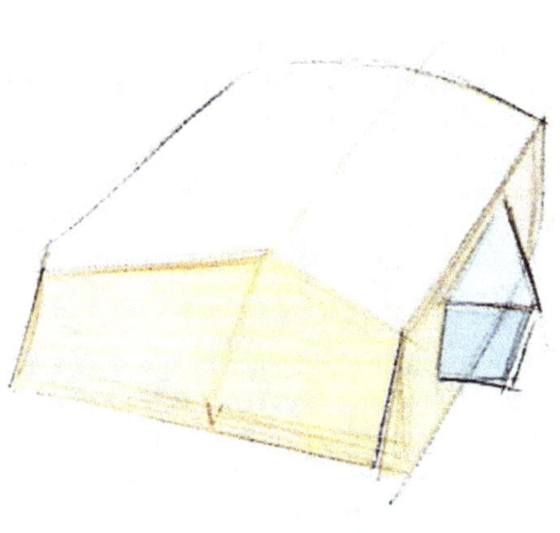

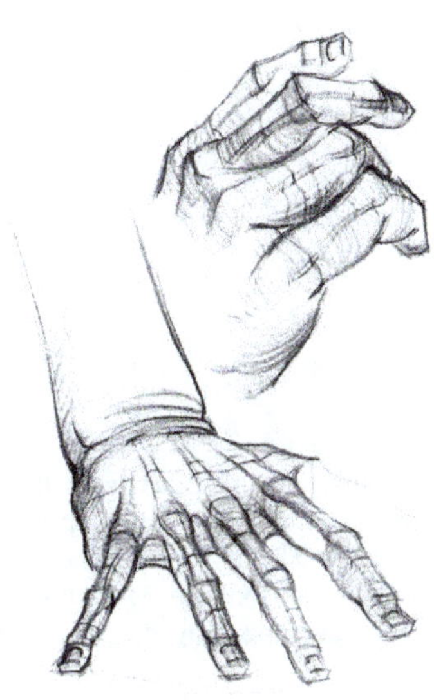

FINGER BONES and KNUCKLES

The drawing to the right shows a breakdown of the bones and knuckles in the finger. Notice that the shape of the knuckles resembles a thimble that has been compressed in the middle. This design allows for the tendon on top of the hand to sit in a groove in the center of the knuckle as it continues toward the fingertip. This is an important feature to develop at a later stage of the drawing. For developing a working process, this can be simplified even further.

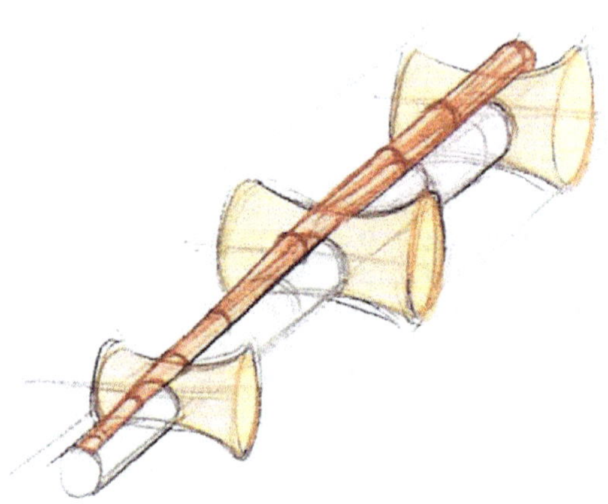

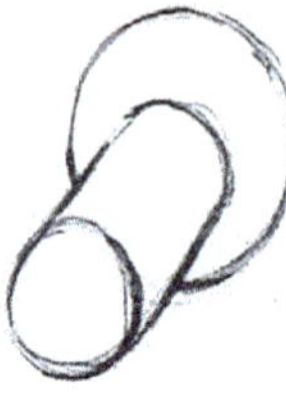

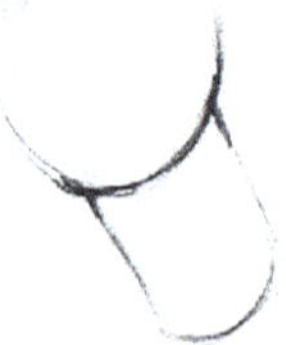

When drawing the fingers, you only need a straight line, a sphere, and a cylinder. The drawings on the next page show both a first and second step.

The first stage in drawing the finger involves using a sphere (for placing the knuckles) and a straight line (to determine the direction and placement of the proximal, middle, or distal phalanx). This approach gives you a great deal of flexibility in positioning the fingers — especially when using this method for figure or character invention.

The second step involves positioning the fingers in space using volumes (note that this still follows our overall process — first gesture, then development of volume). On top of the straight line, a cylinder has been added that corresponds to the direction in space.

The diagram just above details the two types of interaction that exist when using a sphere and a cylinder together to develop the fingers. The example on top shows how the sphere and cylinder will interact when the finger is raised and coming towards the viewer. In this view, the cylinder continues to the inside of the sphere, creating the "T" overlaps. To show the finger pointing away, the cylinder stays outside of the sphere, creating "T" overlaps.

Study the two drawings below to see how this principle has been used.

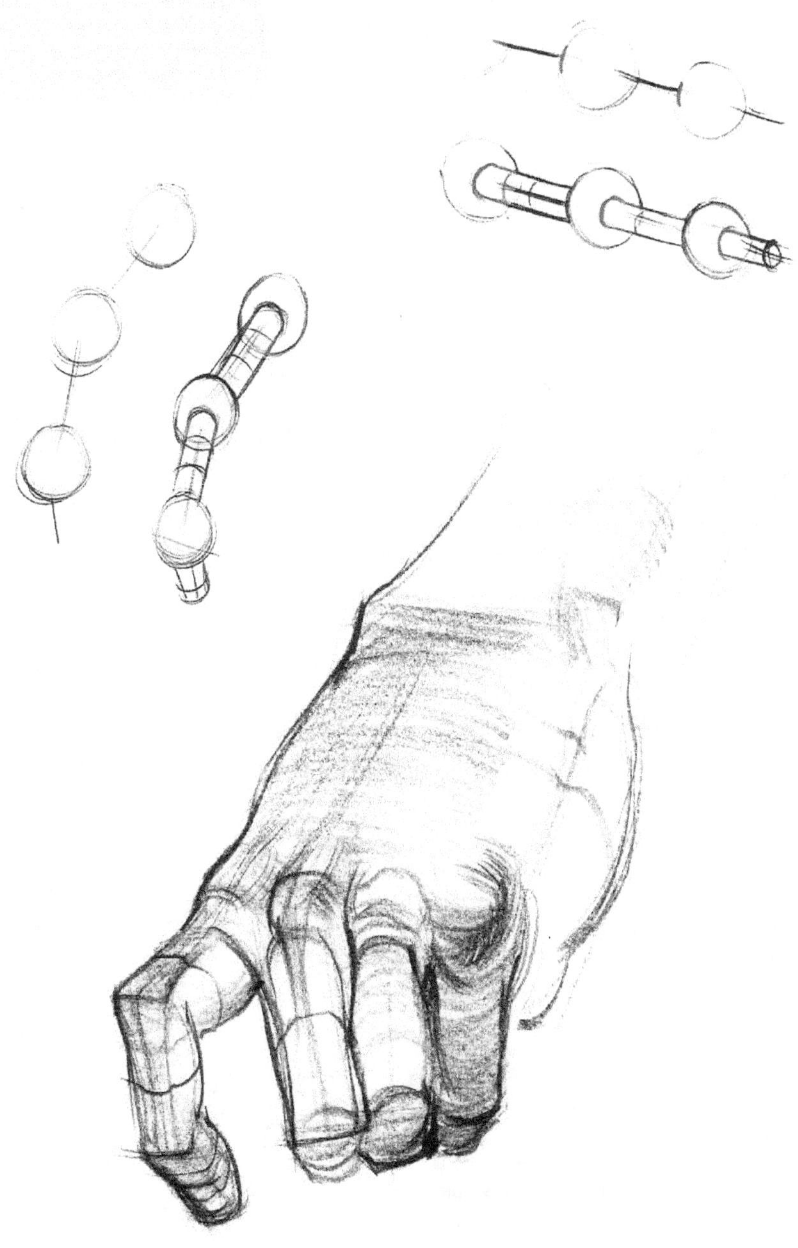

FLESH AND MUSCLE

Having developed the underlying structure, the next step in drawing the fingers involves adding the flesh or muscle. It is important to work through the previous stages before creating a finished line. Going directly to a contour risks creating a drawing with no feel for the skeletal structure or perspective. However, when you have had enough practice in working through the construction method, you should be capable of creating a finished volumetric line without working through the construction.

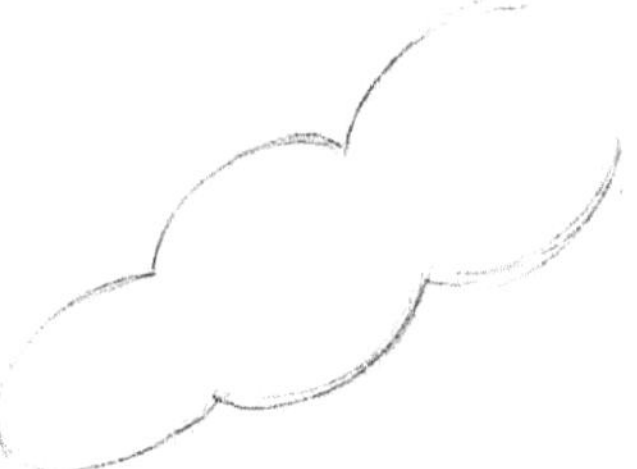

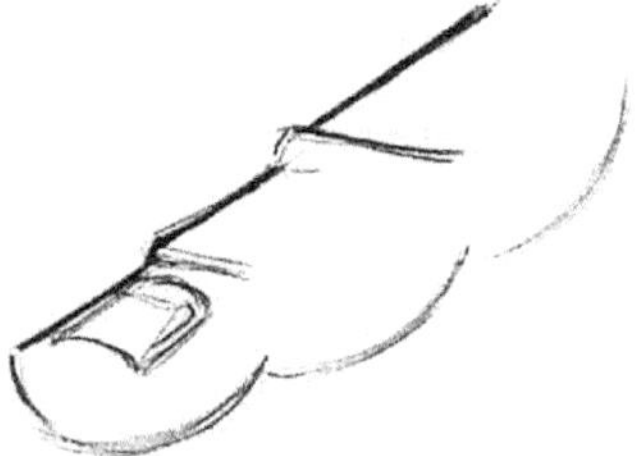

When designing the fingers with a finished line, one common mistake (shown above left) is paralleling forms. This approach gives the fingers an awkward sausage-like appearance, and does not create a fluid experience of the form.

The drawing to the upper right shows a design solution more in-sync with the natural qualities of the finger and hand, keeping a sense of naturalism to the finger by only using straight lines on the top portion of the finger (or palm) and by using a curve on the underside. A straight line associated with a more rigid or abrupt visual mark is fitting for the top portion of the hand, which consists primarily of bone very close to the surface of the skin. A curve, commonly associated with a slower, softer visual experience, is used on the underside of the palm, which is primarily made up of fatty tissue.

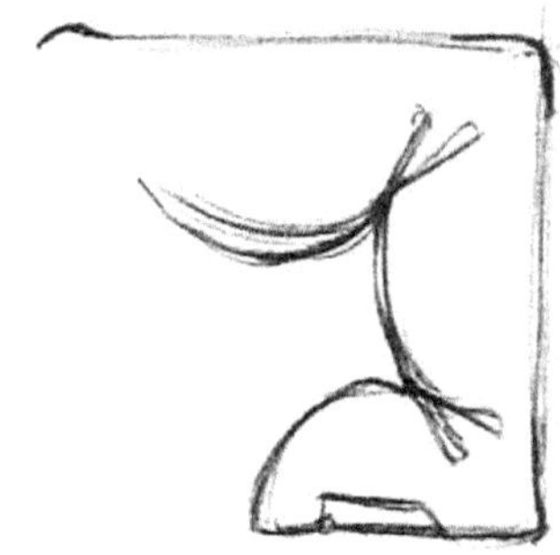

This approach also makes dealing with issues of pinch and stretch in the fingers easier. Notice that the knuckles bend at ninety-degree angles — it is at these points of bend where the curves on the underside of the finger will pinch and the top surface will stretch.

When putting the finished line work on the fingertips, continue to keep straights working against curves and a sense of volume and perspective present. In all of the fingertips shown below, the nail is shown as a flattened plane rolling over the cylinder of the finger. This is an example of how to use an organic element to describe the perspective without turning the finger into a cylinder.

Notice on the finger seen in profile how the shape becomes very pointed at the end, while still having a sense of volume created by the placement of the nail. The drawing to the bottom right shows how to develop a sense of gesture in relationship to the finger's or hand's involvement with an environment. Exaggerating the fatty underside of the finger as it is squashed is a clear way of describing an interaction with an object.

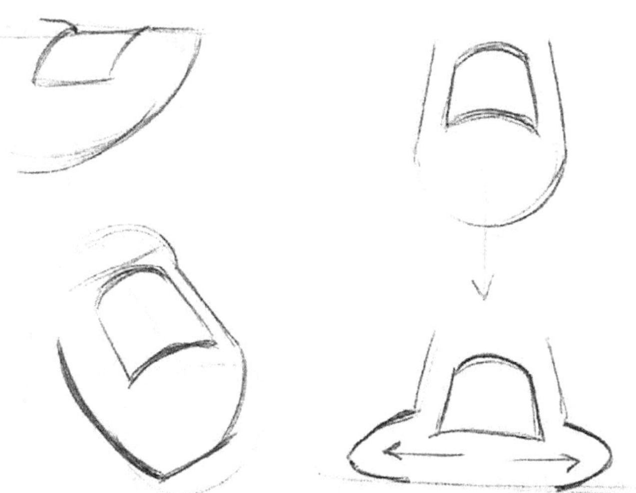

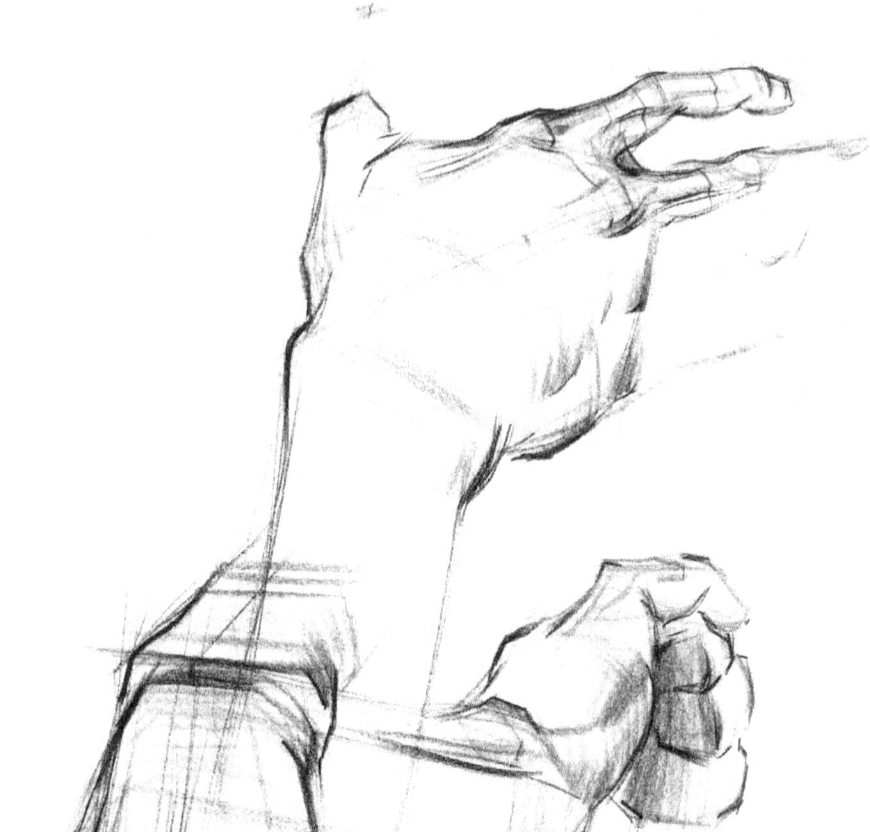

The diagram to the right shows how to put all of this information into a manageable process.

The first step shows the placement of the knuckles and bones only using the spheres and straights. This first step focuses on placement and proportion.

The second step builds upon the first by assigning each finger a perspective using the cylinder and the "T" overlaps.

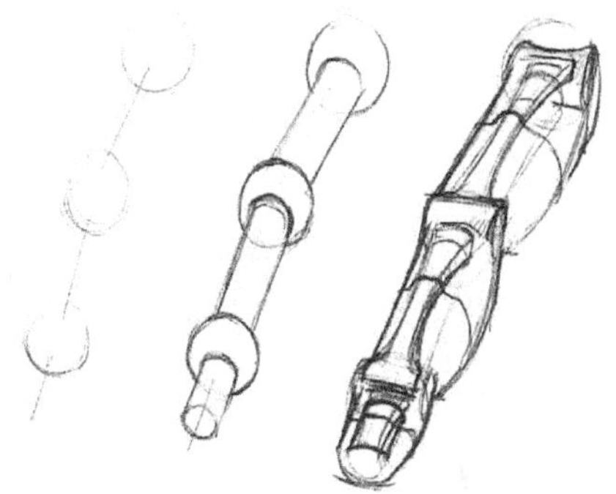

The last step focuses on designing the finger, building on the previous steps. The underside of the finger has been drawn only using "C" curves. A "C" is drawn from the back of one sphere to the center of the next (this curve changes depending on whether or not the finger is pinched). Study the drawing below to see how all of the stages have been used to reach a well-constructed, informed drawing.

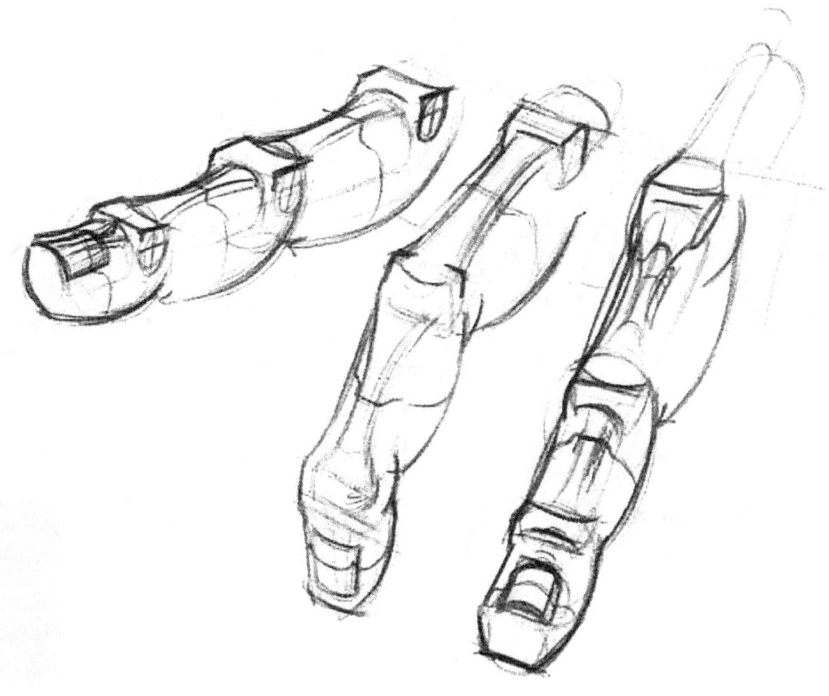

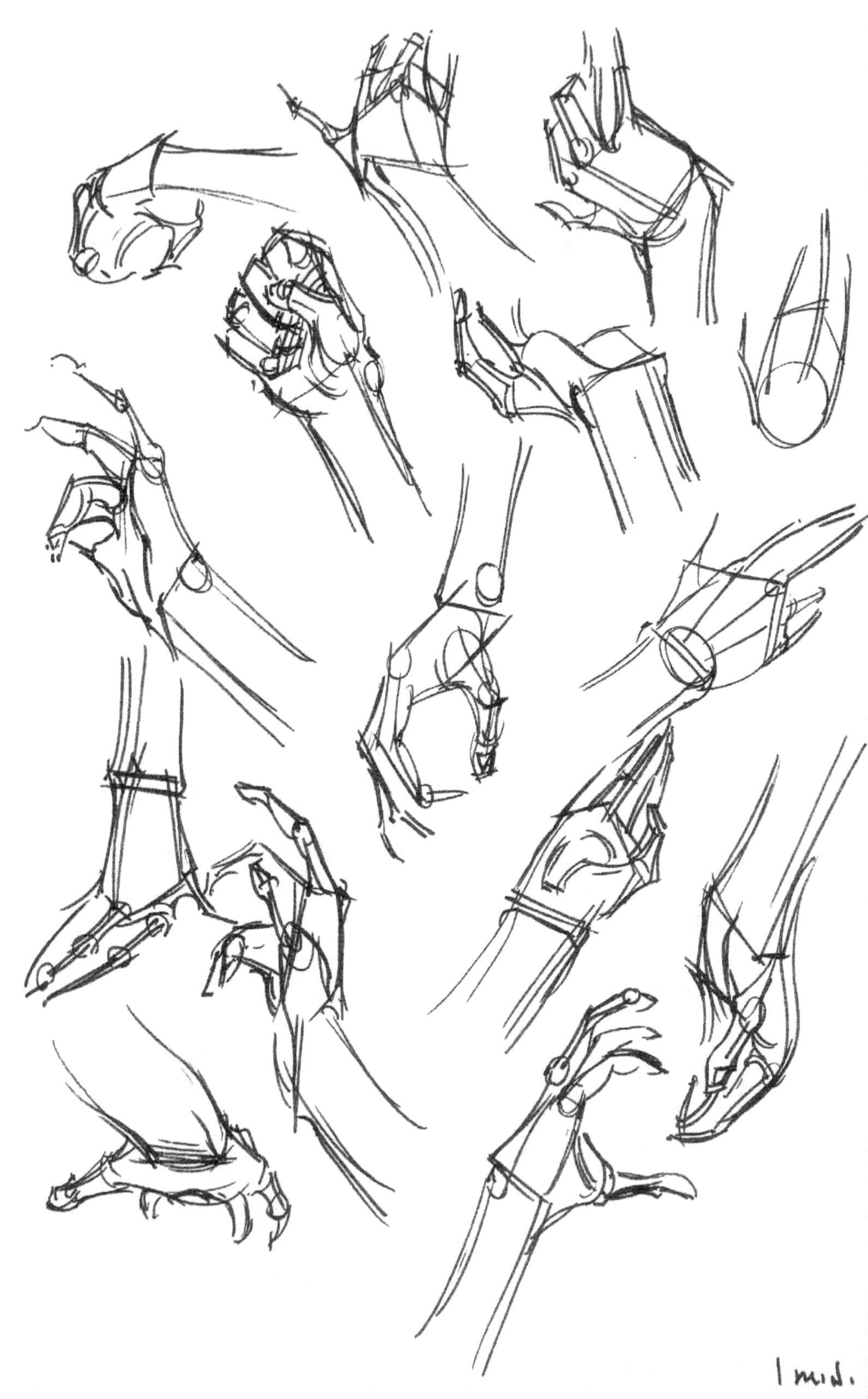

These pages show how to put all of the knowledge from the previous pages into a manageable process. The first drawing shows how to start the hand with a gesture.

The hand is always placed first using the "S" curves to describe the width of the radius and the ulna and the movement into the middle three fingers. A "S" or "C" can be used to start depending on the how the hand is moving in relationship to the wrist.

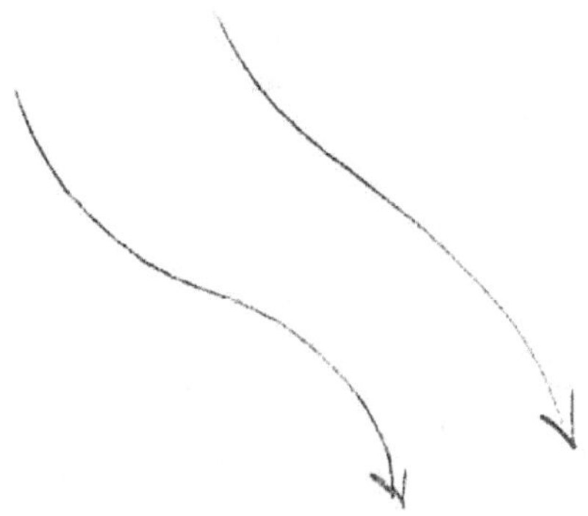

In the second step, a "C" curve is used to place where the fingers end. It is important to use a curve at this stage to begin developing the shape of the carpus group.

Because the hand is based off of the carpus group, everything relates using curves. The fingertips and knuckles will always line up on a curve or arc.

The curve used in this step shows this important element while creating a glove-like shape that places the overall position of the hand.

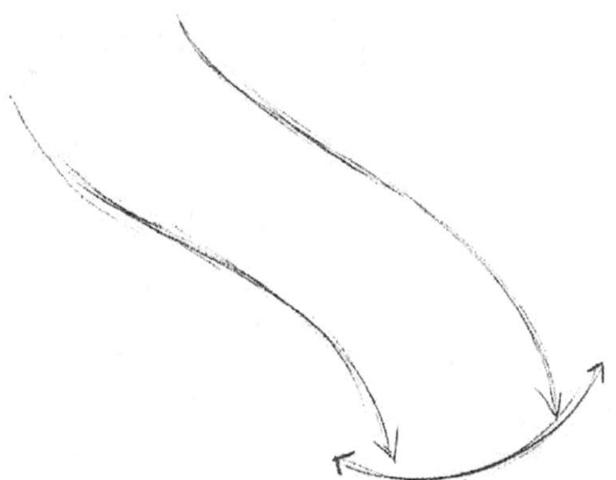

The third step begins with
finding the end of the wrist or
the radius and the ulna. You will
be able to identify this area by
looking for two distinct bones
pushing up against the skin. The
bone of the radius will always
be found on the thumb side of
the hand, while the ulna will
always be on other side in line
with the little finger.

At this stage, a straight line has
been drawn across the top plane
of the wrist. This line should
be thought of as connecting the
radius and the ulna. From the
corner of the top plane, a line
has been drawn straight down
to show the depth of the wrist.
This builds a solid volume to
begin the drawing of the hand.

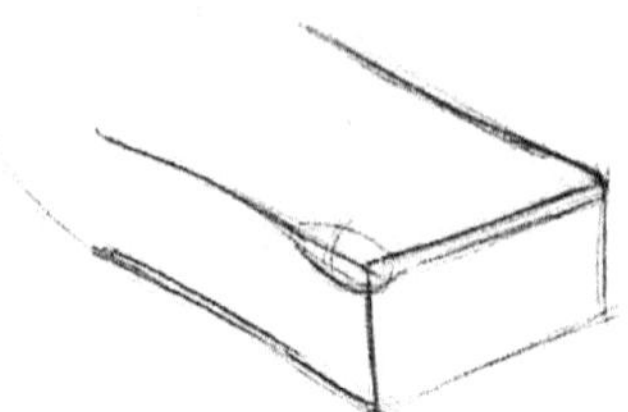

This step shows the developed structure of the wrist
including the box of the palm (discussed previously).
When placing the palm's structure, keep a small amount
of space separating the end of the wrist from the form
of the palm. Additionally, keep in mind the overall
proportions. Remembering that the palm is half the
length of the entire hand helps determine how long to
make this form, based off of the beginning glove shape.

When placing the
structure of the palm,
the corner of the end of
the wrist can often be
carried through into the
palm to help find the
side plane. It is also be
important to keep the
curved surface on the
top plane of the palm.

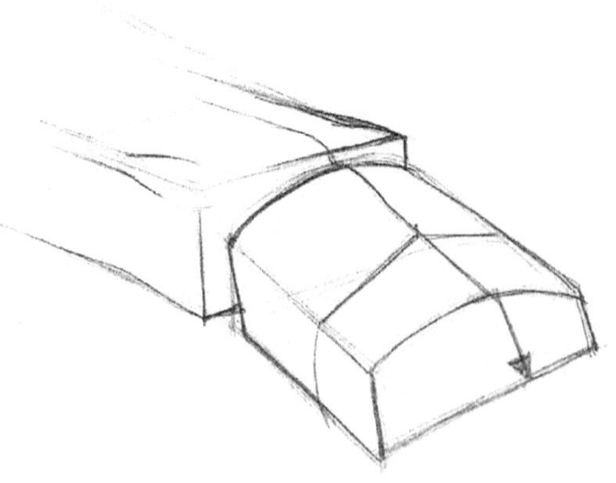

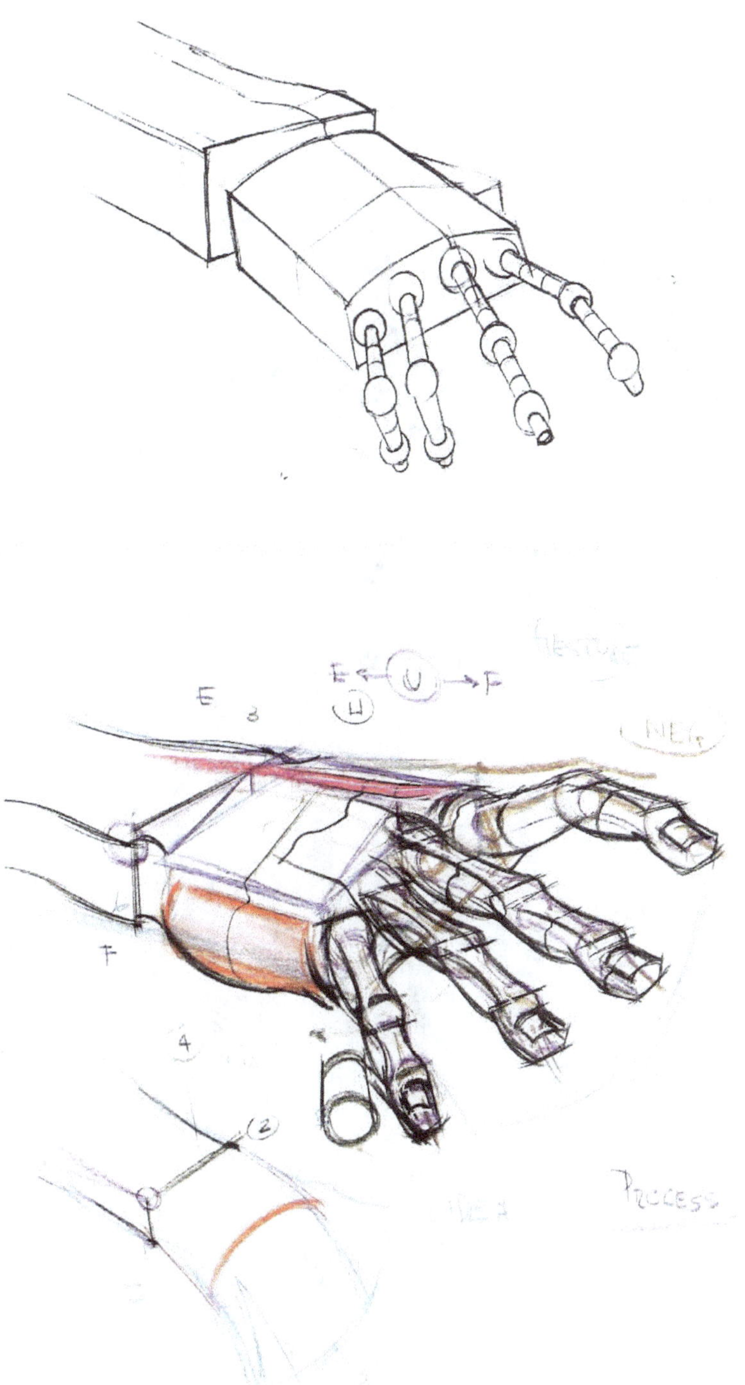
GESTURE
E
U
F
NET
E
F
PROCESS

THE LEG

The process for drawing the legs is very similar to drawing the arms. Both are asymmetrical forms and are represented in space by using the cylinder. Additionally, comparing the shape and function of the arms and legs as you study will give you a definite advantage.

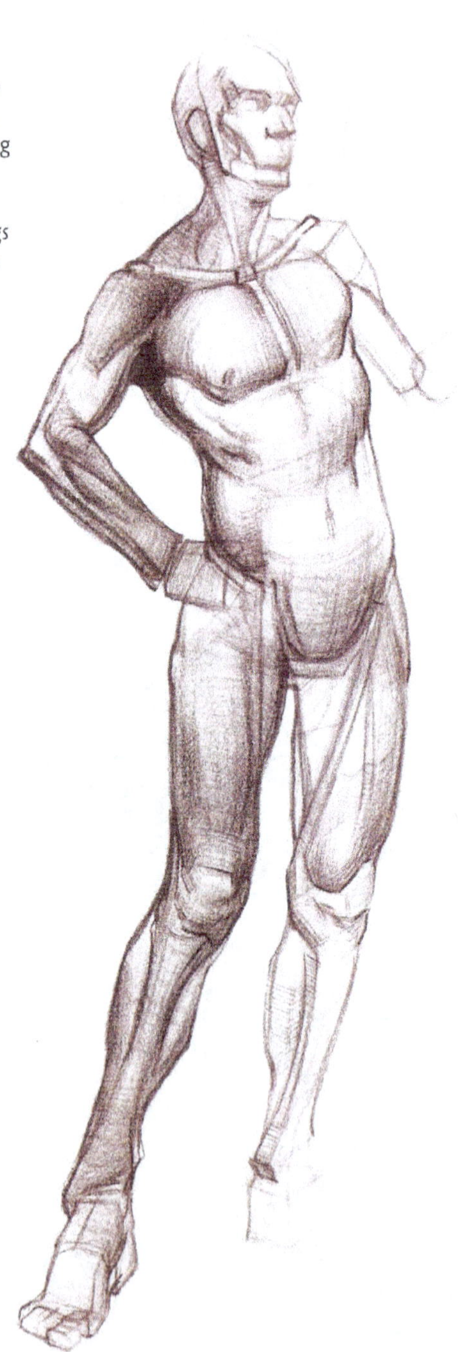

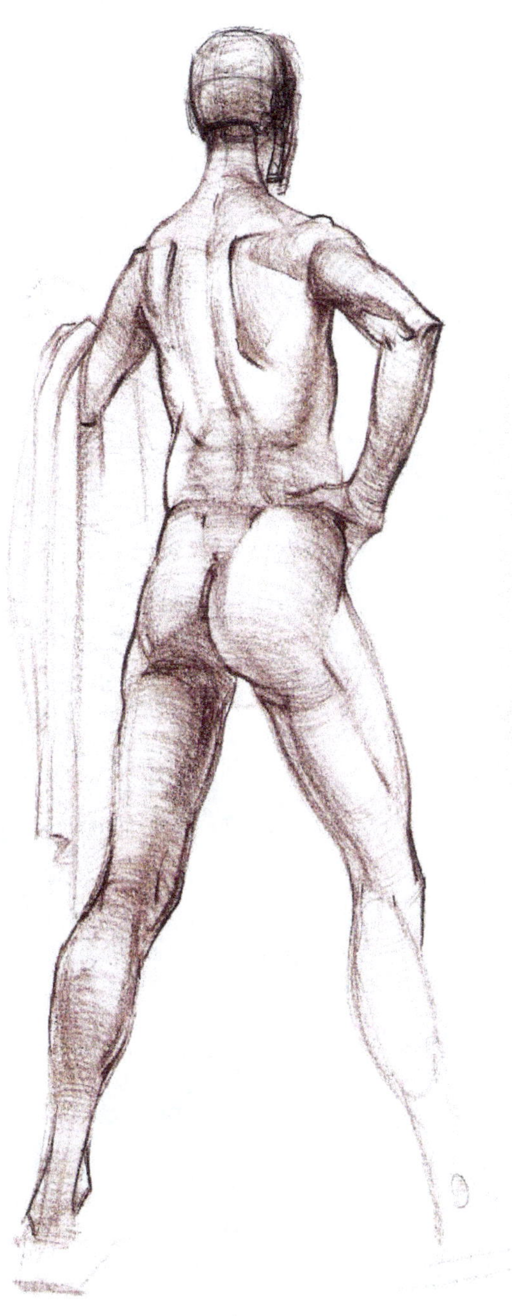 When drawing the
legs, like the arms, the
anatomical shapes always
are placed as ellipses
in an asymmetrical
relationship. For an
effective design, you will
need to remember which
shapes of anatomy are
active and which are
passive.

The following pages will, in
a very general way, point
out the function of the
anatomical shapes in the leg
for use in the drawing process.
Remember that the emphasis
is on understanding the active
and passive relationships of
anatomy in order to design a
compressed or stretched shape
on the perspective of the
cylinder.

The challenges of drawing the leg are similar to drawing the arms. You are working with cylinders, and having to wrap shapes around that much trickier surface. However, keeping in mind the overall process should help make this difficult form more manageable.

The two drawings below represent the stage your drawing should be at before you begin with the anatomy. Gesture, landmarks, and the perspectives of the pelvis and legs should all be finished before proceeding.

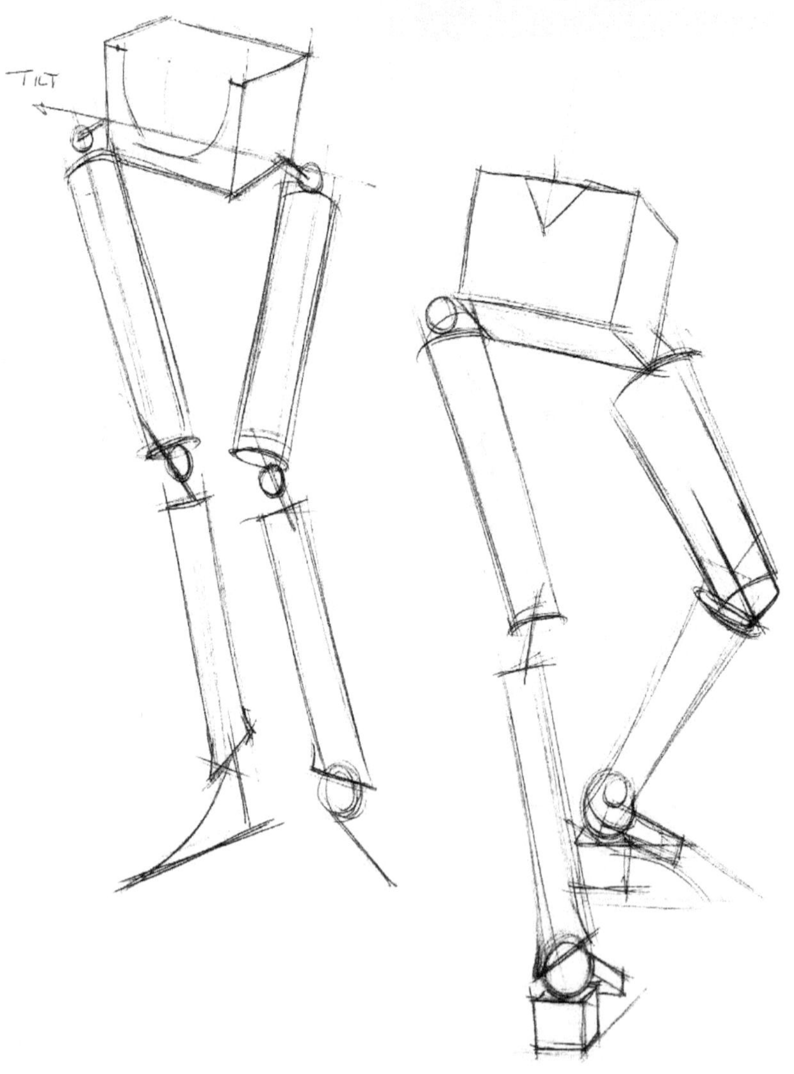

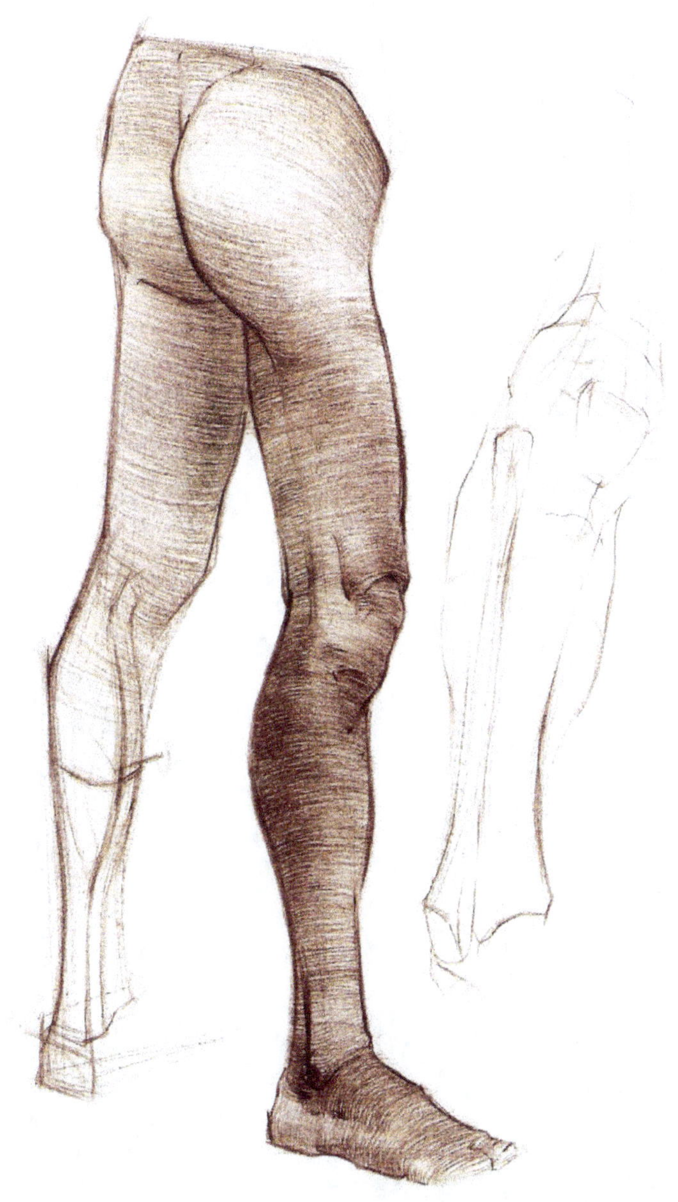

This diagram shows a simplified approach to organizing the muscle shapes in order to make these complex ideas more practical for the drawing. Remember, the goal of knowing the muscle's activity is so you can represent that state in your drawing.

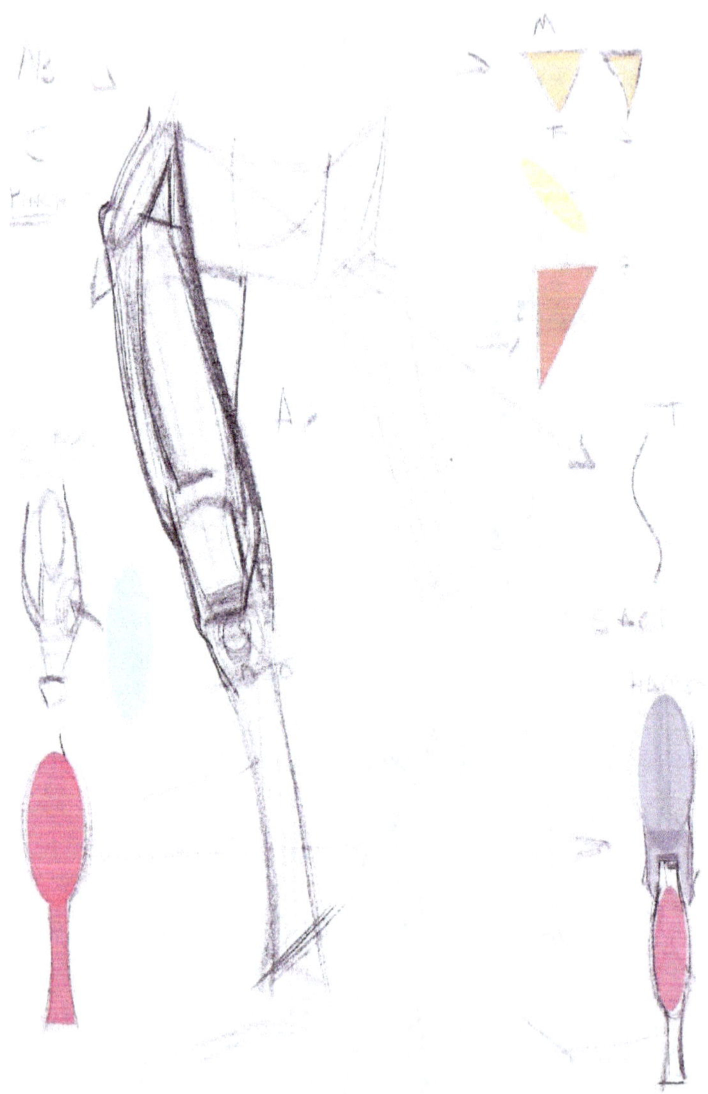

The majority of shapes in the leg are variations of an ellipse. To make remembering the anatomical design of the leg a little simpler, envision and memorize the way these elliptical muscles of the leg tumble down the form.

Throughout the chapter, refer back to this page for simplified views/conceptions of muscle shapes.

The first group of muscles begins on the hip and pulls toward the great trochanter of the femur. The focus when studying these shapes is to understand abduction versus adduction. The first shapes introduced are ones involved in abduction. Keep in mind (once again) that I am simplifying the muscle function in order to emphasize the integration of these ideas into the drawings. These muscles also help rotate the thigh, stabilize the knee, etc.

These first two muscles originate under the iliac crest:

The tensor fascia lata. This muscle abducts and medially rotates the thigh. It is represented as an elliptical shape.

The gluteus medius. This muscle can be represented as a triangle (very similar to the shape of the deltoid).

The shapes of these muscles are compressed during abduction (moving the leg away from the body)

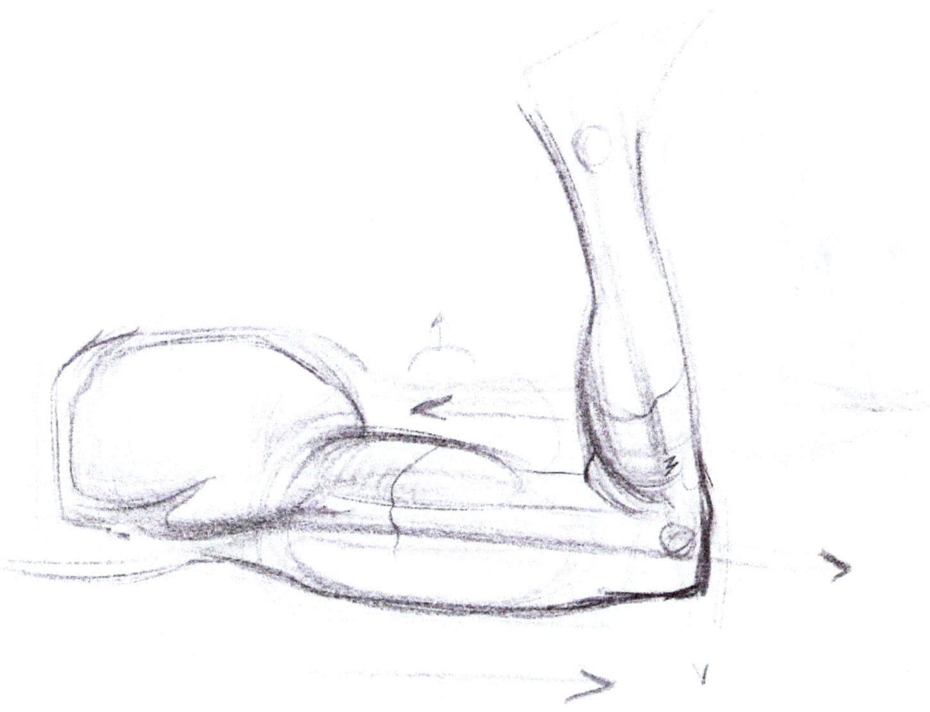

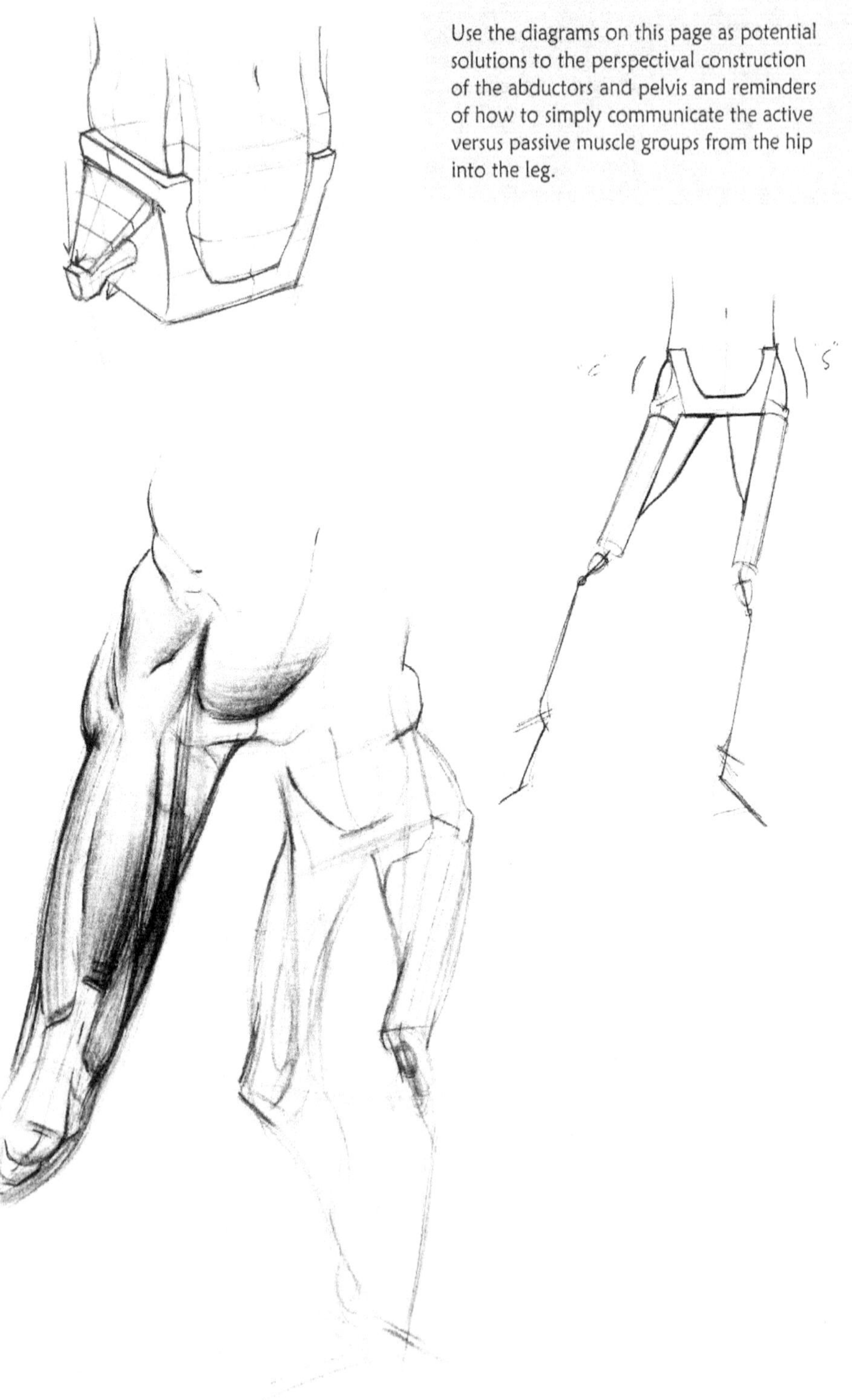

Use the diagrams on this page as potential solutions to the perspectival construction of the abductors and pelvis and reminders of how to simply communicate the active versus passive muscle groups from the hip into the leg.

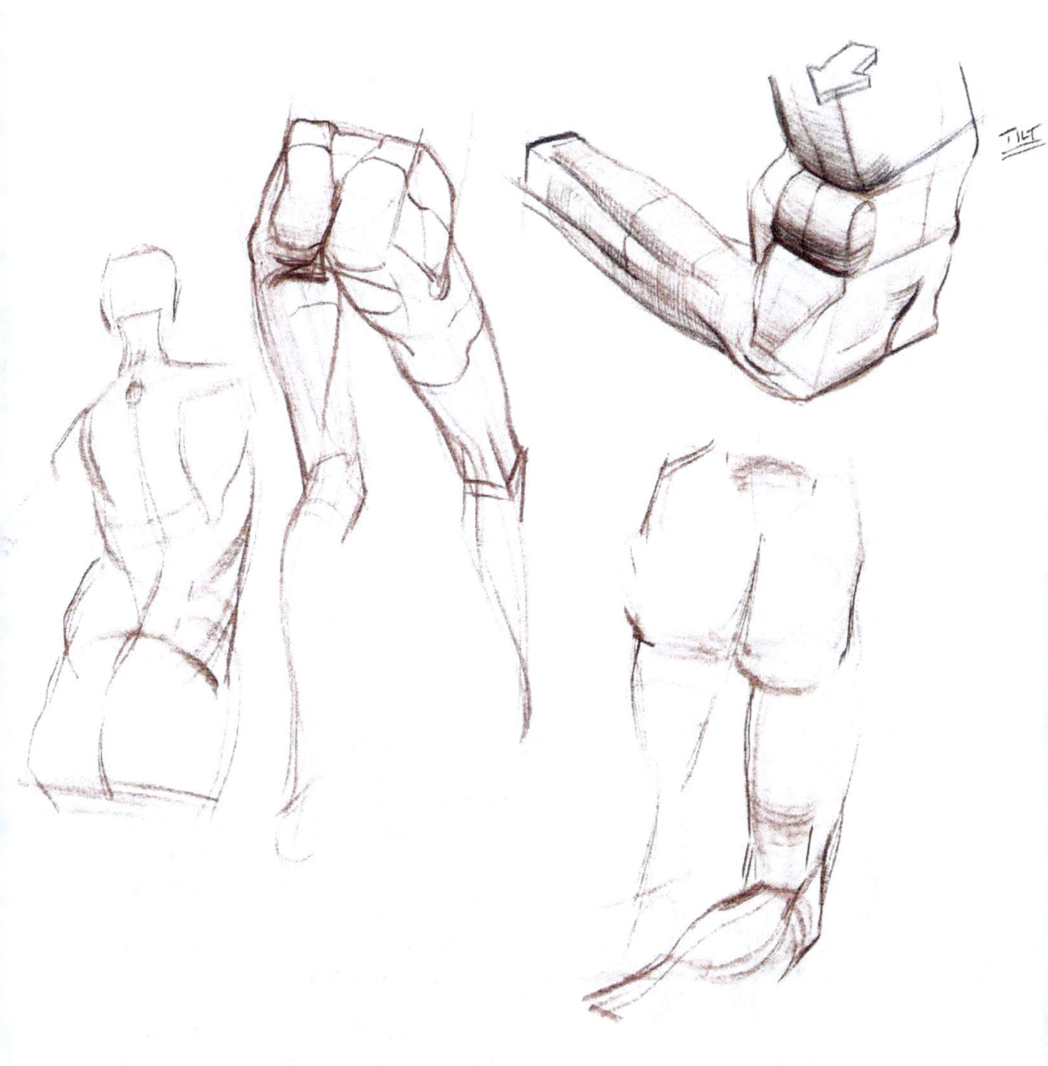

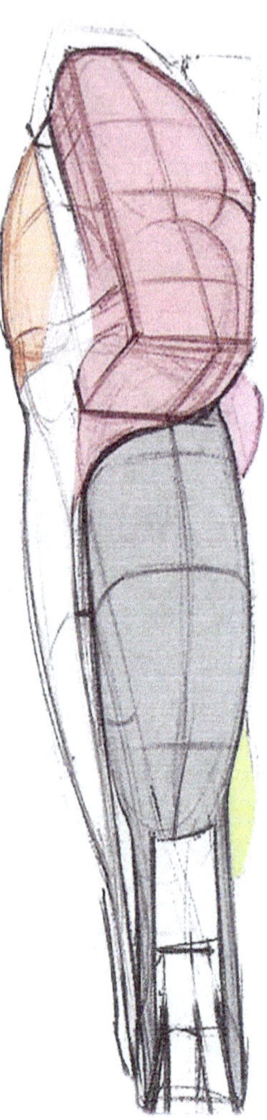

The gluteus maximus is a muscle that can be involved in adduction and abduction.

The primary shape of the adductors (longus and magnus) is a triangle. The base of the triangle extends along the back of the femur, and the tip ends toward the anterior portion of the pubic bone/bottom of the pelvis.

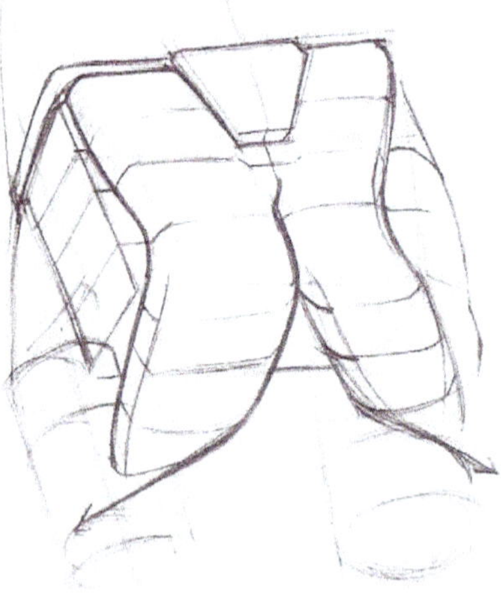

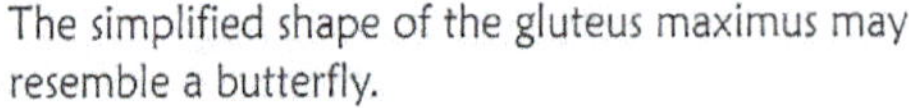

The simplified shape of the gluteus maximus may resemble a butterfly.

Notice the design that resembles a tire at the bottom of the opposite page. This is a beneficial way of thinking of the muscle moving from the hip down to the femur. Remember, for the purposes of trying to emphasize approach, and the practical creation of a drawing, all of the anatomical ideas have been highly simplified into more memorable, basic ideas. The abductors of the hip obviously don't look like a tire on the side of our pelvis, but, as an idea, it may help you simplify what you see as you begin to think about organizing simple plane divisions.

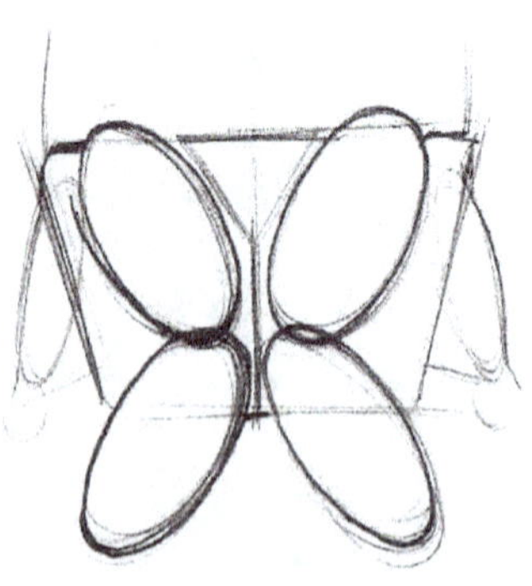

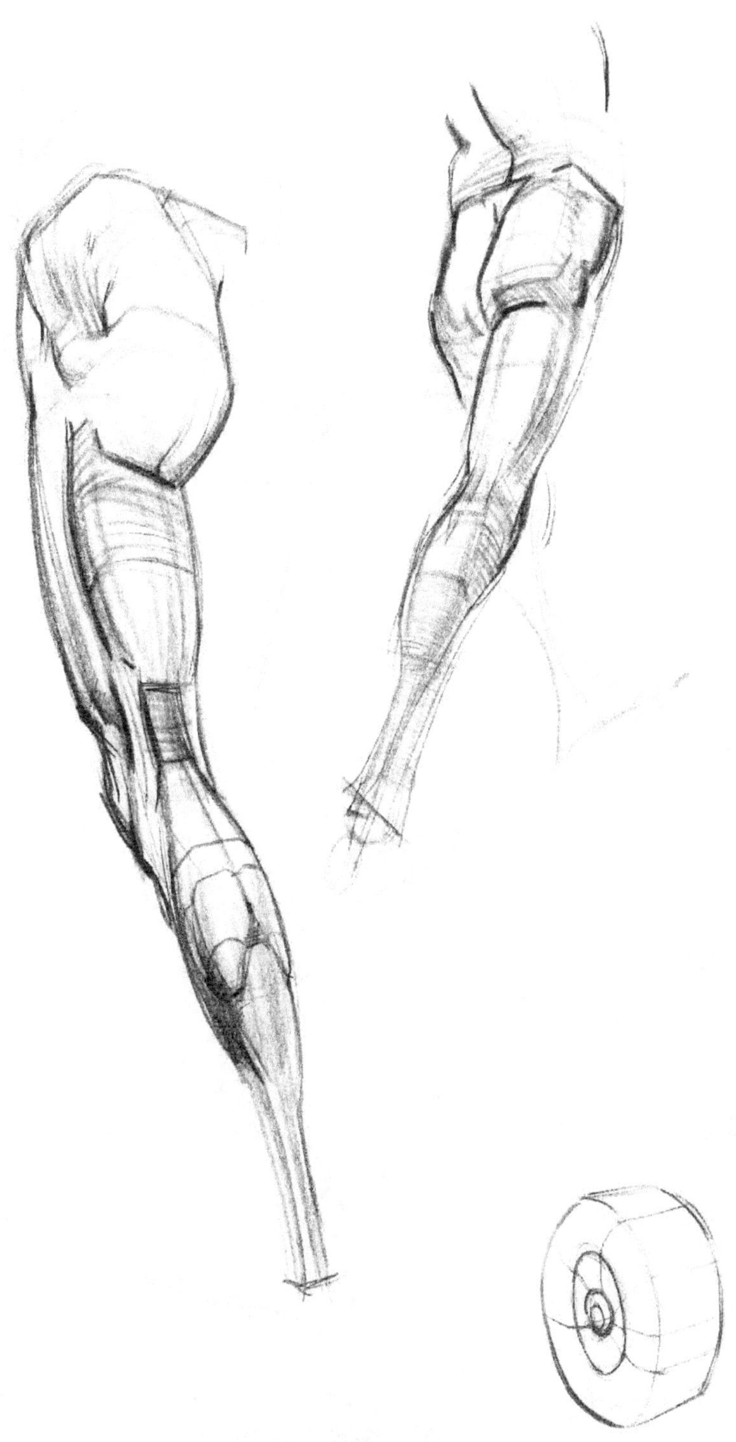

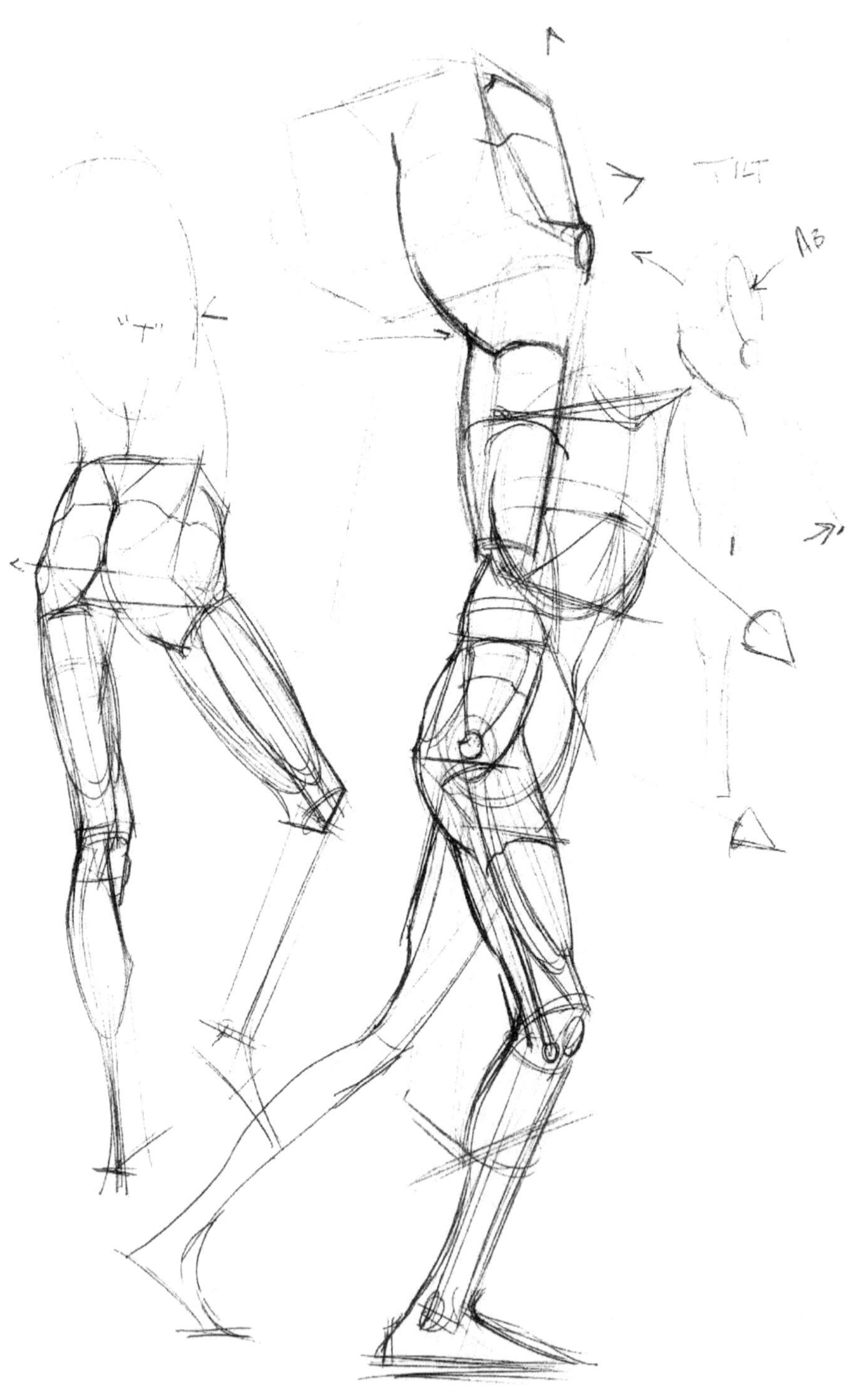
TILT
AB

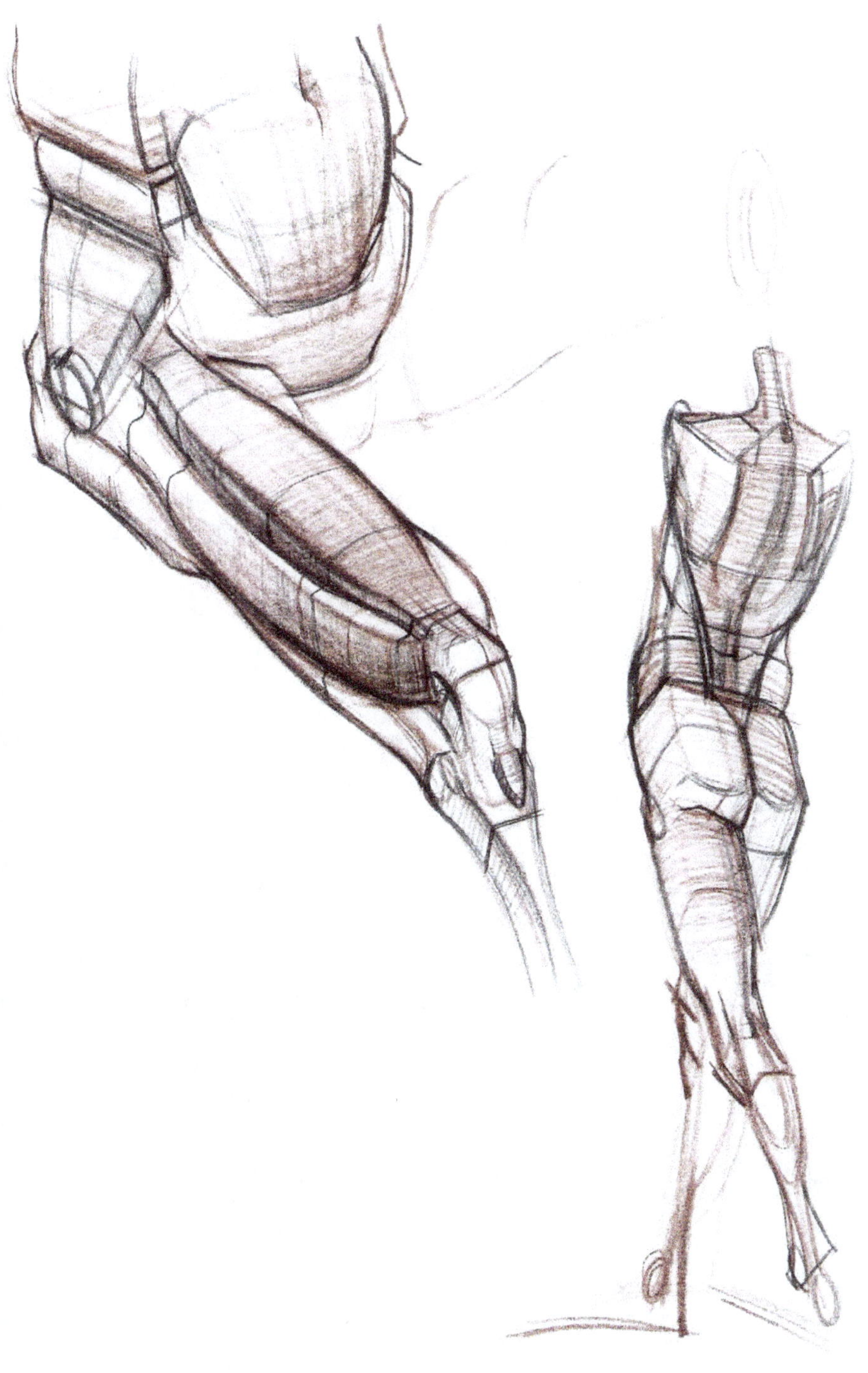

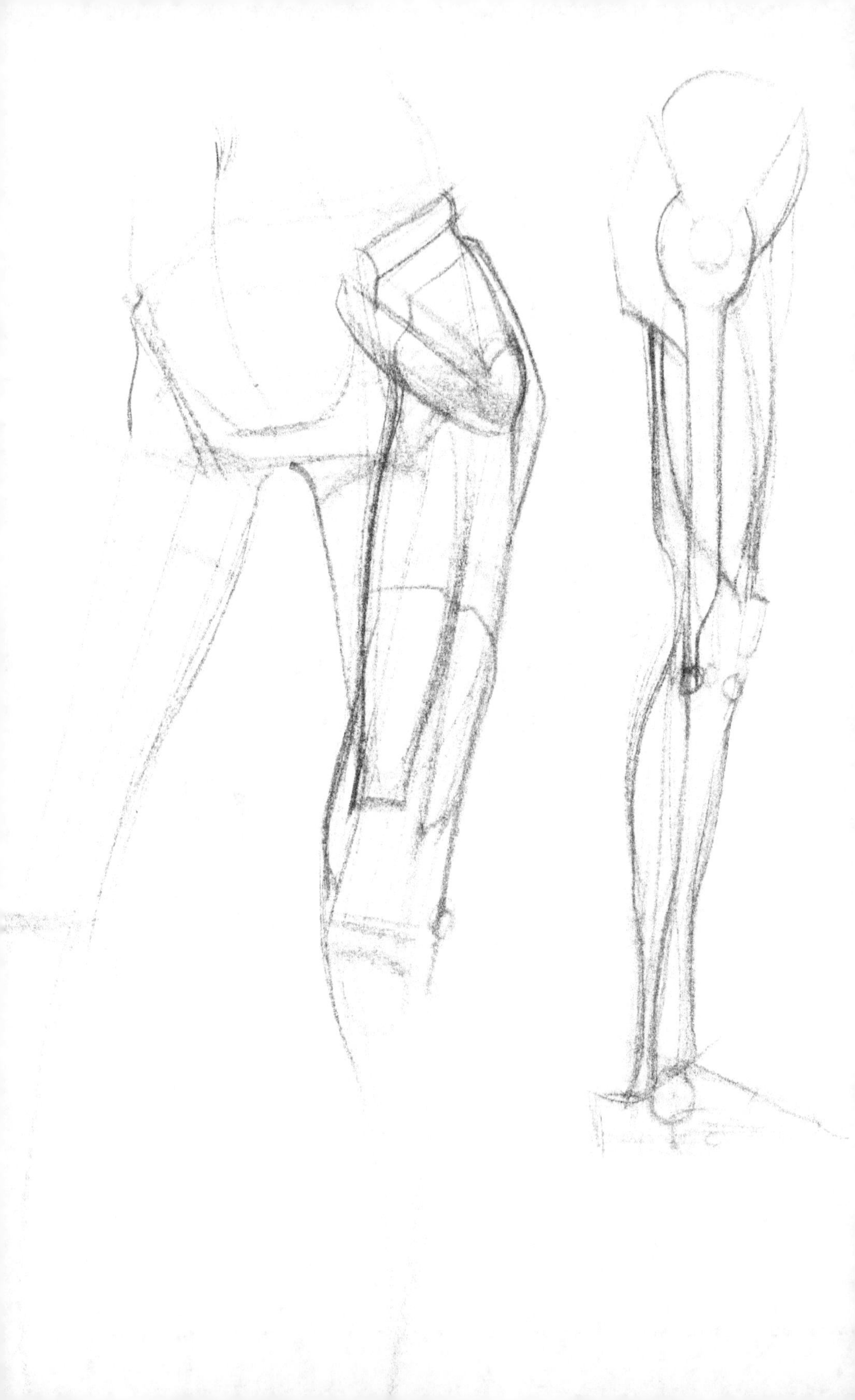

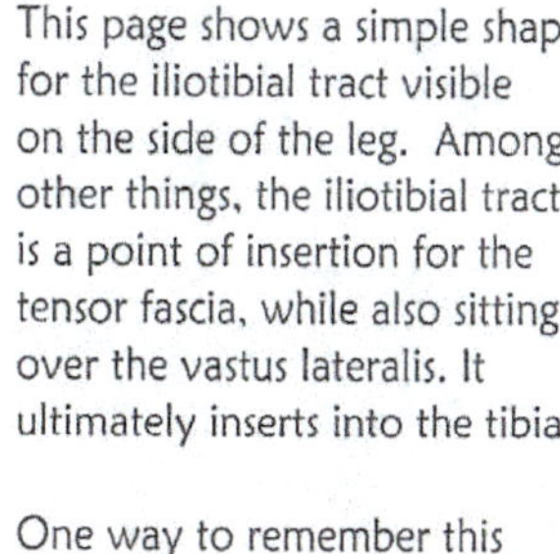

This page shows a simple shape for the iliotibial tract visible on the side of the leg. Among other things, the iliotibial tract is a point of insertion for the tensor fascia, while also sitting over the vastus lateralis. It ultimately inserts into the tibia.

One way to remember this shape is to envision a wrench grabbing the head of the femur, where the bottom or end of the wrench continues down to the lower leg. This shape will be visible on the outside of the leg, in some cases, as a lower arc pushed into the vastus lateralis. Depending on the lighting, this arc may be seen in halftone.

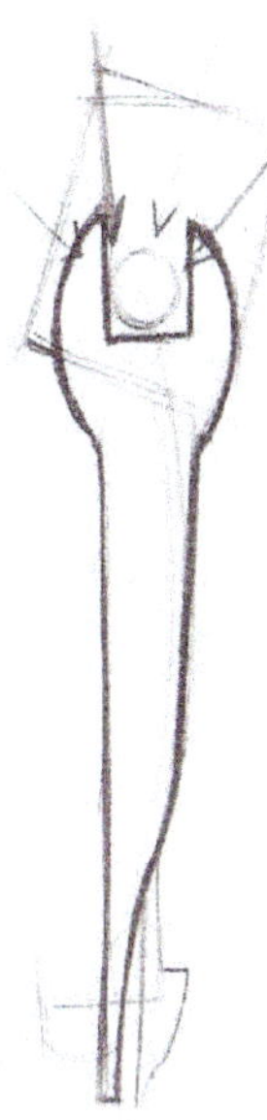

Sartorius – this muscle aids in flexing the lower leg and rotating the leg medially. It can be described as a long "S" curve, beginning at the end of the iliac crest and inserting into the inside of the tibia.

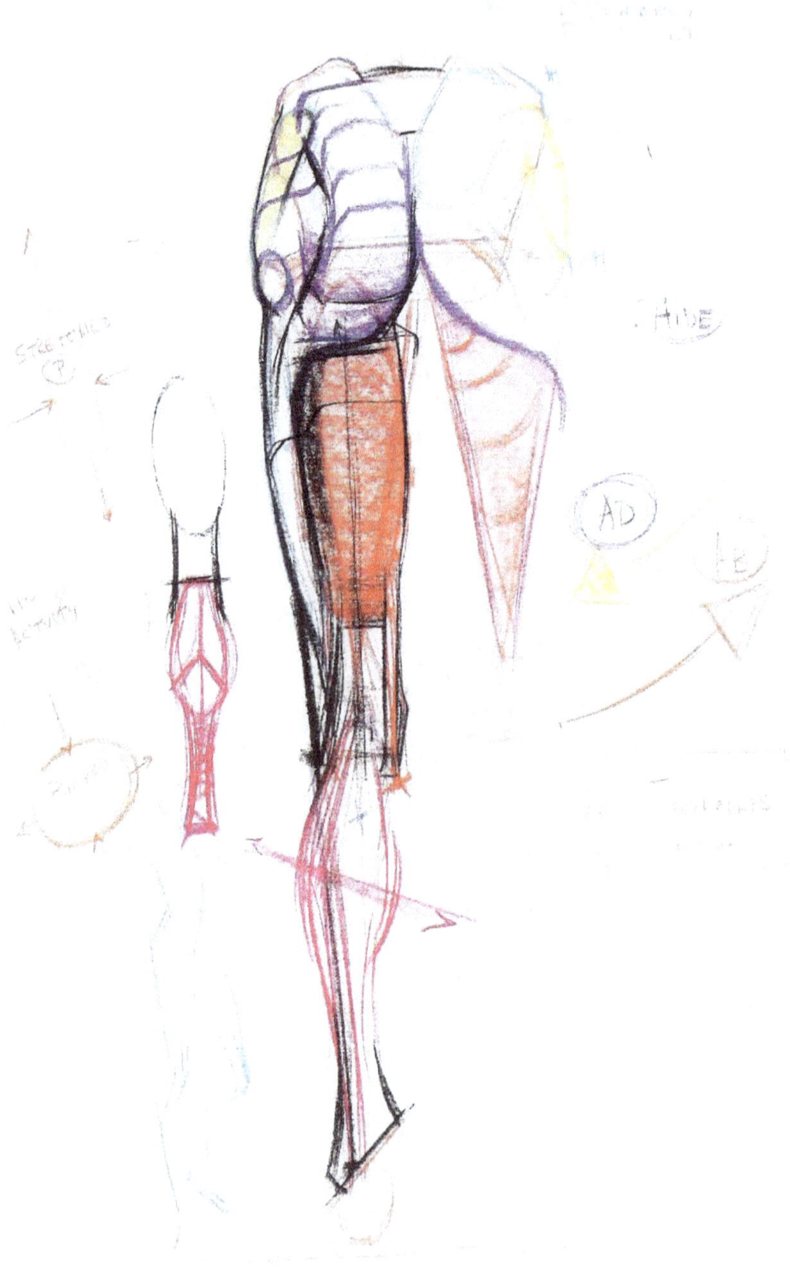

The hamstrings group is a collection
of three large muscles on the back
of the leg (semimembranosus,
semitendinosus, and biceps femoris).
The shape used to represent these
muscles is a ellipse with two legs at
the bottom (legs representing the
tendons which end in the lower leg).
Among other things, the collected
shape of the hamstrings flexes the
knee joint.

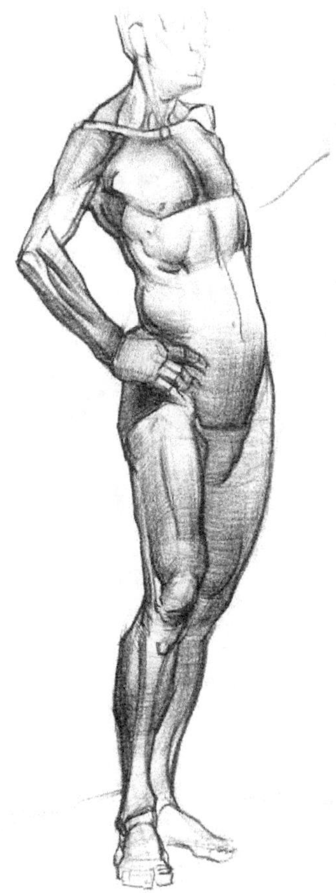

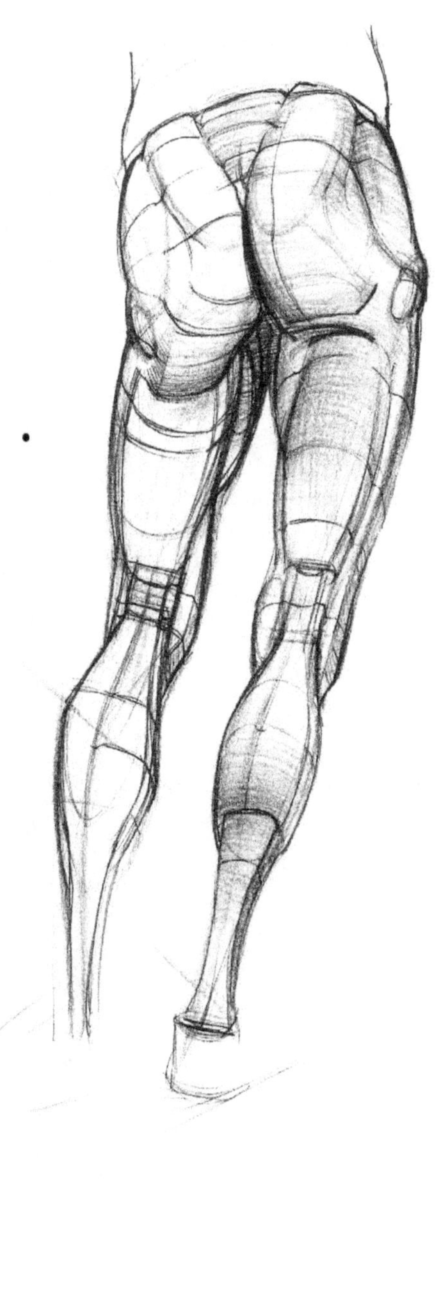

The quadriceps are a large
group of muscles situated
on the front of the femur,
continuing onto the outside
of the thigh. Notice that the
shape used to represent these
muscles, when grouped, is a
large ellipse.

Within the larger shape are
four separate muscles: the
vastus medialis, vastus lateralis,
rectus femoris, and the vastus
intermedius. All four connect
to a compound tendon that
inserts into the tibia. The
quadriceps group flexes the hip
and extends the knee.

The large mass of the
quadriceps should first be
wrapped over the cylinder of
the leg. Second, determine
whether the shape is active
("C") for extending the knee, or
passive ("S").

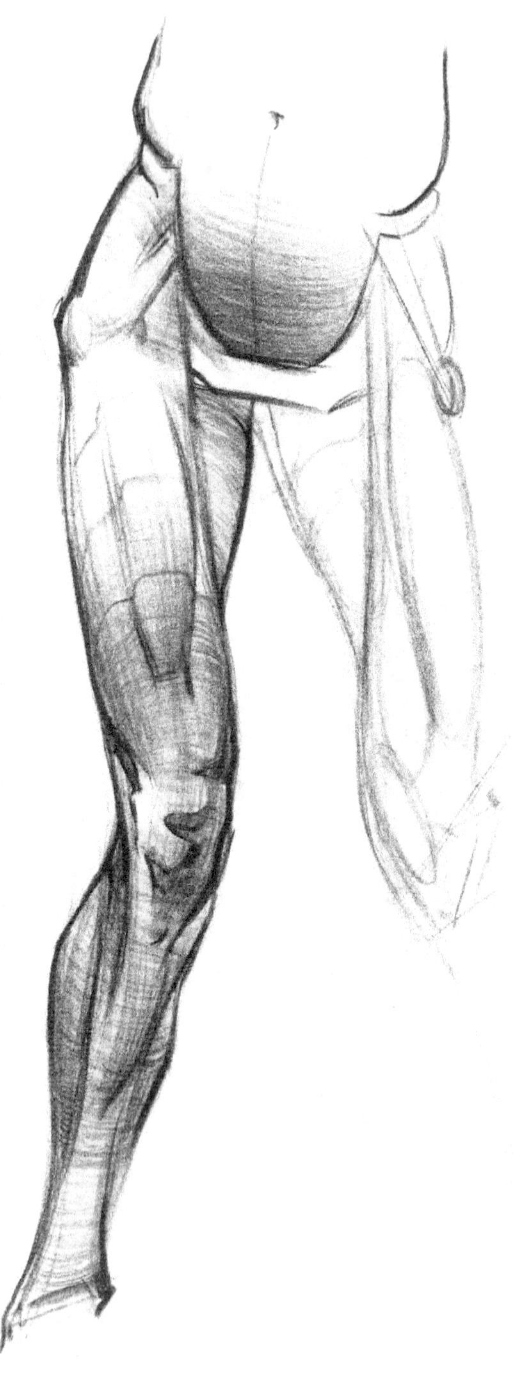

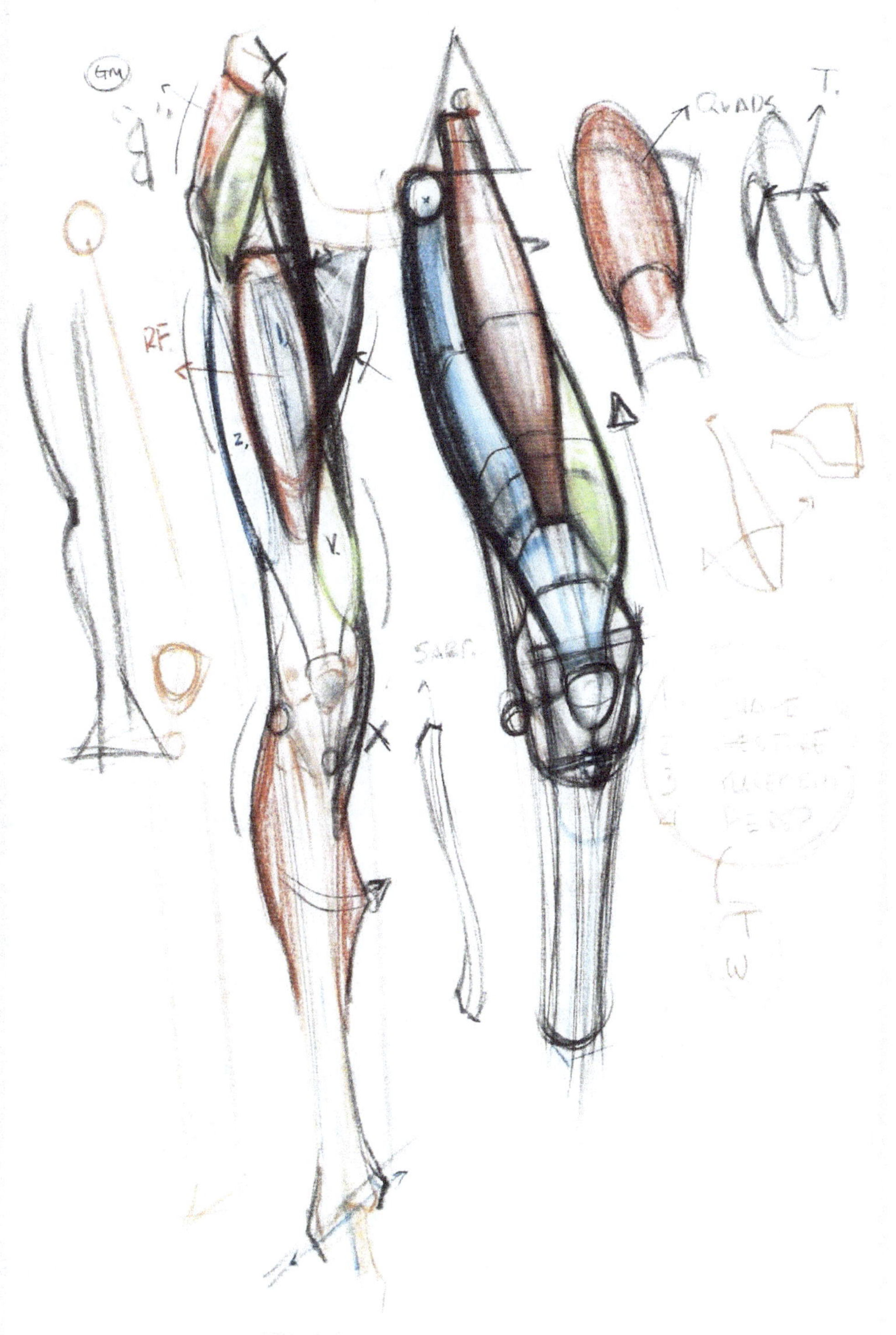

GM
RF.
SART.
QUADS
T.

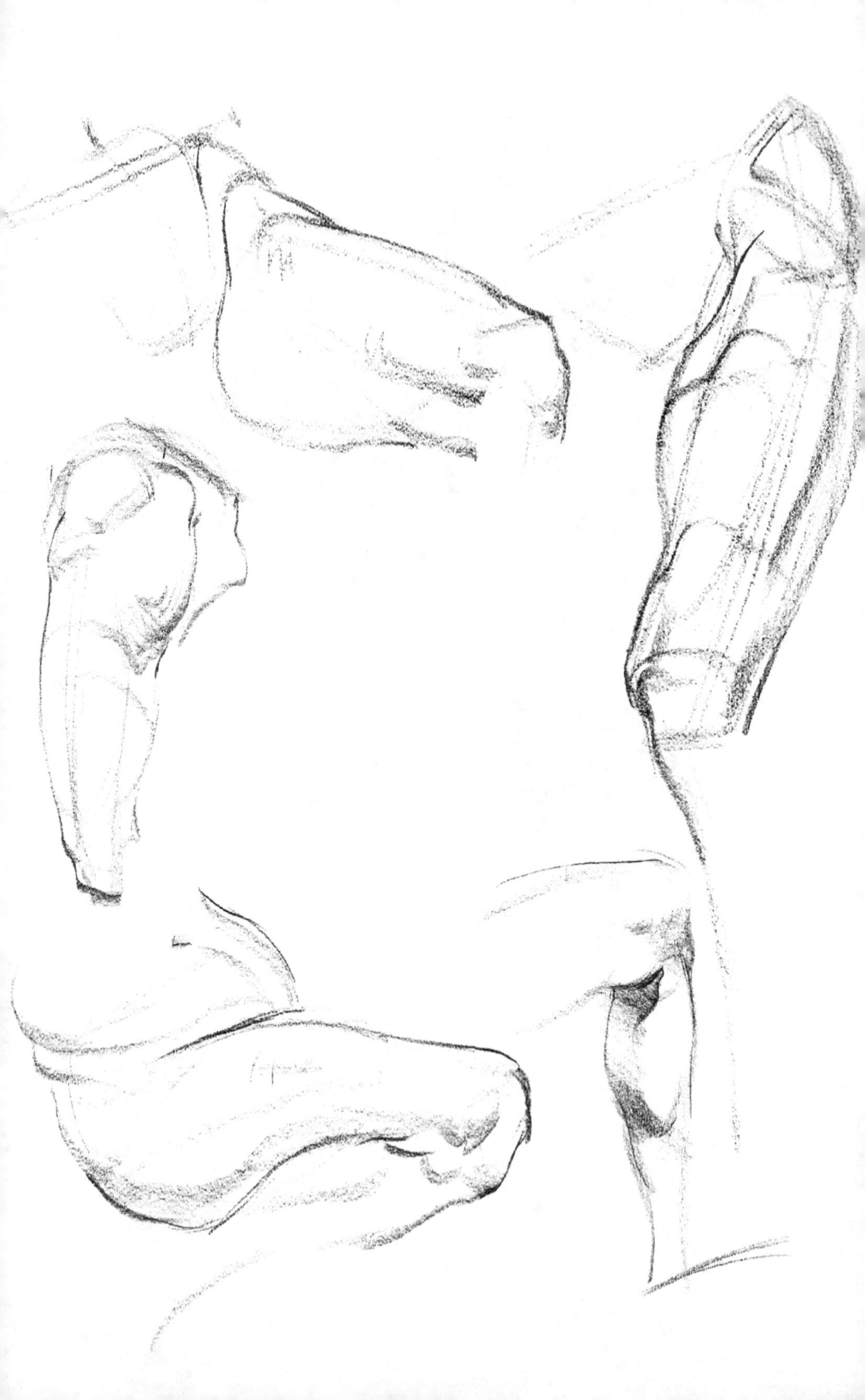

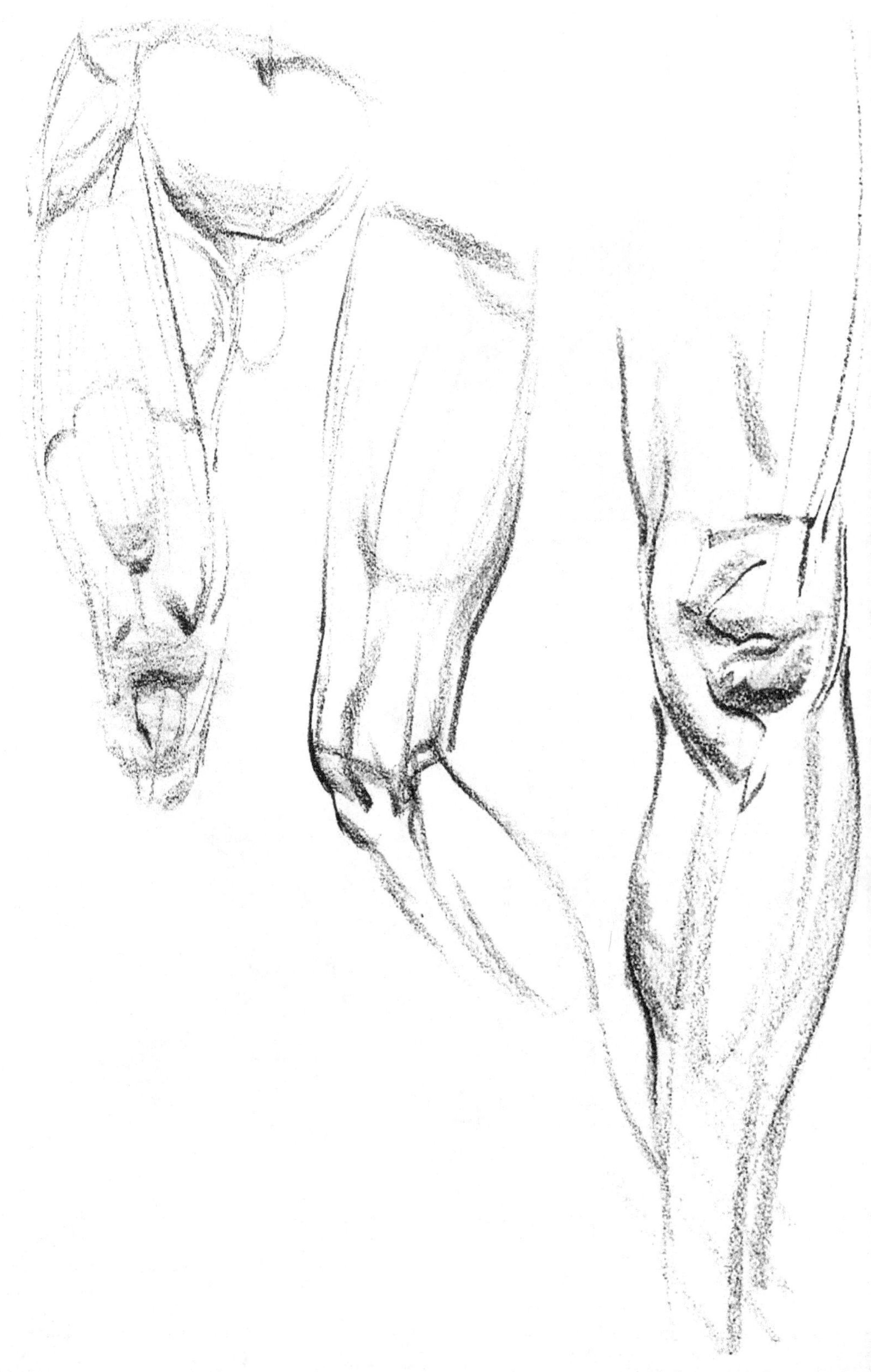

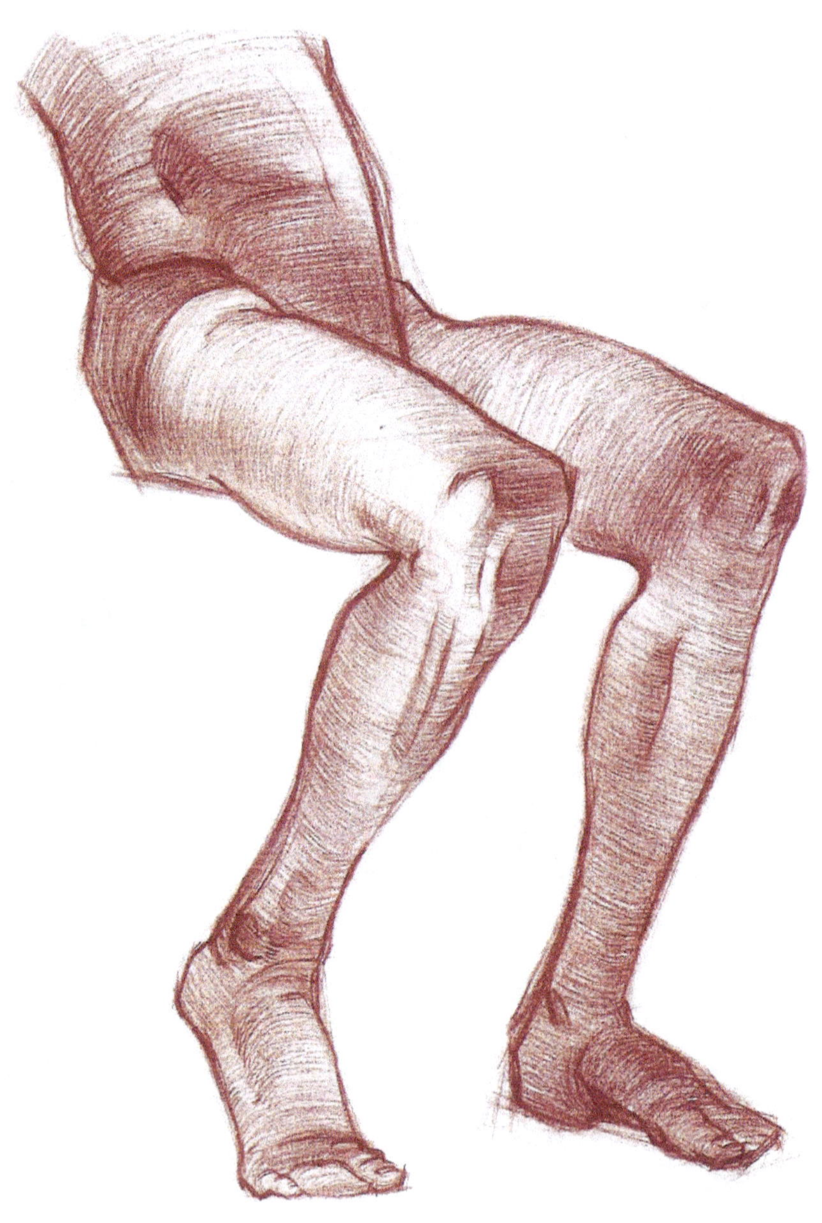

The calf muscles (here, the soleus
and gastrocnemius are grouped
into a common shape) raise the
heel and prevent the figure from
falling forward from a standing
position. The calf muscle has its
origin at the fibula and inserts into
the calcaneus bone/heel block.

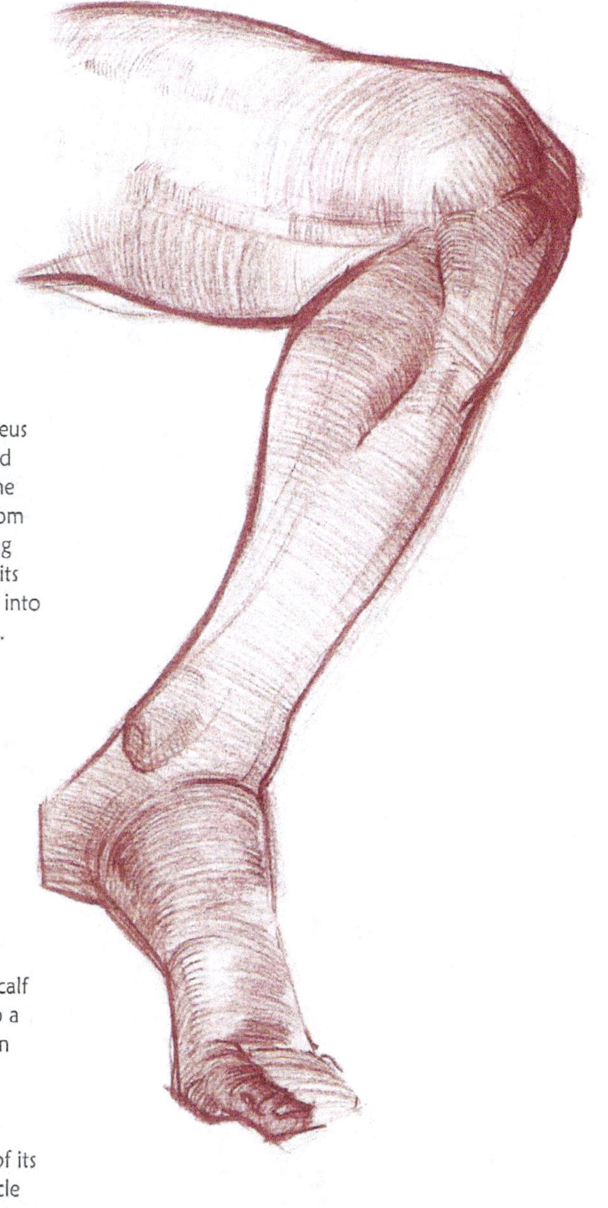

The shape used to design the calf
is a large ellipse that trails into a
more square/block-like bottom
(Achilles tendon).

Try to remember the muscle
through a simple abstraction of its
shape. For example, this muscle
may look like a lollipop, corndog,
or simplified tree shape.

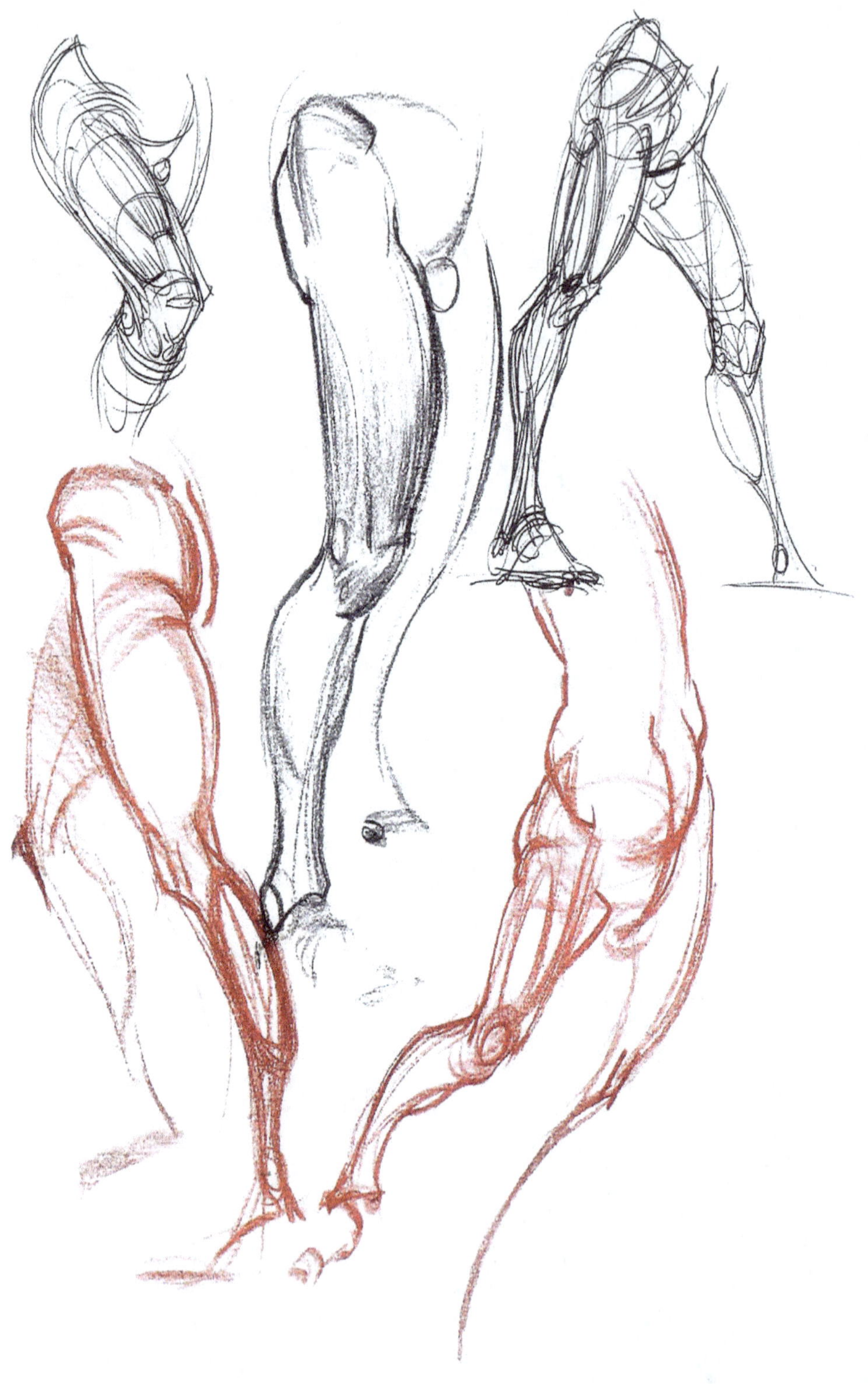

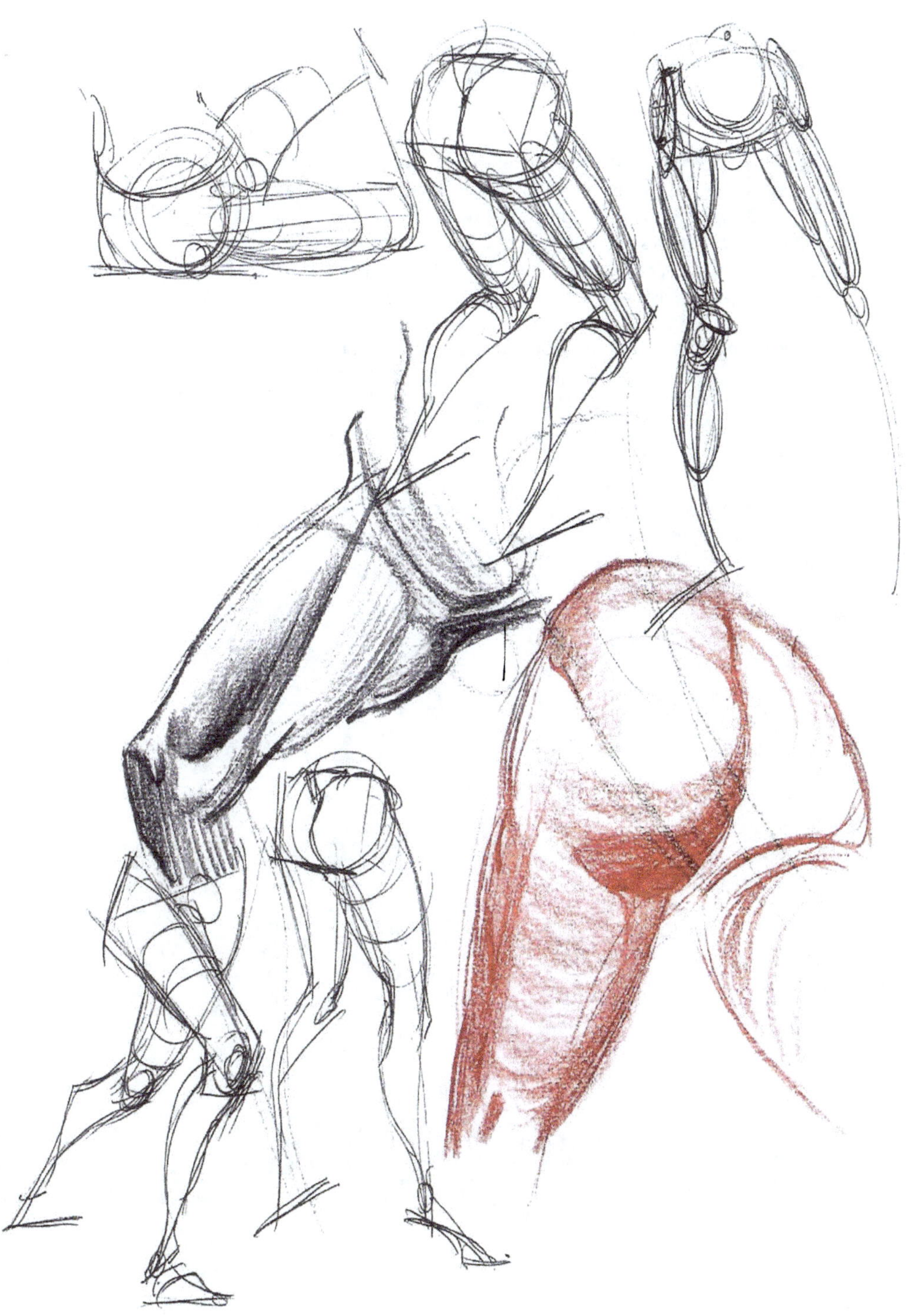

THE FOOT

The process of drawing the foot begins by studying the design of the bones, looking at the placement of the anatomical shapes, and then using this information to design volumes and line. The main design focus of the foot is that it provides stability for and supports the weight of the body. The design of the foot is similar to a shock absorber, in that it can support the body's impact as it walks, runs, or jumps. It also can work as a lever to push or help propel the leg (and body) forward.

Before beginning our study of the foot, remember that whenever you take the time to study a part of the figure's anatomy, there is always much more to learn and get out of it than just the remembrance of muscle names, textures, and origin/insertion points (which have mostly been omitted in this study of the figure). For example, the foot, hand, and rest of the body are perfect lessons in the idea of form matching function. In other words, in addition to being a lesson in how to draw the foot, studying the foot will instill in you the natural principles of how a form (the foot) is designed in order to match its needed function (supporting/absorbing weight, movement, etc.). Thinking this way will sensitize you to creating characters, crafts, architecture, and so on, with the correct functional aesthetic.

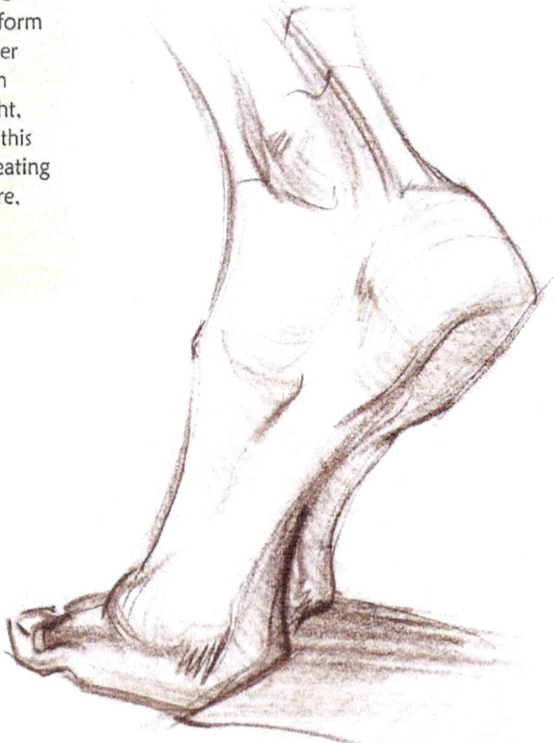

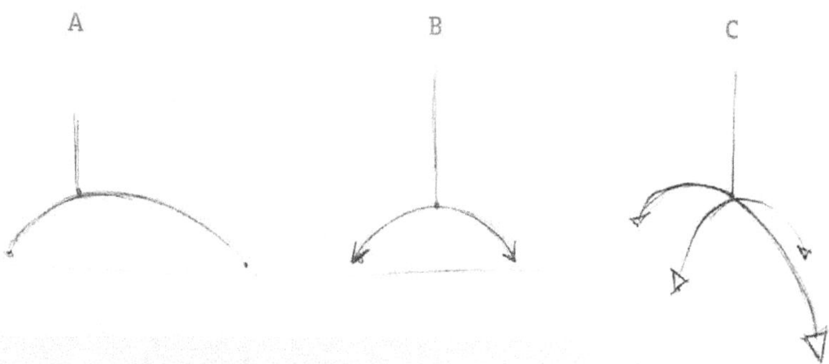

The design of support in the foot is built on two major arches: the longitudinal arch (front to back – example A) and the transverse arch (side to side – example B), and example C shows an idea of how they work together.

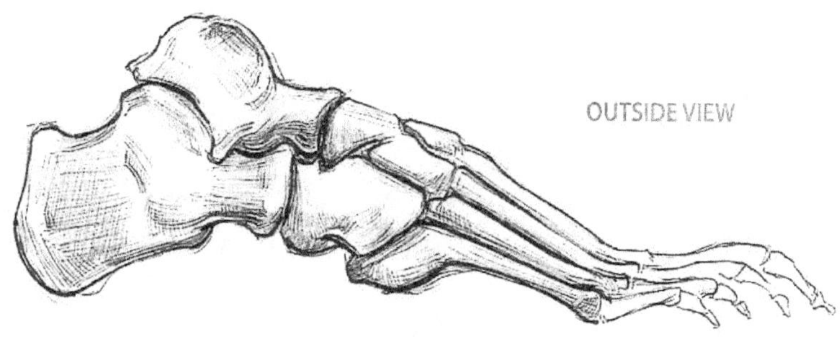

The bones in the foot that create these arches include the tarsus, metatarsus, and phalanges. The tarsus group is the largest collection of bones (taking up roughly half the foot), followed in size by the metatarsus, and then the smallest group, the phalanges. In comparison to the hand, the proportions are essentially reversed. Instead of moving from small to large (which allows for more dexterity and movement in the hand), the proportions of the foot progress from large to small (in order to provide support and stability).

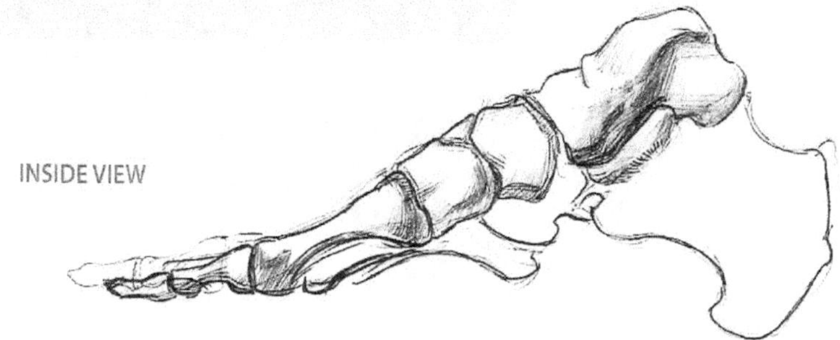

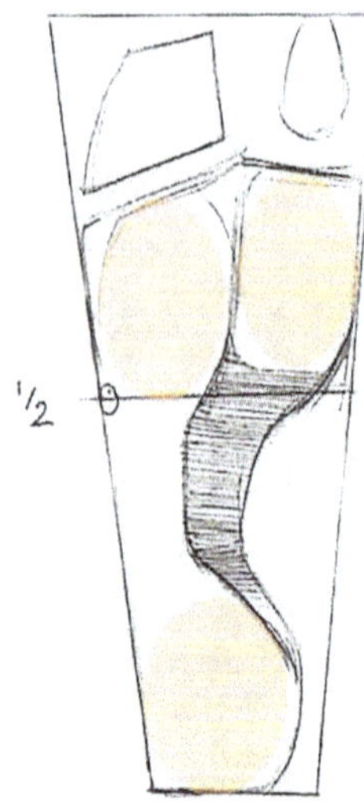

The diagram at right shows a view of the foot seen from beneath. Note that the half-way point is the beginning of the fifth metatarsus (little toe). Additionally, this diagram includes the main areas of padding on the bottom of the foot, which give us the recognizable footprint shape.

Use of padding is a great resource when describing gesture. The gesture here is consistent with the principles used in every chapter so far — squash vs. stretch. The foot has qualities that we have been studying all along, mainly hard vs. soft. Notice that the top part of the foot is primarily made of bone, while only the bottom is heavily padded.

This arrangement makes perfect sense for our design, and the study of form and function, because the bottom of the foot is responsible for supporting so much weight that it is given a significant amount of padding. It is here that the squash and stretch can now be used for the presentation of these ideas in the actual drawing (remember the hand has a similar design with some variation).

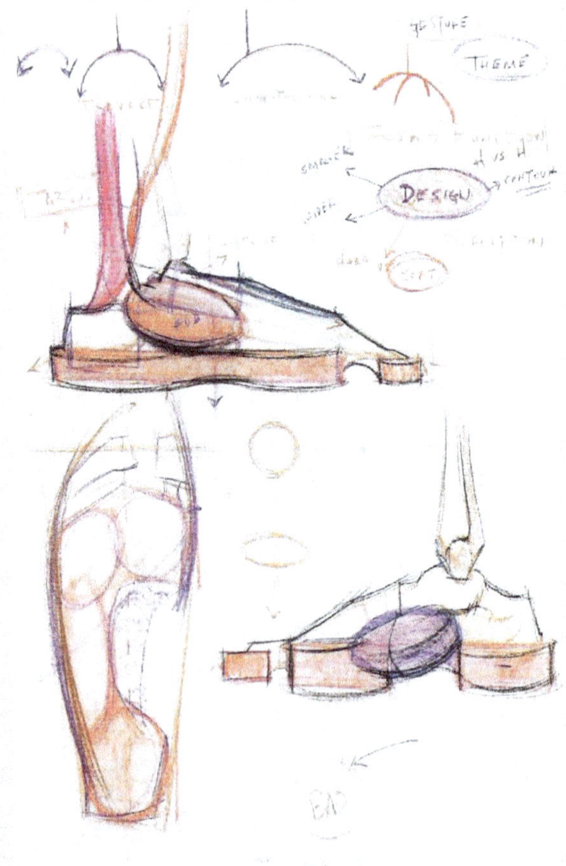

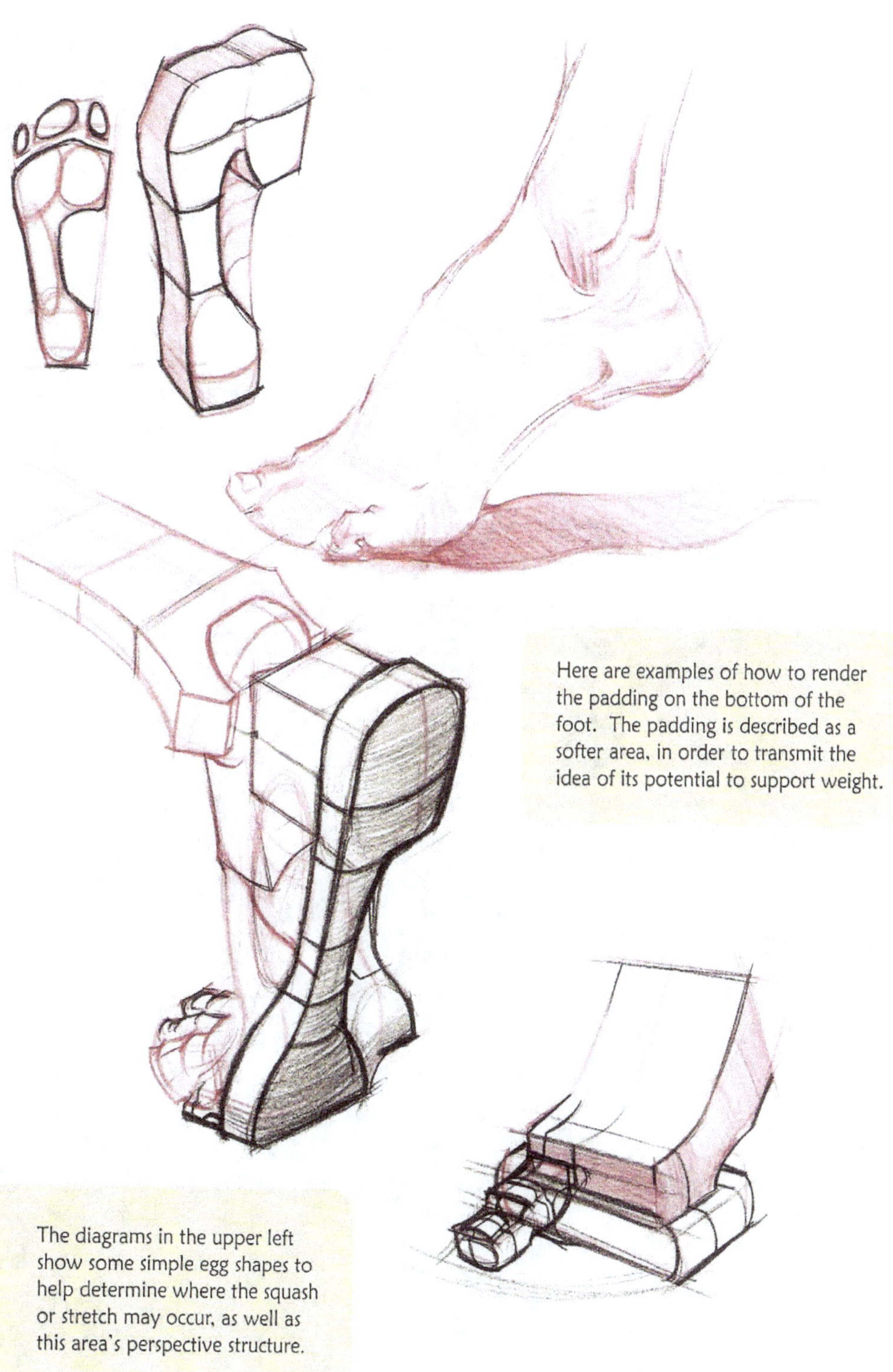

Here are examples of how to render the padding on the bottom of the foot. The padding is described as a softer area, in order to transmit the idea of its potential to support weight.

The diagrams in the upper left show some simple egg shapes to help determine where the squash or stretch may occur, as well as this area's perspective structure.

When drawing the foot, or analyzing a difficult form in general, I start by taking the lessons, ideas, or major themes from the bones and interpreting them through a combination of the major three volumes discussed earlier (cylinder, box, sphere).

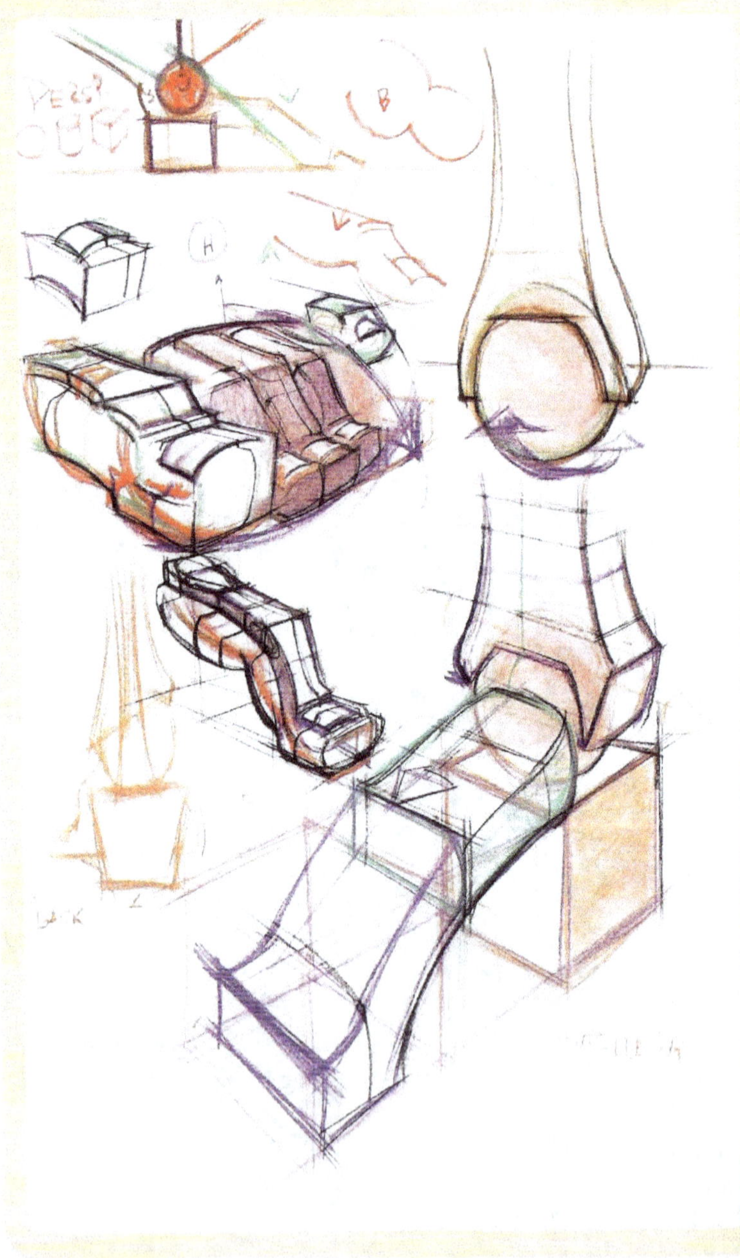

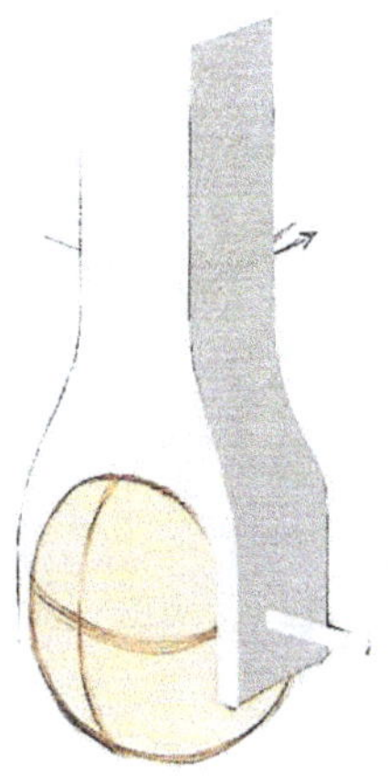

We will begin by designing a structure to describe the connection from the lower leg to the beginning of the tarsus group (again the largest collection of bone in the foot — requiring three different perspectival structures). By manipulating a box, expanding it at the end, and then attaching a sphere to it, you can describe this area of the foot as a wrench grabbing a ball.

This description is also a match in terms of function. Here, the wrench describes the lower leg ending, while the sphere is a reduction of the bones in the upper portion of the tarsus group. Additionally, as demonstrated in the diagram, you can add a pin through this structure, to help remember the movement. For example, if a wrench was to be attached to a sphere, and they both had a pin through them, that sphere would only be able to move/spin front to back.

This wrench grabbing a sphere is then stacked on top of a cube. The cube is a representation of another bone in the tarsus group, the calcaneus or heel block (shown in light blue). From the sphere, which now rests on top of the heel block, the goal is to explain the longitudinal arch as it moves forward (or away) in space. Begin building this bridge from the last portion of the tarsus group as a volume equivalent to a deck of cards (shown in light orange at bottom right), coming out from the sphere at a slightly declined angle.

The last part of the foot to be added (before the toes) is the metatarsus. As shown in the diagram to the right, the metatarsus is the last element within the bridge of the foot (connecting the tarsus to the ground plane). When designing this shape, you can think of a box that is built from an "S" curve, similar to a slide, that connects the tarsus and curves down to the ground plane. When the metatarsus ends on the ground, the perspective form should be left with a flat plane for the inclusion of the toes.

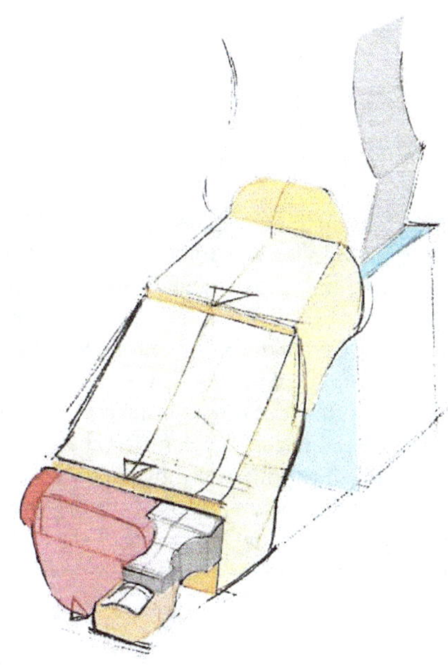

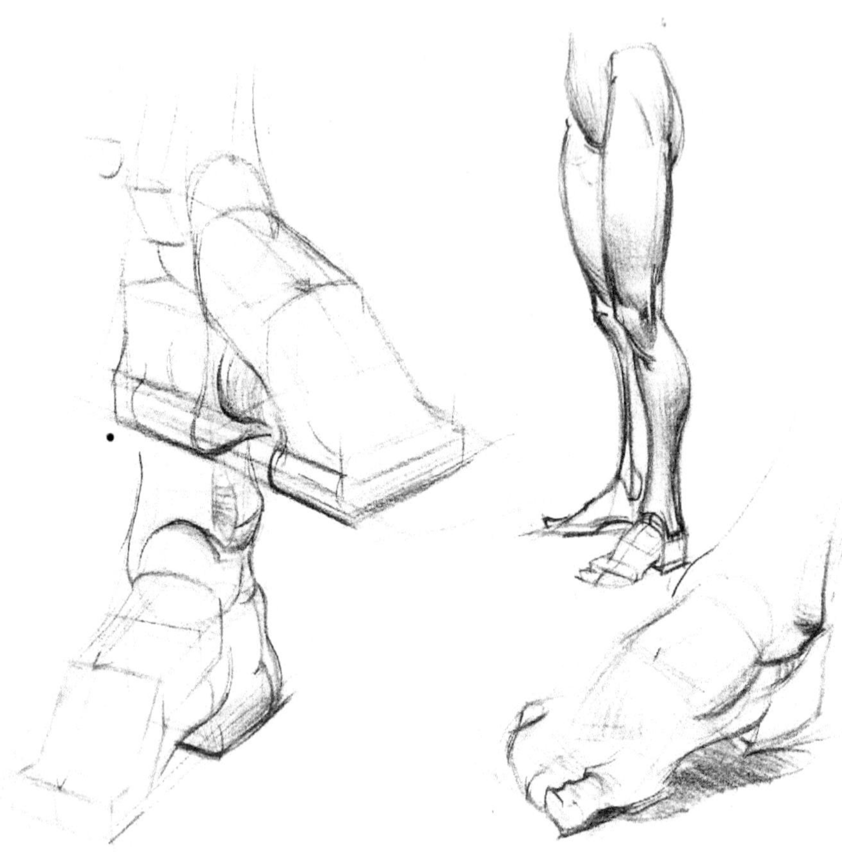

The last part of the foot involves the toes, and this
process begins from the skeleton outwards, adding
on the padding and muscle. The smallest four toes
can be drawn in exactly the same way as the fingers
were drawn in the hand chapter, with some small
differences. First, all of the toes have a natural angle
as they arc from the foot down to the ground plane.
Second, when defining their volumetric character, it is
better to define the forms as slightly wider to indicate
their weight-bearing nature. Look at the drawing on
the previous page for how the four smaller toes can
be summarized in a single shape (shown in red).

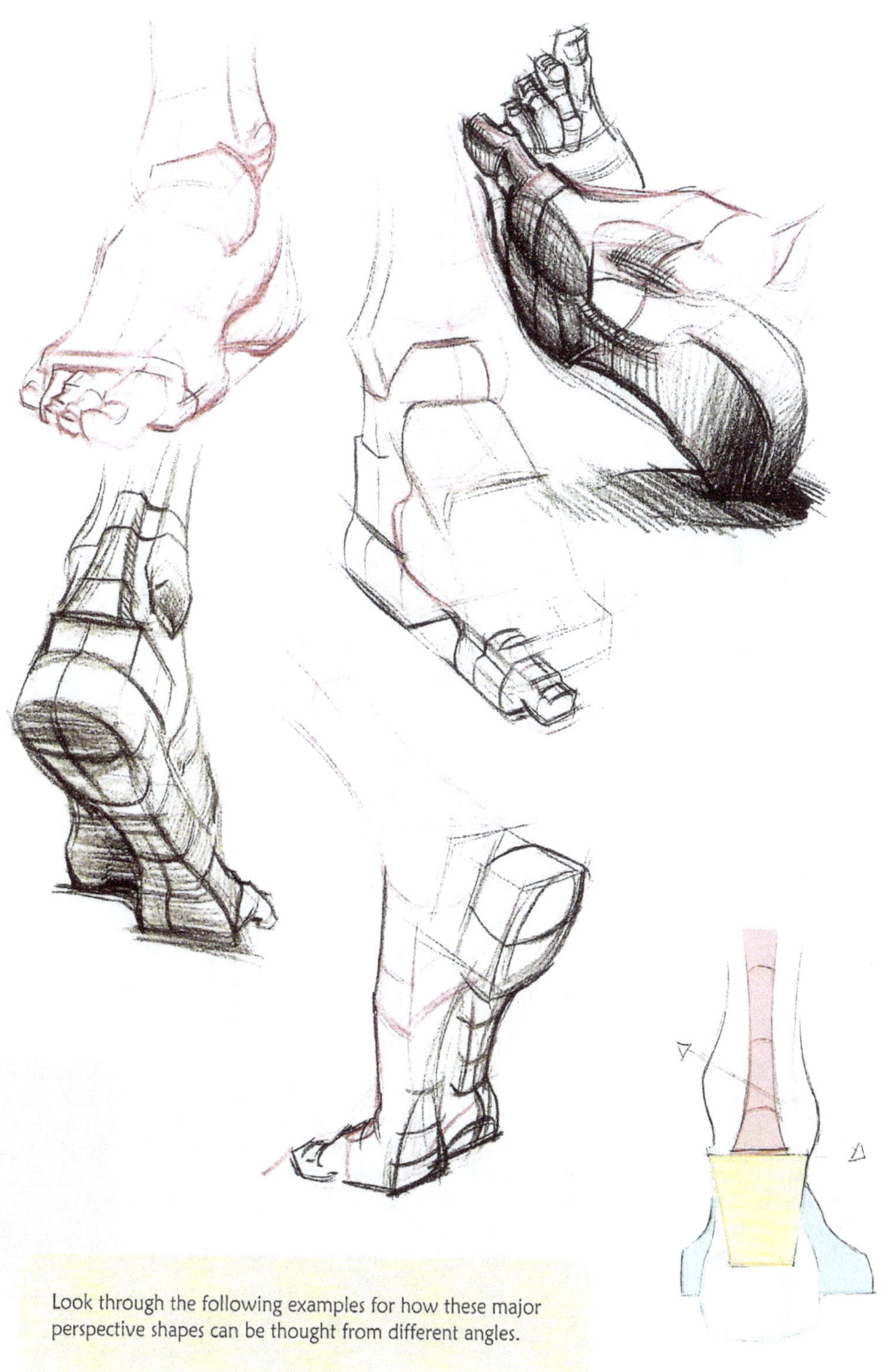

Look through the following examples for how these major
perspective shapes can be thought from different angles.

The big toe is handled somewhat differently. Instead of three
joints, like the smaller toes, there are two. The quality of the
bone is much flatter and closer in perspectival description to
a box, and it has a separate orientation of placement. While
the smaller toes point downwards, the big toe faces upwards,
towards the sky.

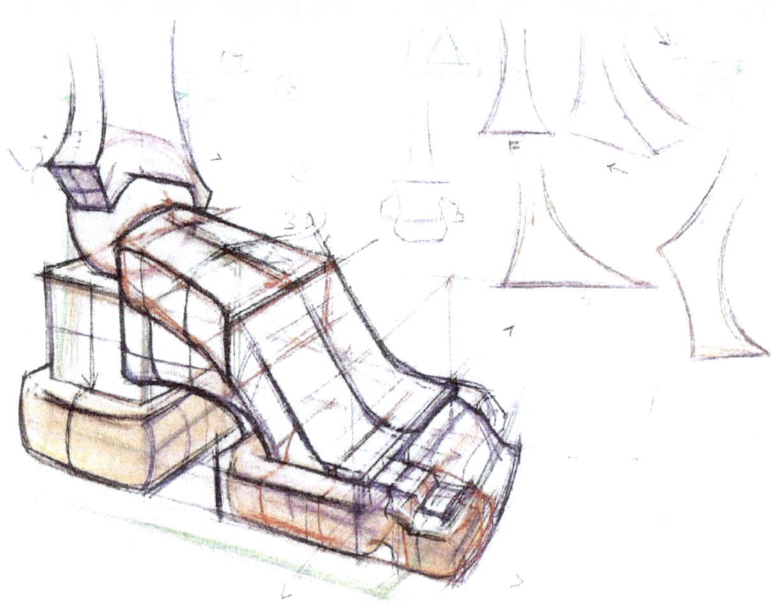

PROCESS

The process for drawing the foot can be thought of in a similar way to the one used for the
hand. When beginning the drawing of the foot, start with a very simple idea for an envelope
— think of the foot with a sock still on — as this highlights the big activity while deterring you
from focusing on contour or details (look at the drawings in the upper right of the diagram
above). This step is done to show simple action, and also makes describing movement and
weight much easier. Also, notice at this point that the enveloped shape for the foot is, in most
views, very similar to a triangle. As the triangle is one of the more visually stable shapes, it
reinforces the idea that the foot is a form built for support.

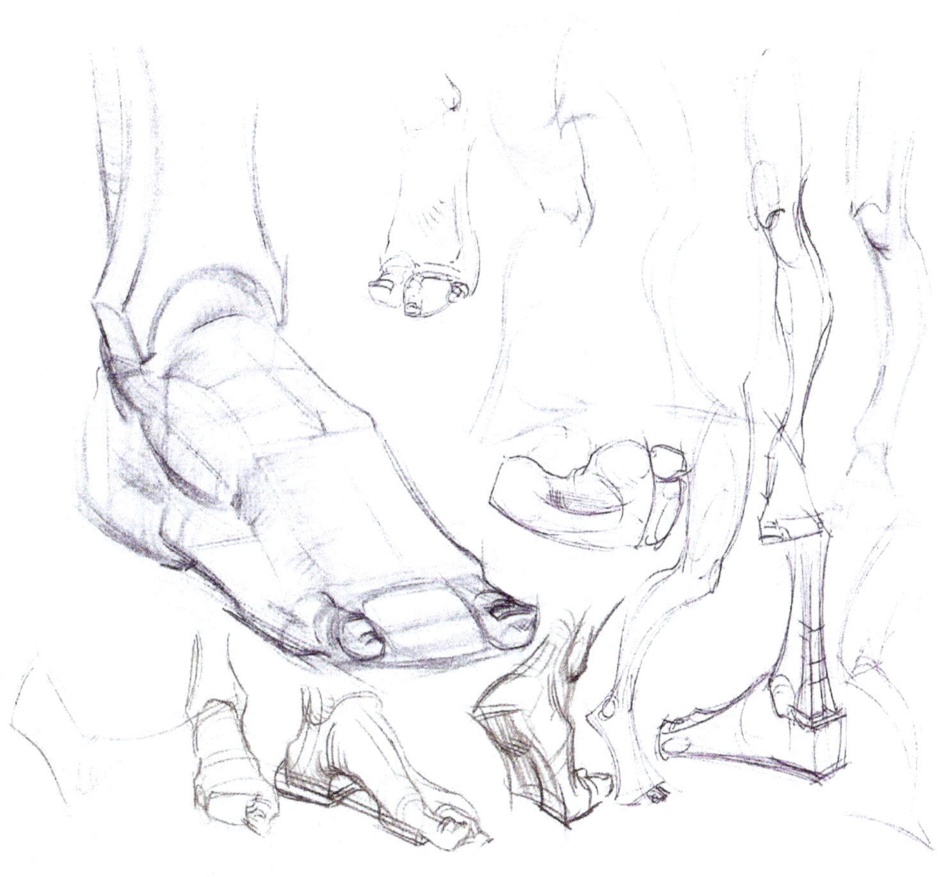

Having established the envelope, just as with the process for drawing the hand, the next step is to break up the shape with one or two proportional measurements, and then proceed with the distribution and construction of the volumes previously discussed, ending with the toes.

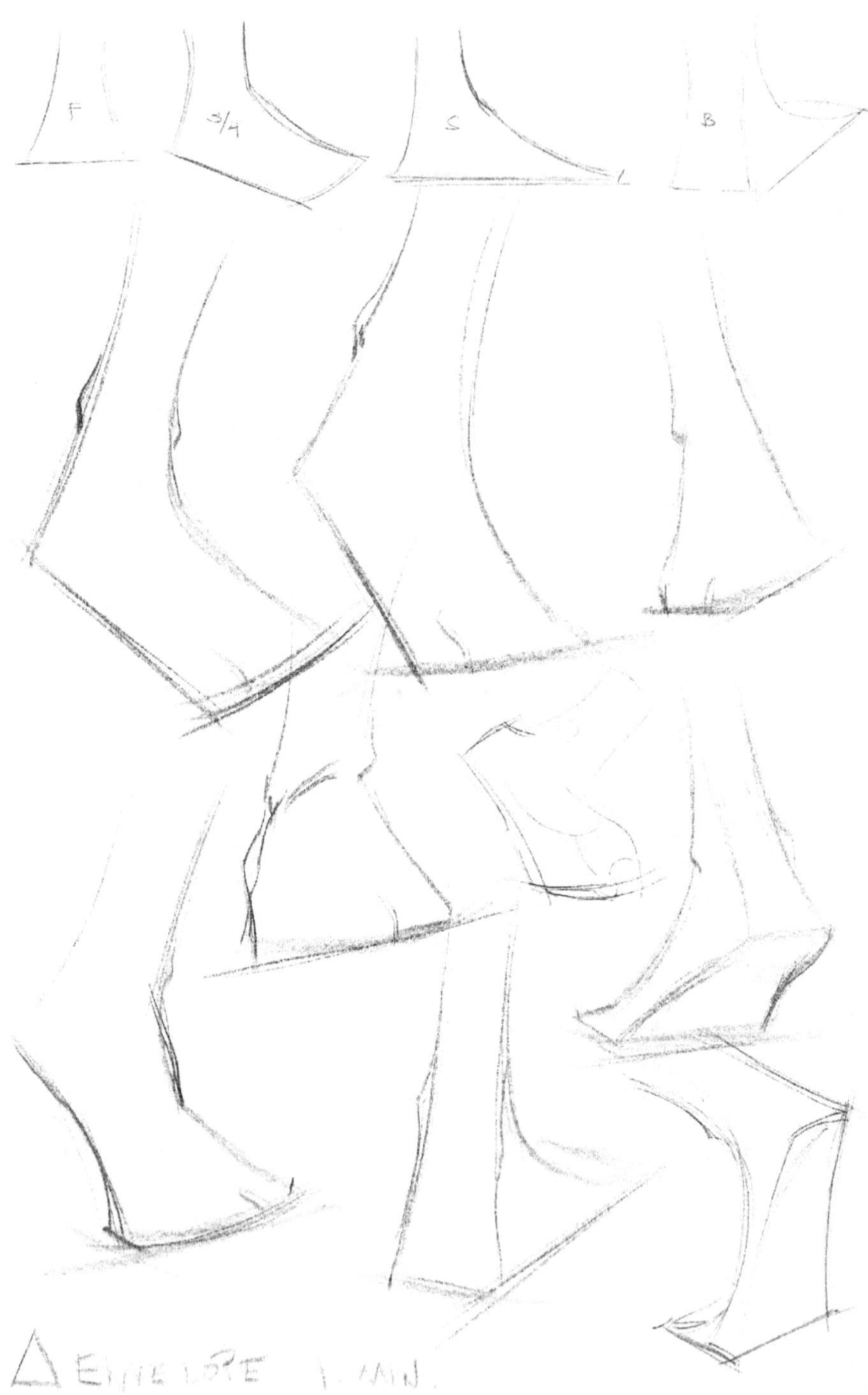

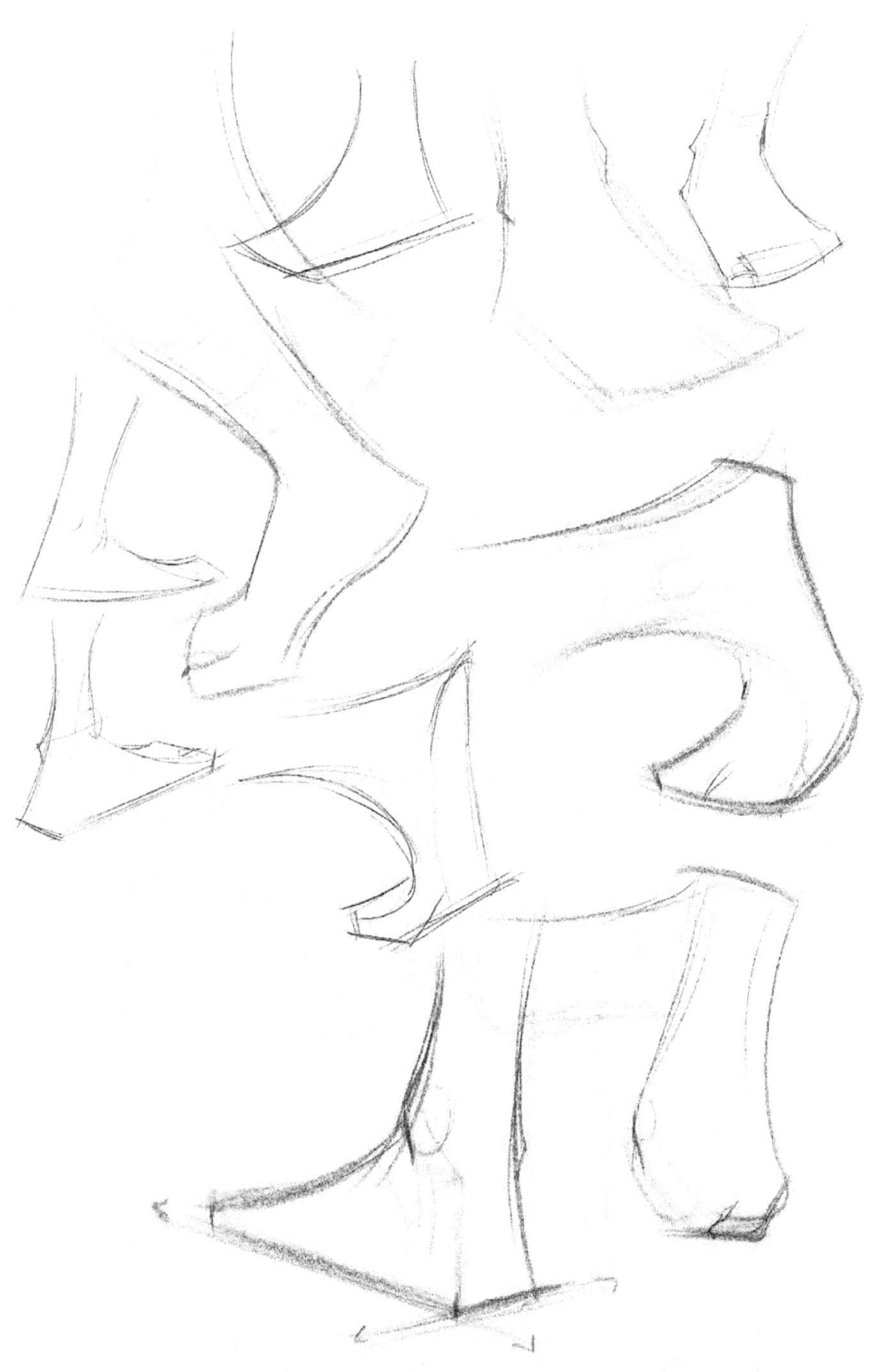

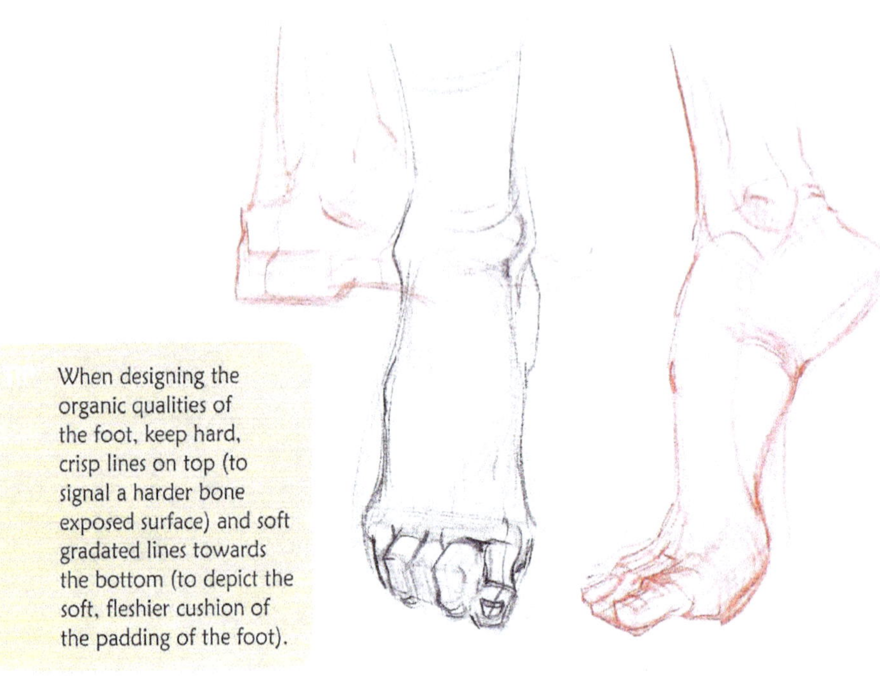

When designing the organic qualities of the foot, keep hard, crisp lines on top (to signal a harder bone exposed surface) and soft gradated lines towards the bottom (to depict the soft, fleshier cushion of the padding of the foot).

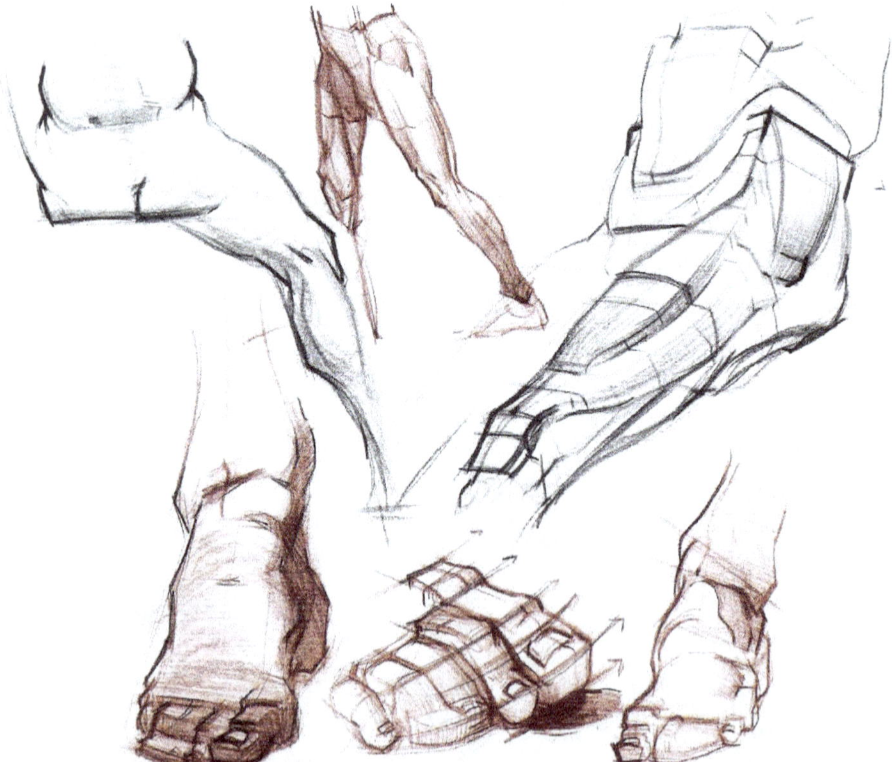

Additionally, keep the line work in the foot as economical as possible; too many lines drawn to describe bumps or smaller details can destroy the overall design.

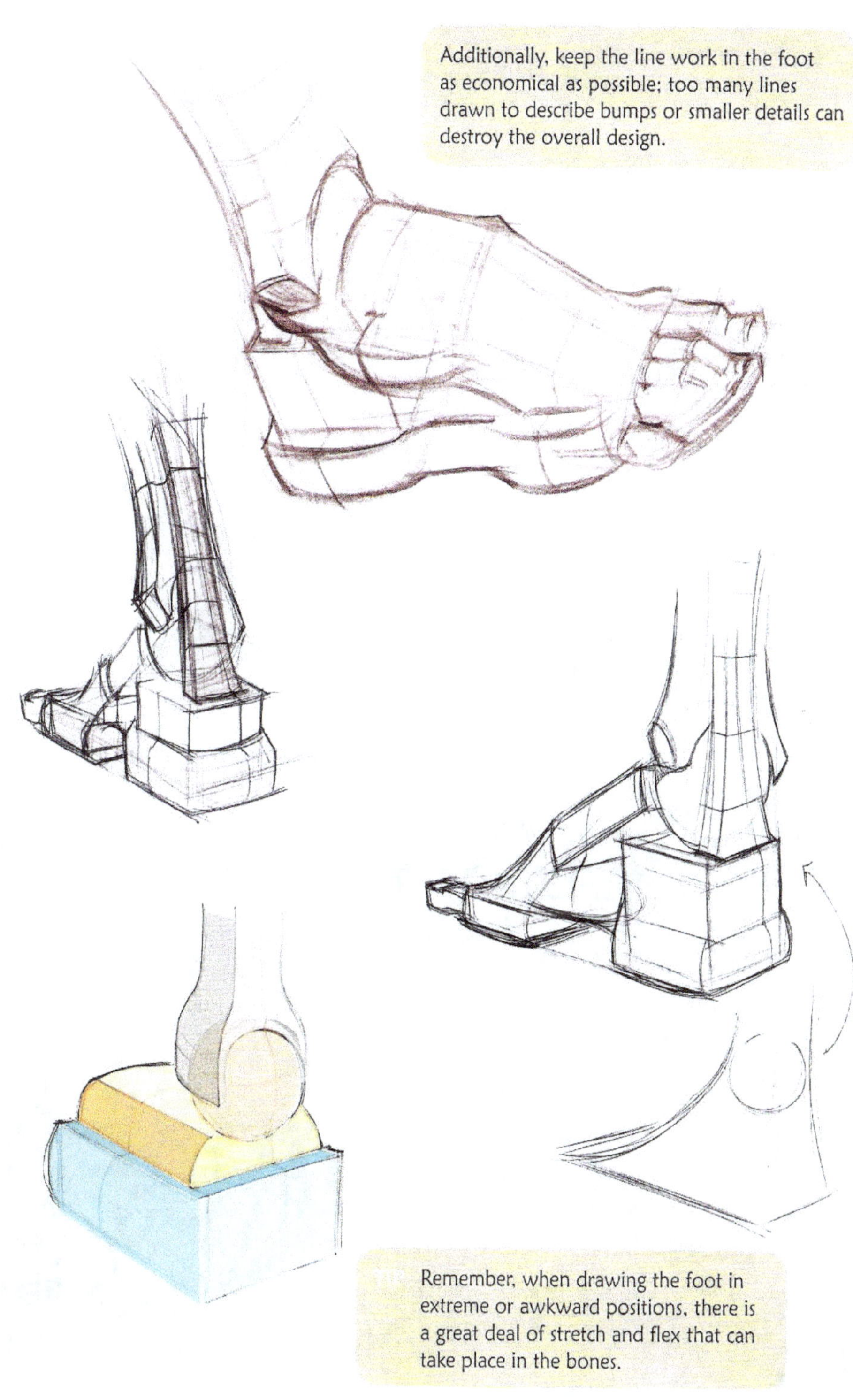

Remember, when drawing the foot in extreme or awkward positions, there is a great deal of stretch and flex that can take place in the bones.

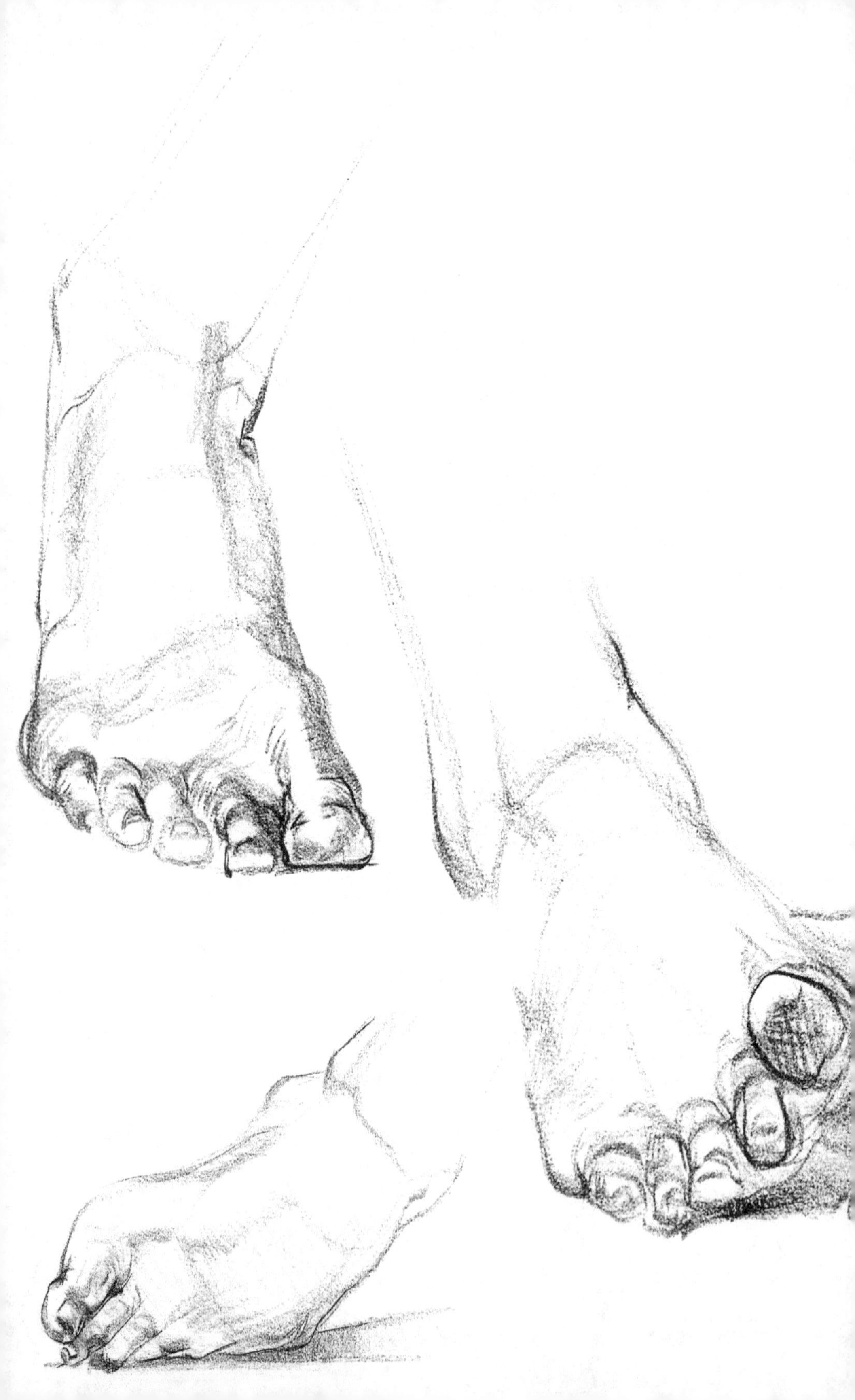

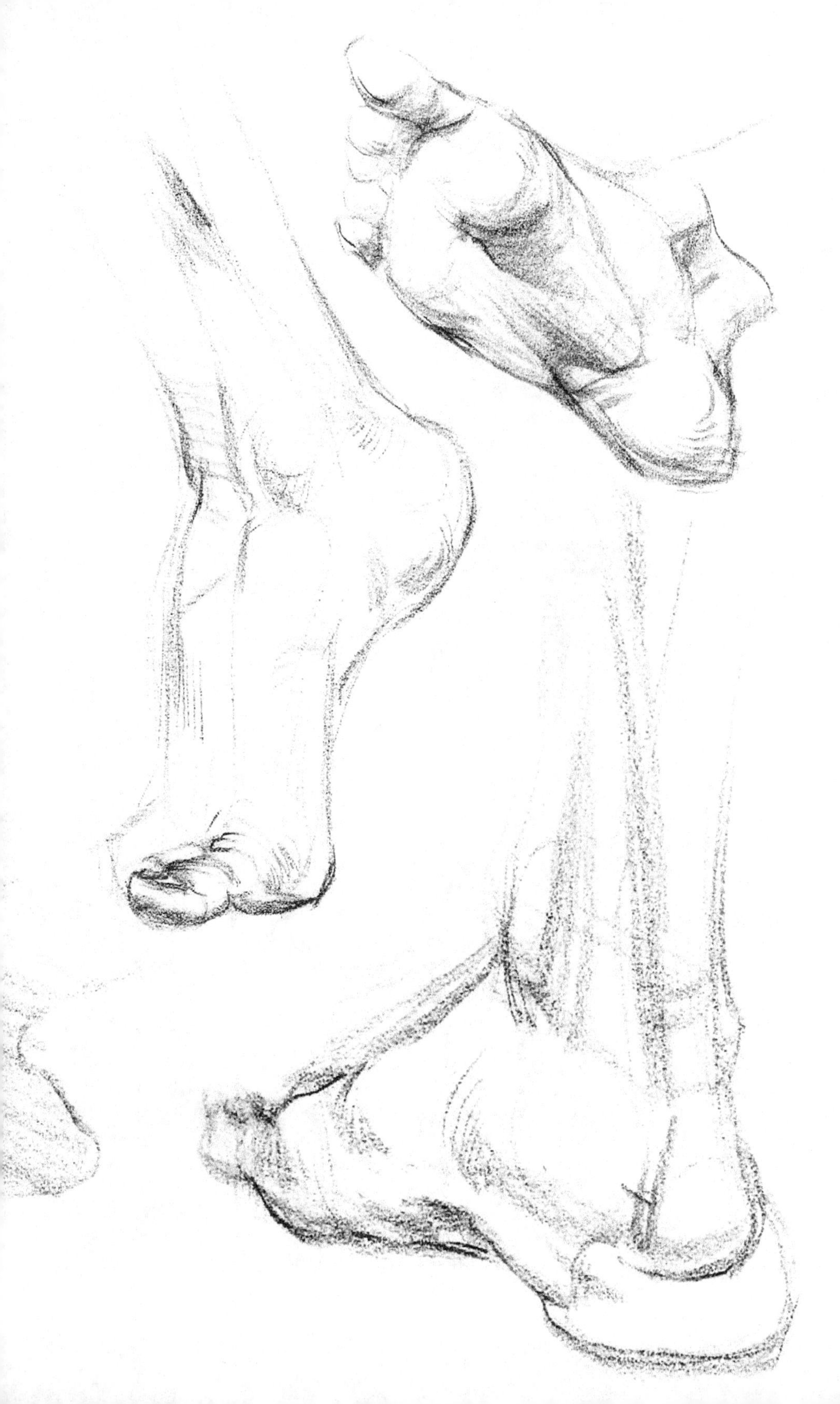

DRAPERY

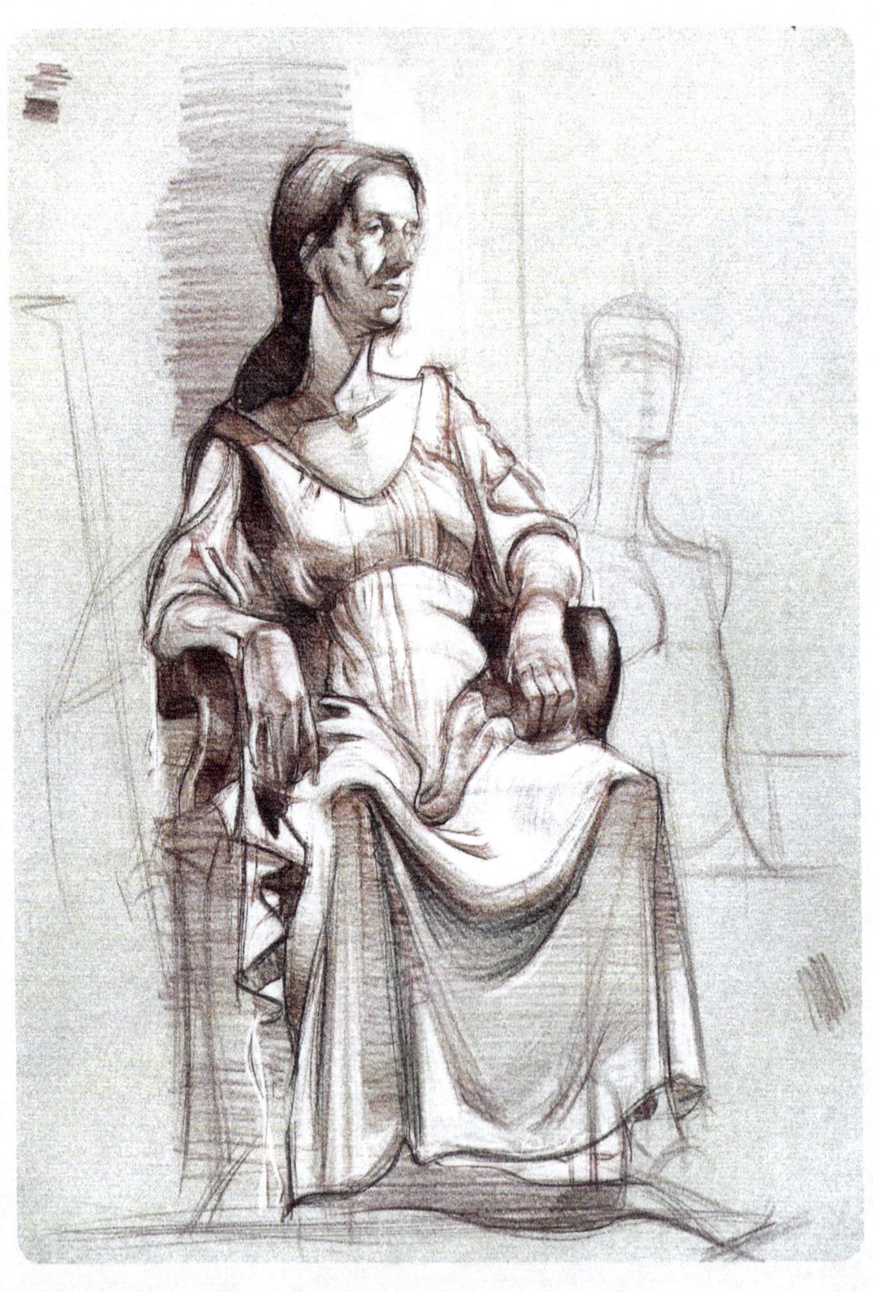

When drawing drapery, all of the previous principles are used in a way to encourage your working process and hopefully simplify a difficult subject. Again, keep in mind that all the shapes discussed can and will need to be manipulated in order to accommodate the myriad possibilities for different fashions/looks from the scope of history. This process should set you up with a foundation for depicting simple fabrics that move with and wrap around the figure, that can later be expanded upon.

To begin, there are only a few things that need to be addressed as being completely new to this area of study. First, in most cases, the gesture of drapery is consistent with gravity, and has a descending linear direction. However, this changes in any type of pose where movement or external conditions (for example, wind) are present. Second, different types of drapery (silk, leather, denim, cotton, etc.) have totally unique qualities in the way they respond to movement and form.

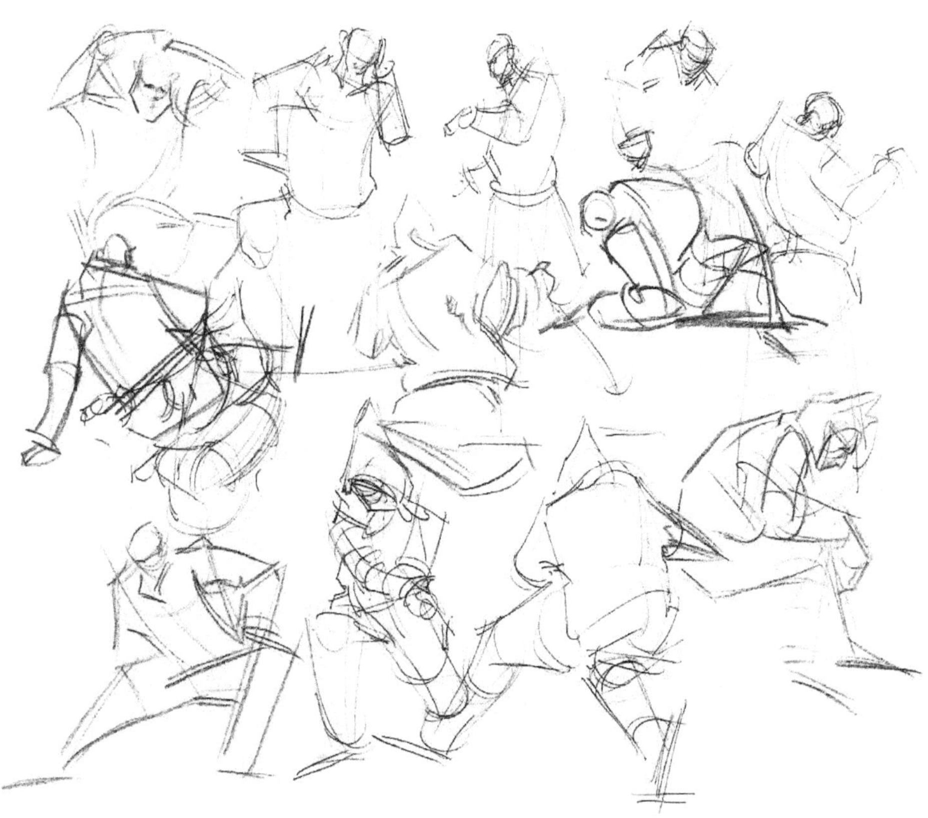

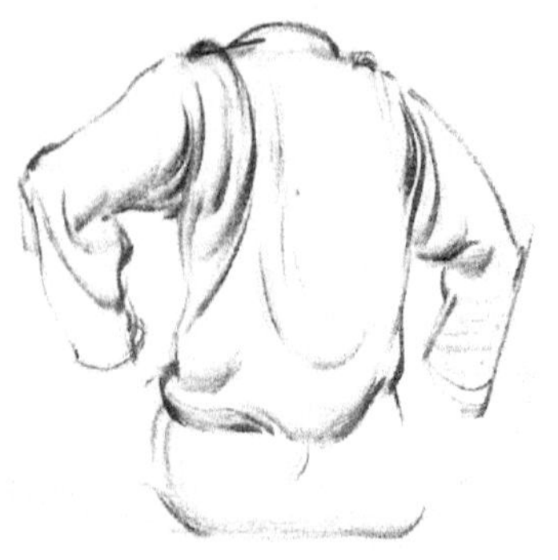

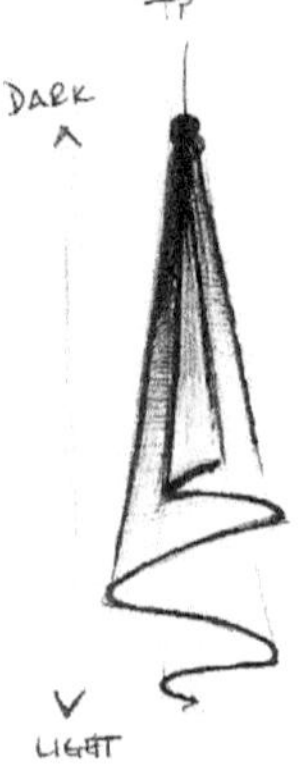

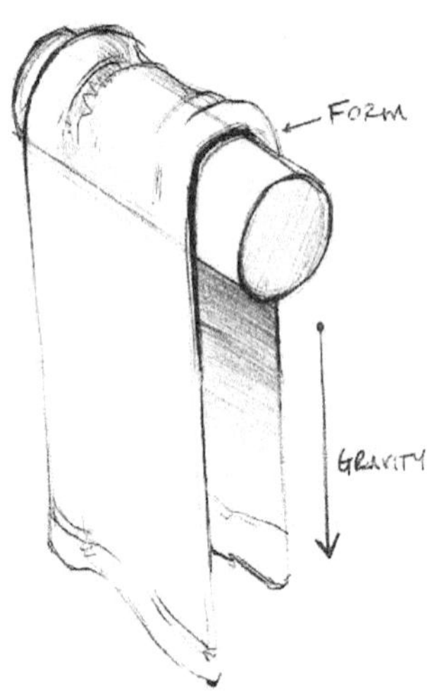

- DRAPERY SHOULD ALWAYS RESPOND TO THE UNDERLYING PERSPECTIVES

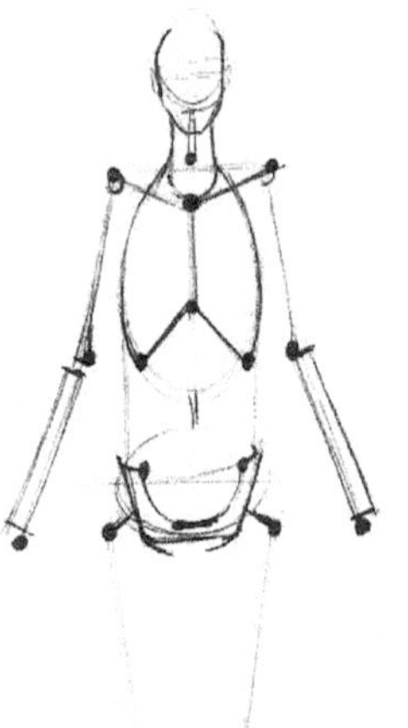

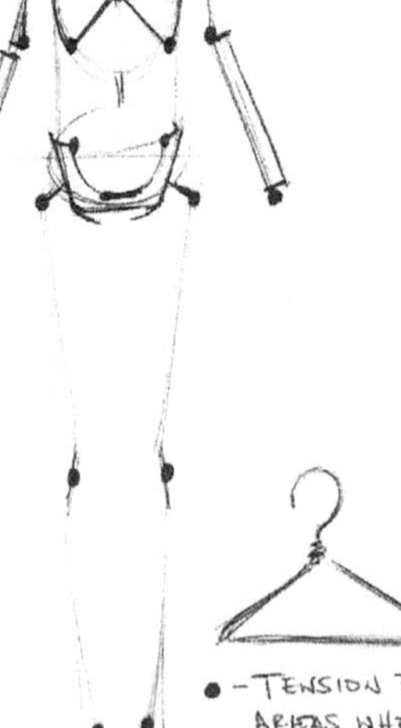

HUMAN WIRE HANGER (SKELETON)

● — TENSION PTS = AREAS WHERE FABRIC WILL HANG OR GATHER

DRAPERY

- TENSION PTS.
- EXTERNAL CONDITIONS
- CHARACTER OF FABRIC
- 6 ACTIVE FOLDS

↳ CAN MIX W/ ONE ANOTHER

① — "T" OVERLAPS
② — WRAPPING LINES ⎤ — HOW

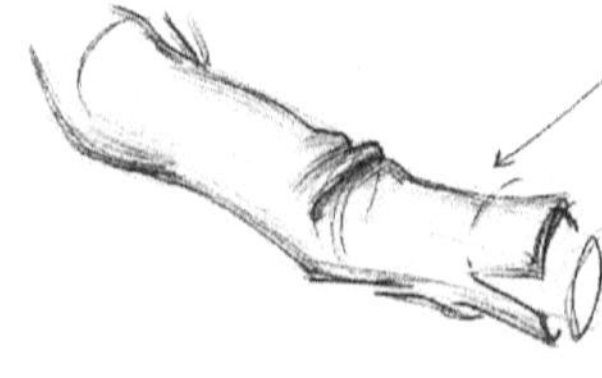

DRAPERY WILL ALWAYS END IN THE SAME PERSP. AS THE FORM BENEATH.

In the study of drapery, we will focus on seven different types of folds that will be intimately linked to tension points located on the figure.

Tension points are the same as landmarks, just renamed to relate them more to the push or grab of fabric. In order to emphasize how these tension points work, try thinking of the skeletal points as being a large, moveable wire hanger. Thinking in this way will make the study of drapery easier, as you will see the seven folds repeatedly occur in the same places.

Additionally, take note that although we are looking at the seven folds in isolation, they have the ability to mix together. The challenge is to simplify and edit what you see for clarity so that the movement and form of the figure have the primary read.

Beyond the new information presented above, the exact same techniques will be used to interpret and show drapery on the figures: wrapping lines, "T" overlaps, pinch vs. stretch, crisp sharp lines vs. softer gradations, and so on.

The first fold is called an *end fold*. The simple idea behind this fold is that when drapery ends, it always describes the perspective of whatever form it is conforming to. In other words, the end fold will always just fit to a wrapping line on whatever perspective it follows.

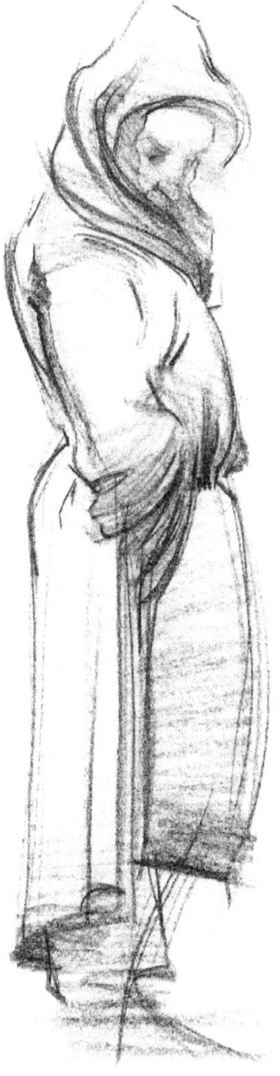

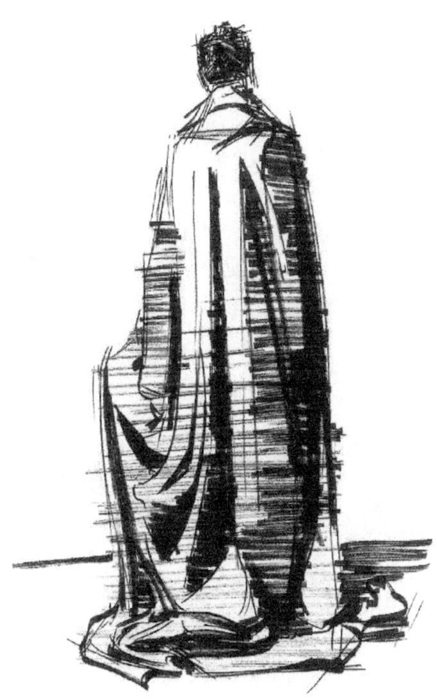

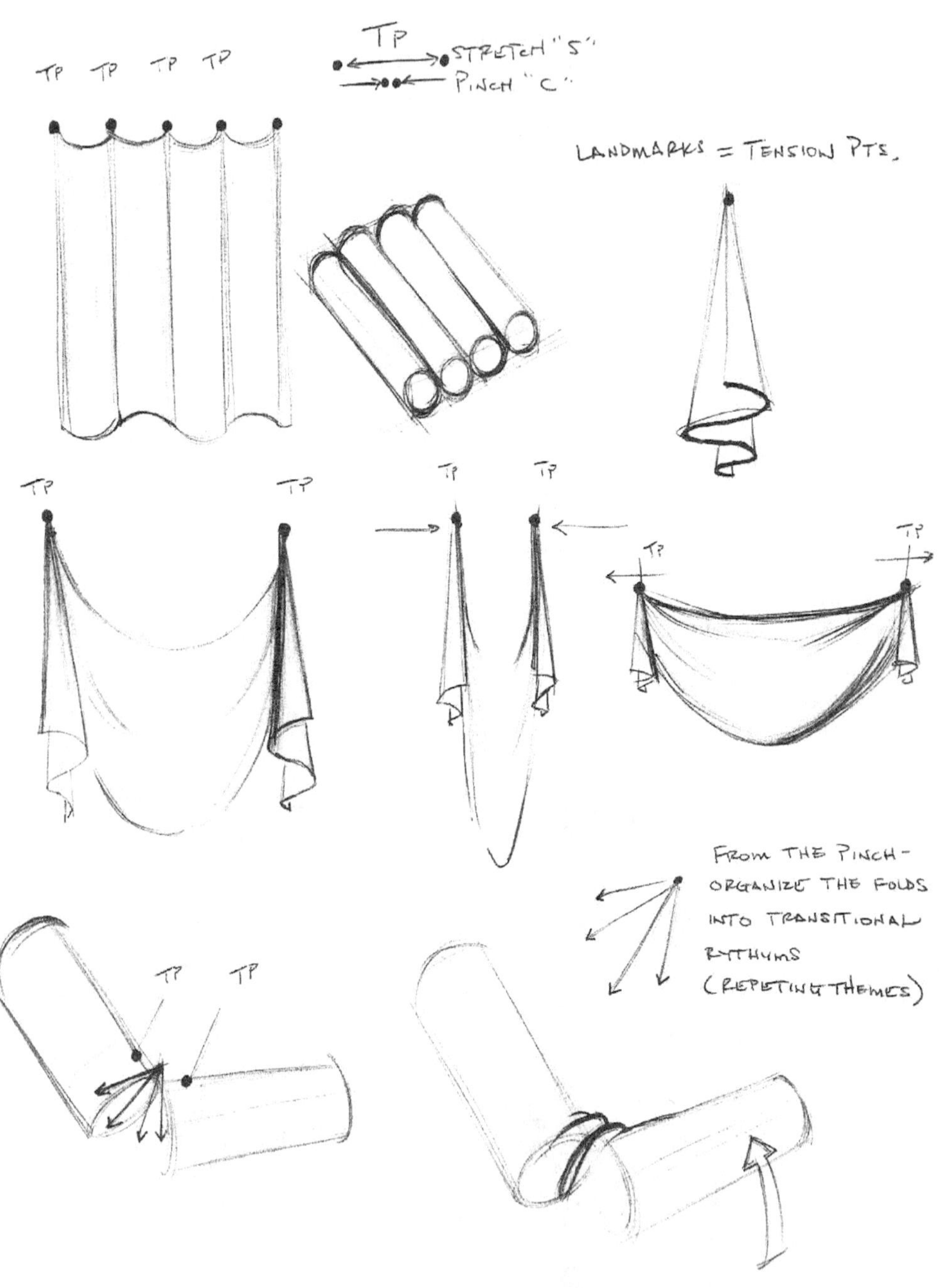

TP TP TP TP
TP
STRETCH "S"
PINCH "C"
LANDMARKS = TENSION PTS.
TP
TP TP
TP TP
TP
TP
TP TP
FROM THE PINCH —
ORGANIZE THE FOLDS
INTO TRANSITIONAL
RYTHMS
(REPETING THEMES)
TP TP

The second fold is called a *cylinder fold*. The cylinder fold is characterized by a consistent placement of tension points. This consistent placement of tension points results in the fabric looking as though there are a number cylinders lined up next to one another. Apart from the figure, you can see examples of this fold in a shower curtain, window curtains, or anywhere else where there is a consistent grab to the fabric. In addition to the fold or fabric as a static form, keep in mind that this fold can change shape through gesture, just like the anatomical shapes. This fold, as well as all the others, stretch or pinch based on how the figure pushes the tension points.

The third fold is a *"U" fold*. Notice the "U" fold is caused by the drapery being suspended between two main tension points. Additionally, notice the squash and stretch of the fold is still designed with "C" and "S" curves. This fold can potentially appear between tension points (landmarks) that have fabric suspended between them.

The fourth fold is called a *pinch fold*. A pinch fold can take place between any two tension points in close enough proximity to pinch fabric between them — for example, the bend of the arm or leg, between the neck and shoulder, the rib cage against the pelvis, etc.

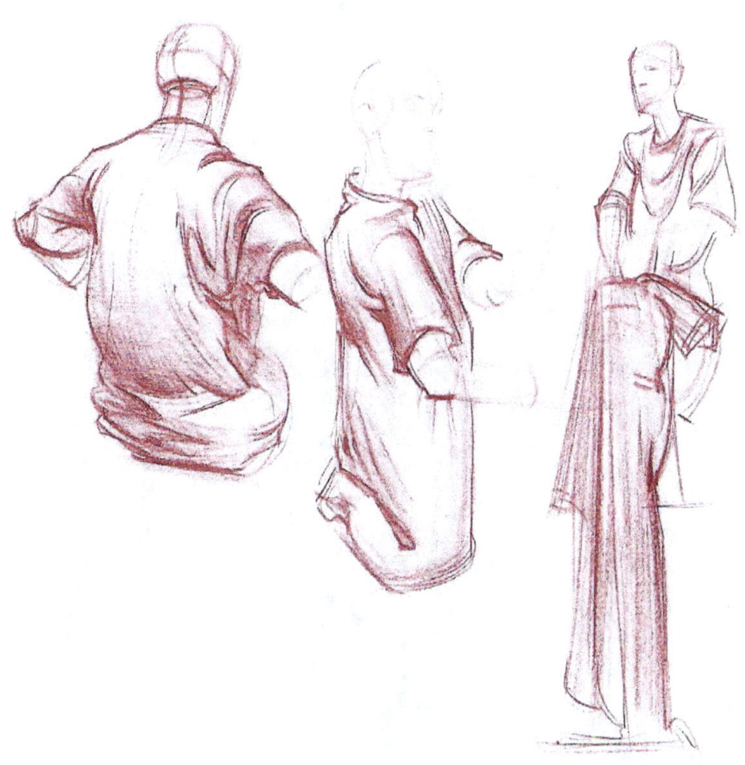

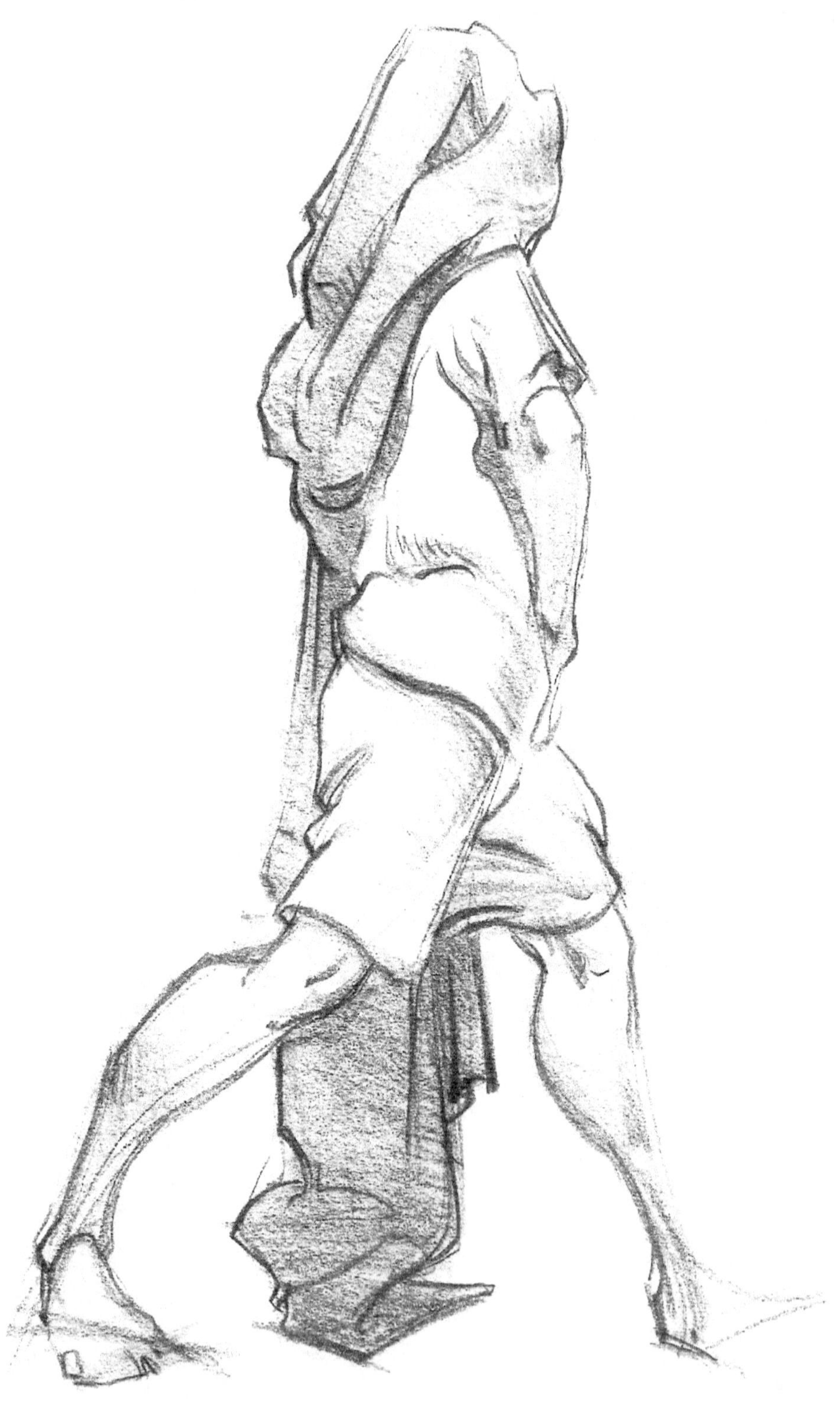

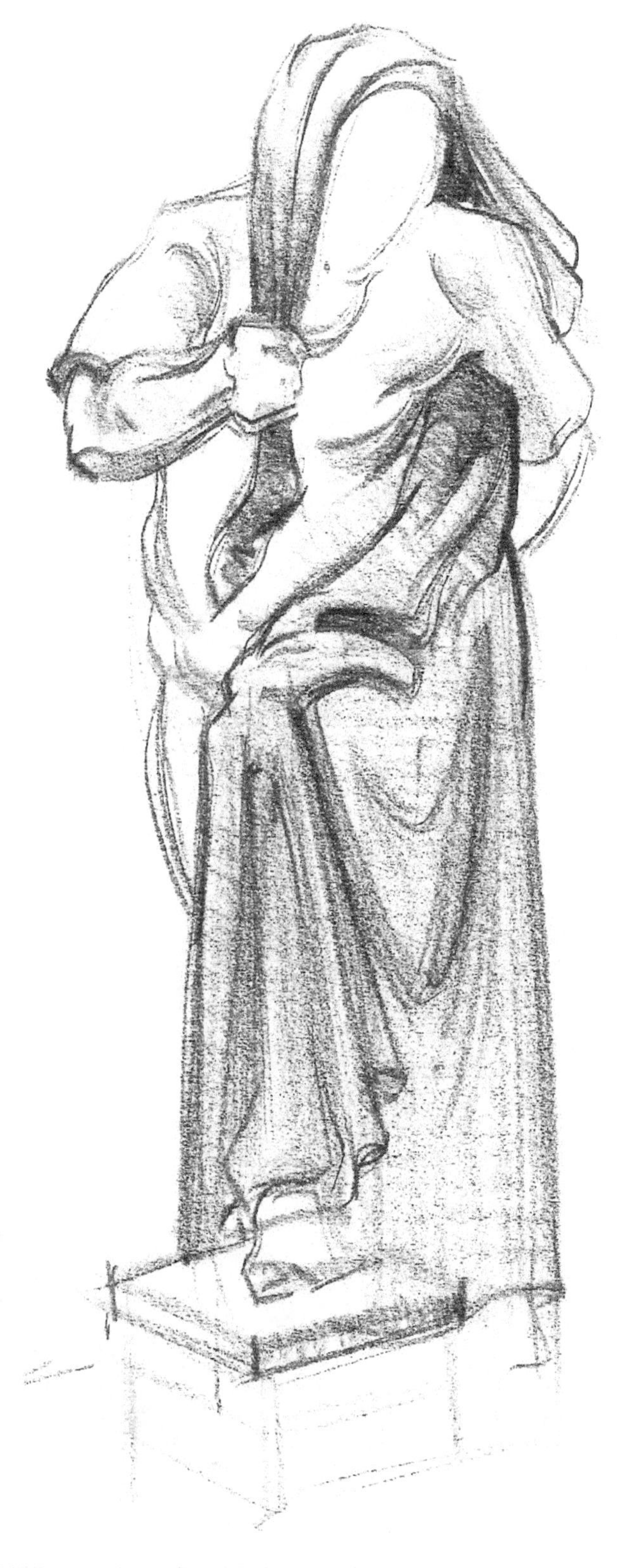

The fifth fold is called a "S" or *spiral fold*. The "S" fold represents two tension points twisting the fabric in opposing directions. This could take place on the arm, or, in the case of longer draping clothing, from head to foot (in the case of a twist of the whole body). Notice that the main design of the fold is first described as the "S" connecting two points, then wrapping across the two volumes, and ending behind the forms in a "T" overlap.

The sixth fold is the "Z" *fold*. The "Z" fold is based off the letter to help remember the asymmetrical compression of clothing as a result of gravity and excess fabric. This fold is most likely seen at the bottom of the pants/leg. In this particular example, remember that you also want to combine the "Z" fold with the end fold to describe the compression of fabric that ends with the perspective of the form it is on.

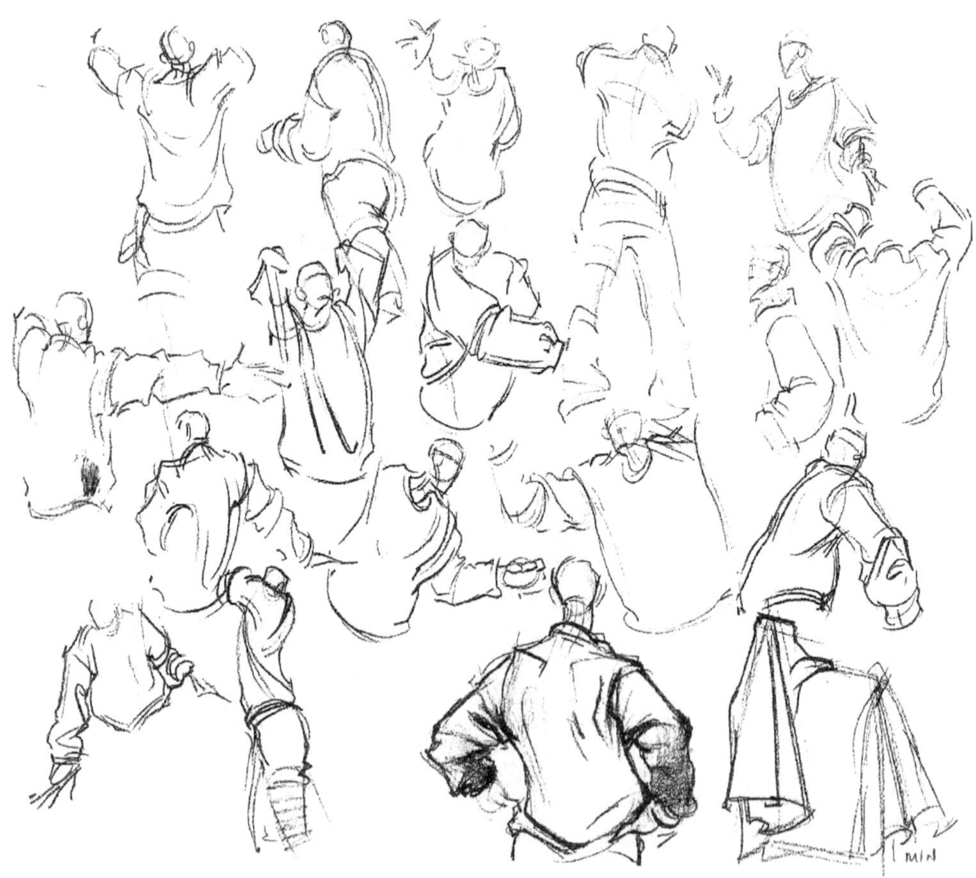

Again, keep in mind that the folds can combine. For example, in the case of the arm, it is possible to have both a bend and a twist, resulting in a pinch and spiral fold happening at the same time.

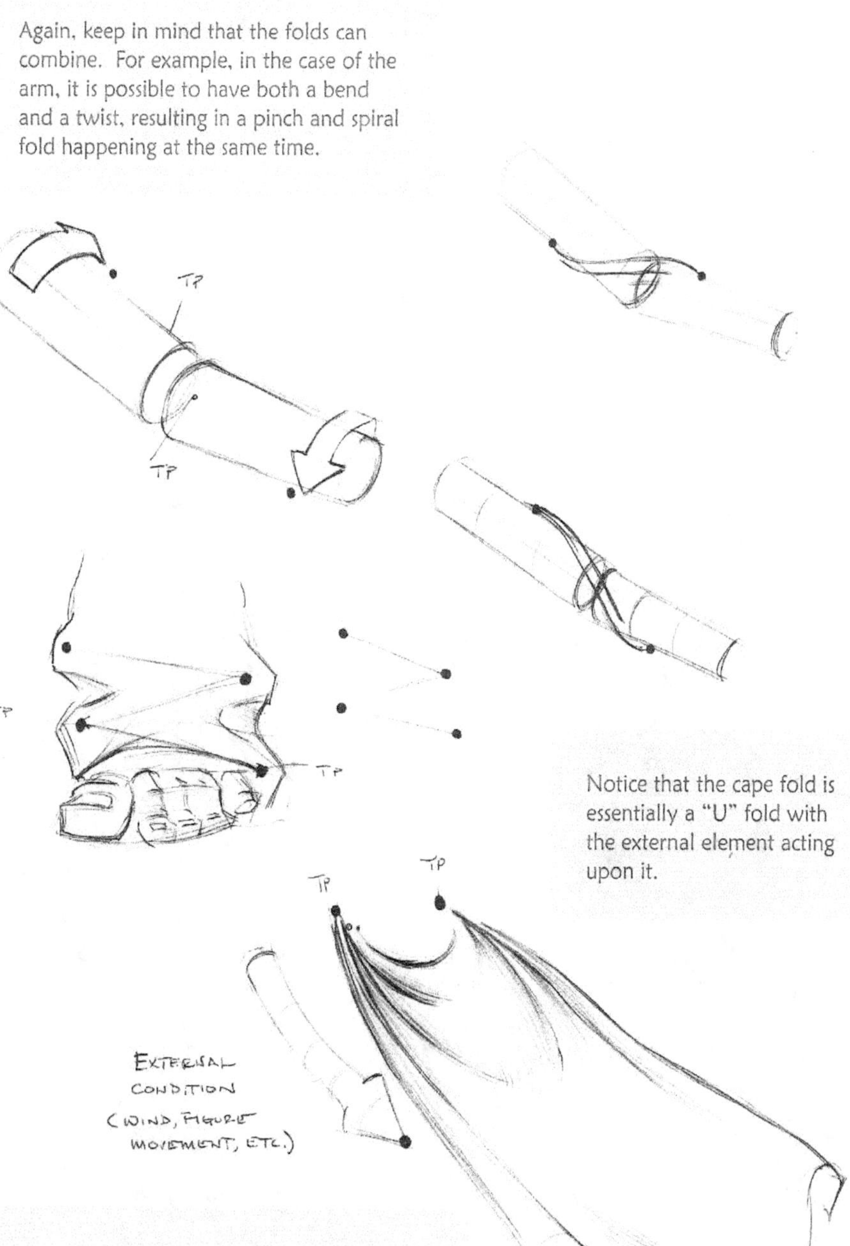

Notice that the cape fold is essentially a "U" fold with the external element acting upon it.

The seventh and last fold is a *cape fold*, and is the only type of fold that really shows an external influence. The cape fold demonstrates the effects on drapery if a figure flies, runs, is in a wind storm — any outside force on the figure.

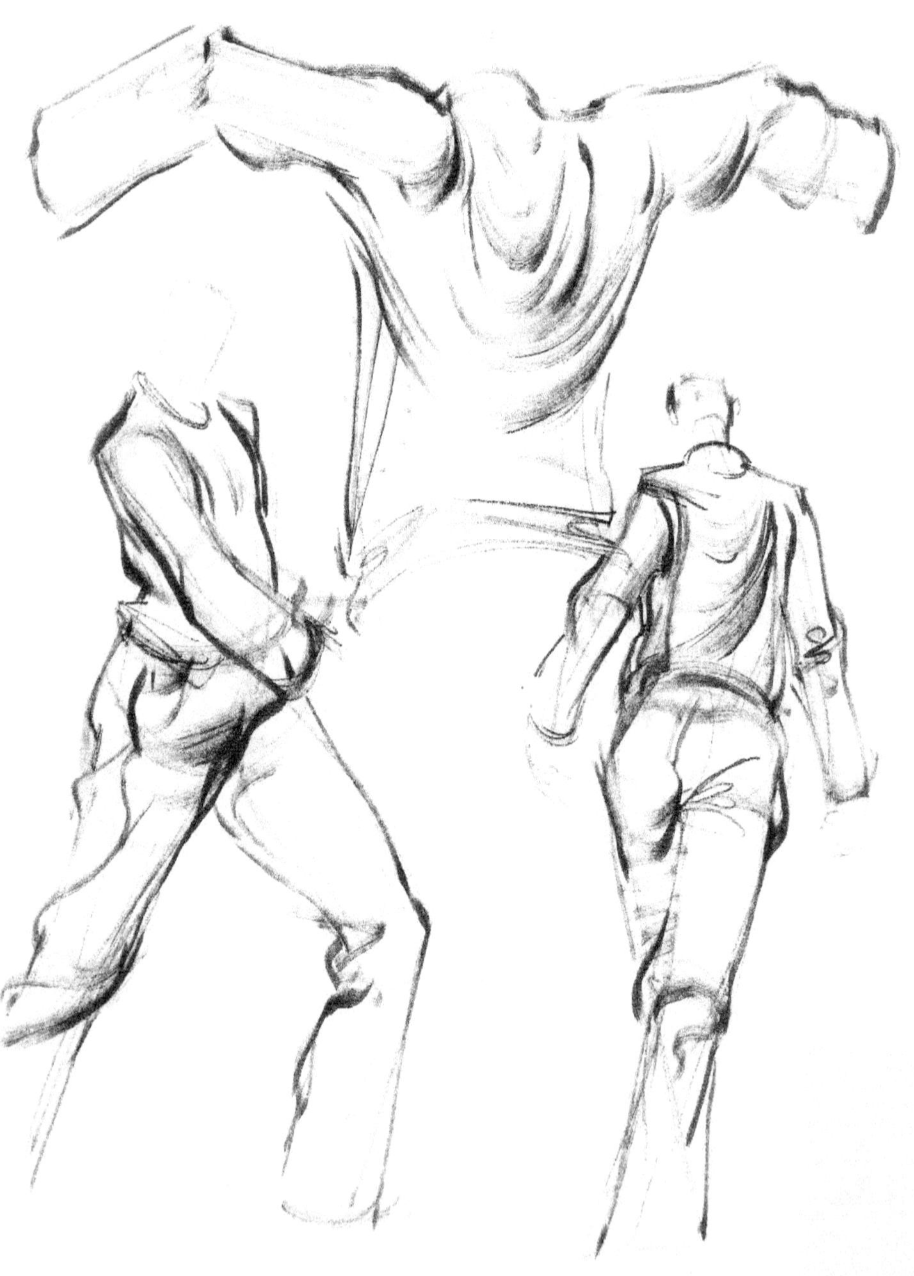

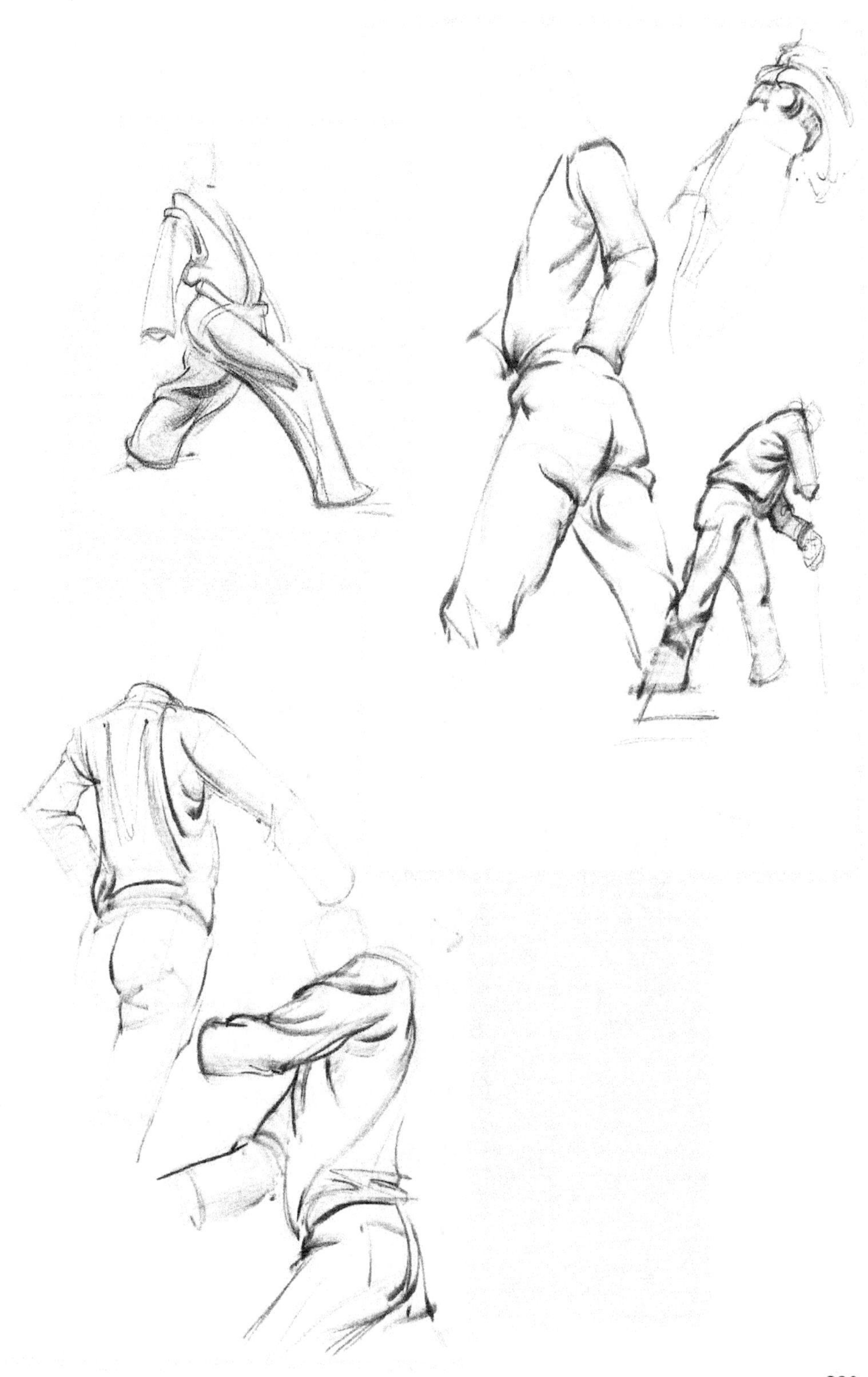

Use the remainder of the drawings in this chapter to study and analyze where these basic
seven folds take place. Additionally, notice that, in some cases, the folds have the same
asymmetrical qualities of line to gesture (Chapter 1). This keeps the folds relating to one
another with a great deal of fluidity. The danger in drawing drapery is that too much
emphasis and attention on any one fold, or separate folds, can break up the movement
and form specific to the figure underneath.

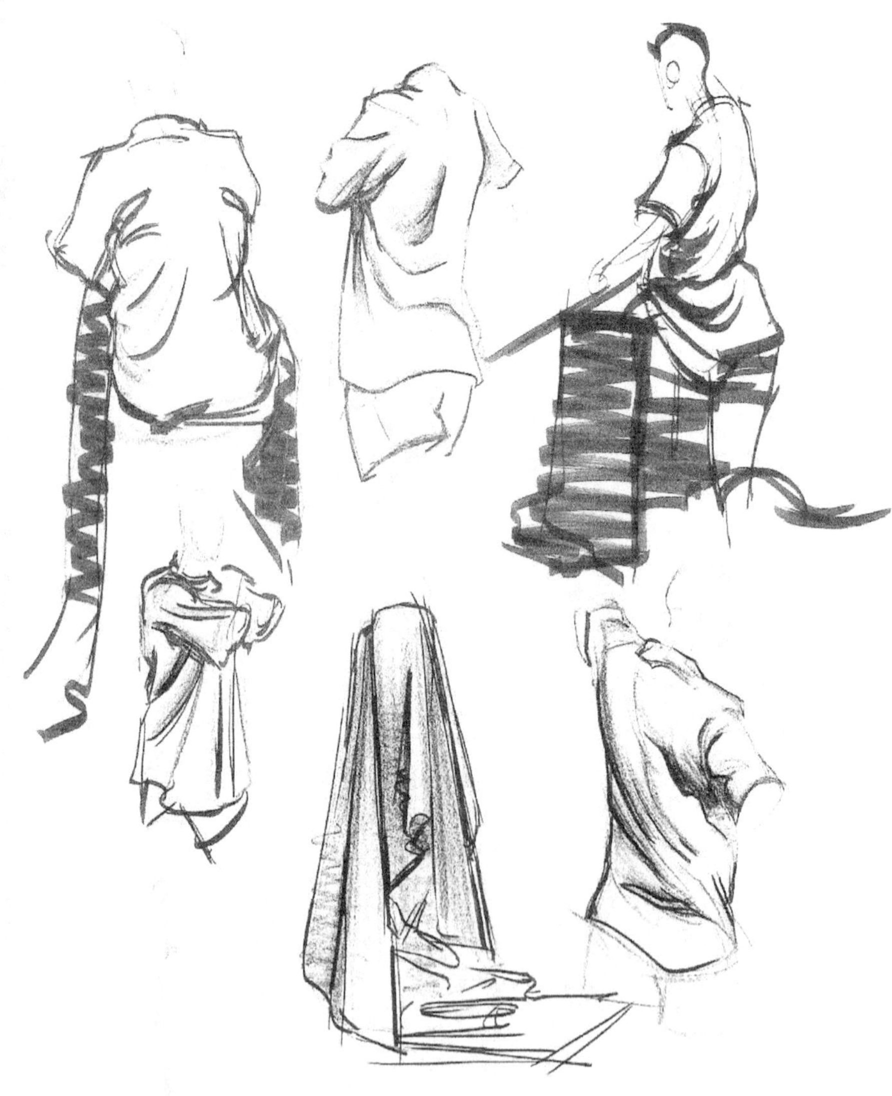

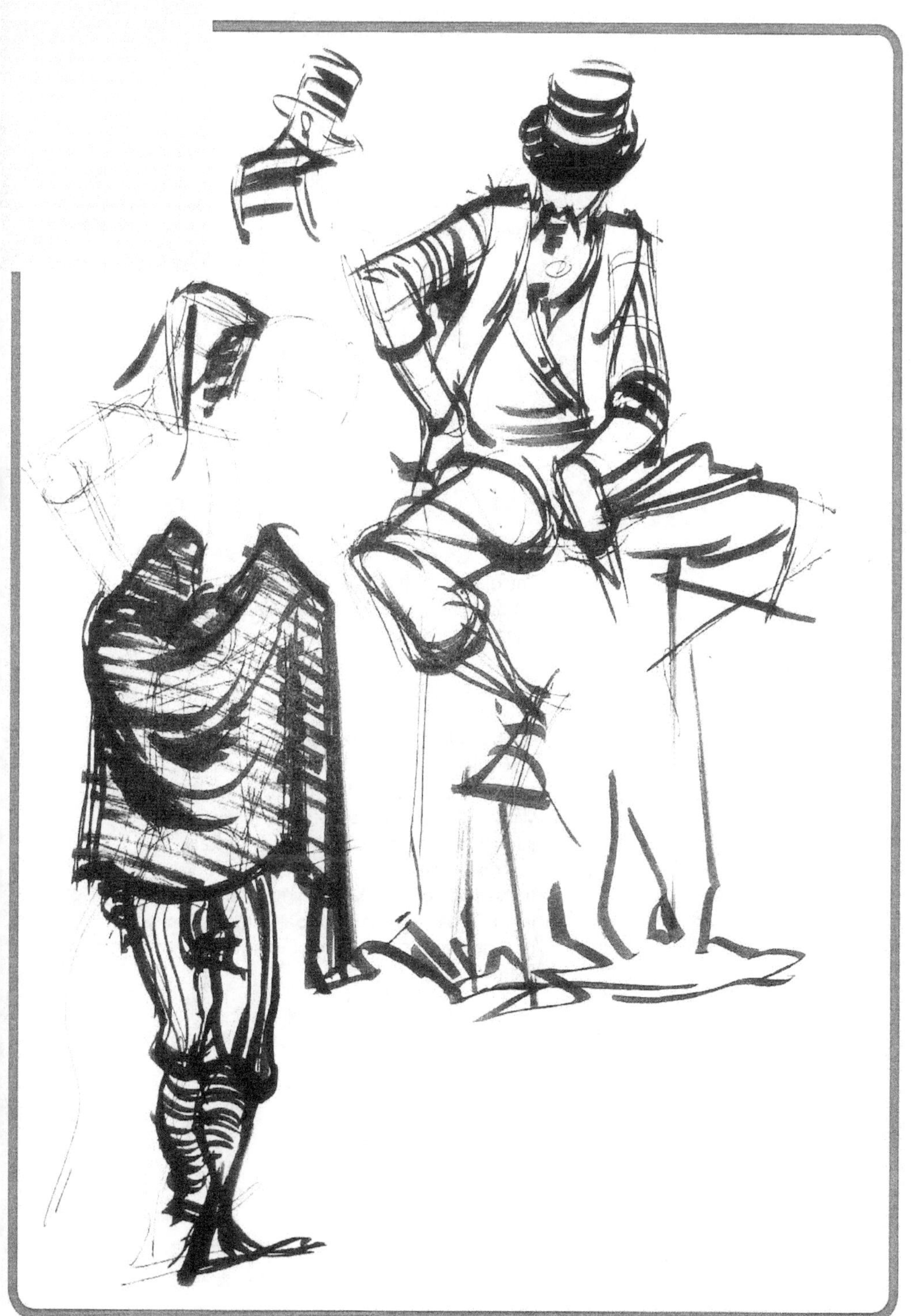

SOME NOTES ON LIGHT AND SHADOW

While the primary emphasis in this book is on the development of form through the use of line, the simple volumes developed will make an easy transition into lighting your figures. By always using some kind of variation on a sphere, cylinder, or box, the job of lighting will become much more organized and hopefully more manageable.

Good luck!

The diagrams on this page show the eight conditions of light and the edges used to integrate them.

Having already studied the figure for its sense of plane, form, and corner, the goal in lighting can be to identify those plane changes and determine if they are slow moving (sphere), quicker moving (cylinder), or abrupt (box). Working from a core shadow, the timing of a form can be softened to more organically describe the surface. Just remember that value shifts equal surface changes. The methodology in this book has been to give an introduction to surface in order to help make a more finished study possible.